Fine Wines
& Fish Oil

Dr Hugh Macdonald Sinclair
DM, DSc, FRCP
(1910–1990)

Hugh Macdonald Sinclair
BSc, LMSSA, MA, DM, FRCP, Hon DSc
Fellow of Magdalen College 1937–80, Emeritus Fellow 1980–90
Director, Oxford Nutrition Surveys 1941–46
Reader in Nutrition, University of Oxford 1951–58
Visiting Professor in Food Science, University of Reading 1970–80
Founder and Director, International Institute of Human Nutrition 1972–90
Sometime Senior Warden and Master of the Worshipful Society of Apothecaries

Fine Wines & **Fish Oil**

The life of
Hugh Macdonald Sinclair

Jeannette Ewin

OXFORD
UNIVERSITY PRESS

OXFORD
UNIVERSITY PRESS

Great Clarendon Street, Oxford OX2 6DP

Oxford University Press is a department of the University of Oxford.
It furthers the University's objective of excellence in research, scholarship,
and education by publishing worldwide in

Oxford New York

Athens Auckland Bangkok Bogotá Buenos Aires Cape Town
Chennai Dar es Salaam Delhi Florence Hong Kong Istanbul Karachi
Kolkata Kuala Lumpur Madrid Melbourne Mexico City Mumbai Nairobi
Paris São Paulo Shanghai Singapore Taipei Tokyo Toronto Warsaw
and associated companies in Berlin Ibadan

Oxford is a registered trade mark of Oxford University Press
in the UK and in certain other countries

Published in the United States
by Oxford University Press Inc., New York

A catalogue record for this title is
available from the British Library

Library of Congress Cataloging in Publication Data

Ewin, Jeannette.
Fine wines and fish oil : the life of Hugh Macdonald Sinclair/Jeannette Ewin.
Includes bibliographical references and index.
1. Sinclair, Hugh M. (Hugh Macdonald), 1910– 2. Nutritionists–Great
Britain–Biography 3. Essential fatty acids–History. I. Title.
QP143.E955 2001 613.2′092–dc21[B] 2001034060

ISBN 0 19 262927 1

Typeset by EXPO Holdings, Malaysia
Printed in Great Britain
on acid-free paper by Biddles Ltd, Guildford & King's Lynn

This book is dedicated to the memory of Hugh Macdonald Sinclair and all those who seek a better understanding of human nutrition.

So soon after the death of a great scientist, no biographer can do full justice to his subject, nor can he assess the full significance of his achievements. The writer can only attempt to describe facets of the personality and assess the scientific achievements as they have affected contemporary developments. He must leave it to future biographers to complete the picture on the basis of future source material and to assess in more detail and more objectively the consequences of the research achievements in the light of later scientific advances.

From *Otto Warburg,* by Hans Krebs, Oxford University Press 1981, page vii

Foreword

David F. Horrobin

Why write a life of Hugh Sinclair? He is not a household name. As a scientist and doctor he did not win any of the great prizes which can come to those who practise in these fields. Yet few who met him ever forgot him and he has left a legacy which as the years pass seems more and more important. He may prove to be one of those people whose long term influence is far greater than ever seemed likely while he was alive.

Hugh Sinclair spent most of his working life as a Fellow of Magdalen College, Oxford, though he made many forays into the wider world, notably during the Second World War when he was involved in planning how the British could be properly nourished and in famine relief in the Netherlands and Rhineland when it was all coming to an end. But his main contributions were intellectual. He recognised the central importance of nutrition to human life and at a time when it had become unfashionable constantly emphasised the importance of the right food for proper health. He made a very particular contribution in identifying the role of essential fatty acids and in a famous letter to the Lancet in 1956, which readers classed either a visionary or lunatic depending on their point of view, he foreshadowed half a century of research on a topic which is steadily increasing in importance.

I was fortunate enough to be appointed a Junior Fellow of Magdalen in 1963 and for three years saw Hugh almost every day, sharing with him and with Brian Lloyd the teaching of medical students. Thus began a life long friendship which sadly ended only when I became one of his executors. I soon learned that Hugh was an extraordinary mixture of contradictions, capable of immense kindness and generosity and of brilliant intellectual insights, but with an uncanny ability to provoke sheer irrational rage in the great, the good, the virtuous and the self-satisfied. At that time my own medical and scientific interests were very different from his but over the years I progressively realised the depth of his insight and for the last two decades much of what I have done in medicine has been following his intellectual heritage. This emphasised the role of nutrition in general and of esential fatty acids in particular in human health, and stressed the immense value of human studies as opposed to laboratory and animal studies. Hugh was one of the last great self experimenters, at various times in his life making himself deficient in vitamins or feeding himself on an Eskimo diet in order to learn for himself the impact of particular nutrients.

In 1937 Australian Jack Eccles (later Sir John and a Nobel laureate) was the tutor for medical students at Magdalen but had decided to accept a position in New Zealand. He wrote to Sir Charles Sherrington, another Nobel laureate and Magdalen Fellow and the pioneer of much of our knowledge of the nervous system, "Who should follow me at Magdalen, who should be recommended to take up this fellowship?" Sherrington responded in a letter dated May 4, 1937, "You asked me to recommend someone to follow you at Magdalen. In my opinion Hugh Sinclair is the best. His knowledge, personality and training all mark him out. He has wide interests outside science which add to his attractiveness. He is not only a brilliant and accomplished man, perhaps more important, he is a charming fellow and a reliable friend."

No one is better qualified to write a biography of Hugh Sinclair that Jeannette Ewin. After obtaining her PhD in human anatomy and biochemistry from the University of Maryland School of Medicine, in Baltimore, Maryland, she was accepted as a Research Fellow in the Department of Nutrition, at Harvard Univerisity School of Public, where she worked with eminent nutritionist, D. Mark Hegsted. Her work there concentrated on the metabolism of essential fatty acids. After years in the academic community, she took early retirement to work as a writer. Her writing skills and her knowledge of nutrition make her the ideal biographer. She has produced a warm, stimulating and appropriately critical account of this brilliant, charming, difficult, flawed, provocative and influential man.

Stirling, Scotland, June 2001

Prologue

Fine Wines and Fish Oil began as a book about essential fatty acids: simple molecules required for normal human growth and health. However, as research on the subject progressed, the story of the life of Hugh Macdonald Sinclair dominated the work. A brilliant, eccentric, and sometimes foolish scientist, Hugh Sinclair shocked the scientific community in 1956 by sending a long letter to the British medical journal, the *Lancet*, proclaiming the importance of essential fatty acids. He warned that diets deficient in these nutrients result in degenerative illnesses such as coronary heart disease, certain cancers, skin conditions, and degenerative diseases of the central nervous system. At that time Sinclair's views were derided by many. Today, however, scientific evidence strongly supports his claim.

Emphasis on Hugh Sinclair's interest in essential fatty acids overshadows another of his contributions: the belief that human nutrition is of great importance and must be studied as a subject in its own right, rather than as part of biochemistry or medicine. Good nutrition is the foundation for human well-being and long life. This is the story of a brave man who — despite professional doubt and ostracism (something of his own making) — never weakened in his commitment to championing this single concept.

Blessed with material wealth and prodigious intellectual resources, Sinclair's story should be one of great success, but it is not. Born in 1910, he was a sickly child living under the pressure of unrealizable parental expectations, and grew into an adult unable to trust or work with but few of his own generation. Rejecting careers in the military and clergy, he chose science as his subject. During a lifetime of eighty years, he struggled to convince the world that nutrition holds the answers to human well-being. Unfortunately, although capable of quickly recognizing the important relationships between diet and disease, he often had little will and fewer resources to pursue their truth. Making matters worse, his arrogance and lack of attention to detail offended many people, and his life eventually became one of professional ostracism and loneliness.

This is Hugh Sinclair's story as this author found it. For me, it began among the thousands of personal letters and files stored at Lady Place, the Sinclair home in Sutton Courtenay, not far from the University of Oxford. I had been given a grant by Dr David Horrobin, founder of Scotia Pharmaceuticals, to write Hugh Sinclair's biography. I could not have guessed at that time that this work would take more than four years, nor did I anticipate that Sinclair's story would come to dominate my life like a tenacious ghost.

The drive from the city of Oxford on the A34 towards the village of Sutton Courtenay passes through a patchwork of agricultural land interrupted by the square ugliness of industrial estates spawned from the scientific hustle of local academics. This was the road my husband and I first travelled to Lady Place: it was a journey travelled many times by Sinclair, driving in his Aston Martin, or an old Bristol, or his mother's ageing Daimler. Sutton Courtenay is an English village like many others, with a perimeter of quick-built houses protecting an inner core of chocolate-box gentility. At its heart are a church and cemetery, a village green, some pleasant pubs, and a collection of period houses and barns. Not far away, on that day we first visited, Lady Place stood in quiet decay on acres of fertile land. We turned off the High Street between rusting gates hung heavy with trailing weeds from the previous summer onto a broken drive leading towards the house. Ahead were two signs: LADY PLACE, CENTRE FOR CRAFTSMANSHIP AND MUSIC and DRIVE SLOWLY: PEACOCKS ON DRIVE.

Parking the car, we saw several of these regal white birds strutting along a path by a walled garden partly overgrown with ivy and tumbles of dead nettles. In a window at the front of the house we could see a young woman playing a violin: a sullen looking child sat nearby. This must be the music school, we agreed. A car drove in and parked next to ours. This was my first meeting with Dr Brian B. Lloyd CBE, an old friend of Sinclair, who was to show us around those parts of Lady Place that he thought would be of help in writing the biography. Brian had worked with Sinclair on a series of nutrition surveys during World War II. Their friendship continued, and after Sinclair's death, in 1990, Brian served as Director of the International Nutrition Foundation, the charity Sinclair established during the last years of his life. Although now retired from his position as Director of Oxford Polytechnic (which became Oxford Brookes University), Brian continued work on the survey data he helped collect fifty years earlier.

We were led through the property. Built in 1898 as a private home, Lady Place became the Grange Hotel in 1924. Five years later it was reconverted into a private residence, and eventually sold to Rosalie Sinclair, Hugh's mother, in 1933. Although she had the greatest influence, his sense of humour and enjoyment in the bizarre were evident everywhere. The gardens were studded with gnomes, obelisks, and gaping frogs. At one time his collection of the absurd included a pair of fully clothed human skeletons sitting at a table in front of a window, deeply engrossed in a game of cards. Most of the rooms were overflowing with files and boxes of papers scattered among heaps of rejected jumble. Cheap pottery mugs and bits of wooden-ware shared table space with porcelain saucers orphaned after the sale of Sinclair's belongings. We walked through the massive kitchen area. In a cavernous room lined with shelves I wondered if an alcove was once where Sinclair housed the freezer containing the seal carcasses he ate for 100 days during a life-threatening experiment conducted during the 1970s.

Upstairs we paused in a room Sinclair used as an office during his last years, and I surveyed the scene from the discomfort of his tattered leather chair. The place had been repainted, but windowpanes were obviously held in place by rotting wood. Two large identical prints of Pavlov were propped against a wall. A large kite in the form of a fish stretched over a table. Looking into the hall beyond, I could see that considerable damage to walls, carpets, books, and files had been done by water flooding through some tear or break in the fabric of the building. We left the house and tramped down a curved path partially obscured by layers of rotting leaves. To one side was a low outbuilding. Through a dusty window we could see a man sanding a plank of wood: outside, discarded under a bush was a laboratory bench with its sink and storage cabinets still intact. That bench was a sad reminder of Sinclair's dream to build an international laboratory for research on human nutrition on this site.

Further down the path we came to a single-storey breeze-block building, approximately twenty-five feet wide and seventy feet long. Grey, clumsy and unpainted, it would have been more at home on an army base than on a scientist's fashionable estate. This building was to have been the heart of Sinclair's great plan: a library housing the books, journals, and data needed for nutrition research. Brian Lloyd turned a key in the Yale lock securing the place, and we entered.

Inside was an academic pack-rat's dream. Spread over tables and standing on dozens of racks of shelves was a vast collection of books, scientific periodicals, thousands of carefully wrapped and catalogued reprints, photographs, bits and pieces of materials used as teaching aids (including a box of deformed bones), and massive cases of drawers filled with approximately a quarter of a million punched cards holding data from Sinclair's nutrition surveys. Even the dusty rafters, laced with spiderwebs, held boxed reports and data. Just as we were to leave, Brian handed me two little grey volumes entitled *Malnutrition and starvation in the Western Netherlands — September 1944–July 1945*. 'Here,' he said, 'you will probably find these interesting.' As the reader of the biography will see, I did.

Several months later, we visited Lady Place again. This time we were met by Professor Christine Williams, newly appointed as the first Head of the Hugh Sinclair Unit of Human Nutrition at the University of Reading, Dr Andrew Neil, University Lecturer in Clinical Epidemiology at the Institute of Health Sciences, University of Oxford, and several graduate students. One student was Rachael Huxley, who was anxious to begin research using Sinclair's World War II nutrition database. Results of her work would first appear in 2000, in a paper published in collaboration with Brian Lloyd, M. Goldacre, and Andrew Neil, entitled 'Nutrition research in World War 2: The Oxford Nutrition Survey and its research potential 50 years later. [The paper was published in the *British Journal of Nutrition* (2000) **84**, 247–51.] The long delayed story of Sinclair's war years was beginning to unfold.

While with Christine Williams, we examined Sinclair's hoard of documents including diaries, notebooks, lecture notes, professional correspondence, research proposals, and reports and personal letters. He had even kept letters received from his parents and nanny when he was a child. I was surprised to see that files and papers had been moved about since my first visit. From hand-written labels on archive boxes it appeared that at least two attempts had been made to organize Hugh's papers; one had been very orderly, the second haphazard, leaving some papers in open-topped cardboard wine and spirit cartons rather than in proper archive boxes. In what was once a music room there were stacks of architectural drawings for Magdalen College, Oxford, and Lady Place. Letters, slides, paper tubes containing academic certificates and notification of various awards were scattered about. Alone on a table lay a file marked 'Defamatory Statements'. Shelves stretching from floor to ceiling contained some of his favourite books, including a volume by William Farr on epidemiology, a copy of *Camp and society*, an autobiography by his father, and *Confessions of an optimist*, by his friend Woodrow Wyatt. On the side of the room opposite this collection stood an accumulation of supermarket pasteboard boxes filled with memories of his maternal heritage. Rosalie Jackson Sinclair was from a wealthy family, who were the source of the affluence Hugh enjoyed during his life. He had enjoyed suggesting that his money was inherited from the Sinclair side of his nature: from some Scottish laird, perhaps. As I read the files I could see this could not be true. The money was from the Jacksons. Why had he misled people? This, and many other discrepancies between what Hugh had wanted others to believe and what his files showed to be true, had a deep effect on how I decided to write this life story. I came to rely heavily on written words. Eventually I would interview more than twenty people who knew Hugh Sinclair well. I used information from this work, but because I heard conflicting versions of the same stories, and because it was obvious that some of what was reported was, in fact, a repeat of what Sinclair had told others about himself, I chose to be selective in what I would use in this biography.

Looking around Lady Place, I felt the woebegone collection of papers and memorabilia left by an eccentric old man was interesting, but was he truly a figure so important as to warrant a biography? I asked Christine Williams. A neatly packaged woman with a determined stance and expression, she gave me a withering glance. 'Hugh Sinclair truly understood nutrition', she said. 'For him, nutrition was a whole; it was a legitimate subject for scientific investigation in its own right. Not many people can see that.' I needed no better answer.

Nutrition remains a Cinderella subject in most academic circles. It is hard to know why this is the case, because its complexity requires creative minds to see the problems and ask the relevant questions. To be a good nutritionist you need to know something of biochemistry, human biology, sociology, clinical pathology, anthropology, and even botany. It also helps to have some insight into the commercial aspects of food and agriculture, and understand the basics of food technology. The scope of what one needs to understand is immense. The rapid outpouring and scientific journals filled with new research findings challenges even the most avid student to keep pace.

With all this new research unfolding, it may be surprising that modern scientists want access to the nutritional data Hugh Sinclair accumulated during the 1940s and stored away in the breeze-block building at Lady Place. At the time the data were collected there was no means of complex analysis. Modern computers, however, make it possible to look at these data in new ways and recognize significant facts about the importance of diet. Andrew Neil explained it this way:

'What Sinclair was trying to do and our interest in the risk factors for chronic disease are coming together in a most important way. Over the last decade Professor David D. J. P. Barker, of Southampton University, has developed a hypothesis suggesting that the maternal diet during pregnancy influences her child's risk of developing cardiovascular disease. Put crudely, the hypothesis suggests that intrauterine malnutrition may programme the foetal development in such a way that those babies with the lowest birth weight also have the highest levels of cardiovascular morbidity and mortality.

'We are looking at pregnant women included in Sinclair's Oxford Nutrition Survey who were treated in Oxford and attended the maternity unit at the Radcliffe Infirmary; it is probably about 400 women. We hope to be able to combine his original nutritional data and the birth weights of their children with subsequent levels of risk factors for cardiovascular disease.

'The thing I find most disturbing at Lady Place is that one man devotes his whole life to collecting this vast quantity of information and is unable to do what he presumably wants to do with it. I wonder if he ever felt that devoting such enormous energy to this was (dare one say it) misdirected energy.'

My work as Hugh Sinclair's biographer is finished now, and so his ghost no longer sits beside my desk and prods me on. Having been given the privilege of closely examining his life, I do not believe Hugh Sinclair ever questioned the value of what he did, although I think he grieved that others saw his scientific career as a failure. I believe he would have found the story you are about to read fair, and as accurate as one can be when faced with the complexity of the life he constructed and the scientific ideas he championed.

Acknowledgements

Many people gave me help while I wrote this book, and I am grateful. First and foremost, I thank Dr David Horrobin, without whom this work would not exist. This was a long project, and without a generous grant from David it could not have been undertaken. He gave hours of his time for interviews, and read the various drafts as they were produced. Best of all, he encouraged me each step of the way. One cannot ask for better support.

I owe warm thanks to Dr Brian Lloyd CBE, who also gave me hours of his time for interviews and allowed me to read the diaries Hugh Sinclair kept during and immediately after World War II. He also kindly reviewed the manuscript prior to publication.

At the University of Oxford, I am indebted to Magdalen College President Antony Smith CBE for his time and for valuable guidance as to source materials I might use. Dr Andrew Neil gave his time for interviews, and also provided a spark of enthusiasm for the work that helped tide me over when I was losing faith in my ability to complete it. Dr John Ledingham provided both information and encouragement. I also thank Messrs Terrance Newport and Michael Strutt for insights into Sinclair's life at Magdalen College. Time spent at Oxford left many good memories.

At the University of Reading, the recipient institution of all Hugh Sinclair's papers, and the location of the Hugh Sinclair Unit in Human Nutrition, I am most grateful to Professor Christine Williams for her interest and help in this project, and to Professors Ann F. Williams, Raymond Dils, and Michael Gurr for their interviews and for various papers they contributed to my mass of information on their late colleague. I thank Maria Itta for her willing assistance whenever I asked for help.

Books do not happen without publishers, and at the Oxford University Press I wish to thank Susan Harrison and Richard Lawrence for believing in this book. It was they who introduced me to Eric Buckley, whose skill as an editor greatly improved this work.

Professor [emeritus] D. Mark Hegsted, Ian Whitelaw, Marguerite Patten OBE, Dr Andrew Neil, Professor Christine Williams, Professor Ralph Walker, and John Fields all read this book in preliminary form and offered suggestions that strengthened the story and helped eliminate many of my careless errors. Remaining errors are my responsibility.

I am grateful to: Dr Agnes Mayer Huber; Colonel Charles O'Leary, of the Worshipful Society of the Apothecaries; Colin Maynard, of the Marylebone Cricket Club; and the eminent gastroenterologist, the late Sir Francis Avery Jones, for allowing me to interview them. They each also provided documents that informed Hugh Sinclair's story. I also thank members of Hugh Sinclair's family for their help and interest.

Valuable information was provided by the libraries of the Ministry of Agriculture Fisheries and Food, the Wellcome Foundation, and the Royal Society of Medicine. Of great importance, this book lacked veracity without valuable documents provided by the Rockefeller Archive Center in Tarrytown, New York.

Finally, I thank my husband, Richard, for being there when I needed him.

Contents

CHAPTER 1

The boy who bowed to the King

The memorial service for Hugh Macdonald Sinclair was held at the University of Oxford in the Magdalen College Chapel on 13 October 1990. As sounds of the choir filtered from the stone building along the gardens and paths of the College, aproned butlers and kitchen staff rushed heavy trays from the kitchen pantries and cellars across the ancient yard to the President's residential quarters. The day was warm for the time of year. The staff gossiped as they worked, recalling evenings when Sinclair generously entertained them at dinner in the College; with good food, fine wine, port, and cigars. Always the best from Hugh Sinclair. He was rarely demanding, unlike many Fellows, and he frequently took time to speak with the chefs, butlers, and cleaners alike. He had been a 'gent'. Tall, softly spoken, balding, and bespectacled. Perhaps shy, some said, but always charming.

Staff chatter recounted the confused legacy of truths and half-truths that surround Sinclair's name: 'He was rich, wasn't he?' 'No, I hear he was broke.' 'I heard he had a great row with the University and they kicked him out.' 'No. He left to start that Institute of his.' 'He was the one who saved this country from starvation during World War II by telling everybody to eat whale blubber.' 'I don't know about whale, but I know he ate seal meat while he was at the College.' 'Seal meat? I remember him eating caviar and shrimp when he dined in.'

Following the service Mr Antony Smith, President of Madgalen College, greeted mourners arriving at his door before they were ushered towards tables carefully laid with linen and silver. His wake was Sinclair's final party for his friends, and he had left detailed instructions for everything. Along with his family, his students, and the many colleagues who attended to celebrate his memory, all the College staff were invited. Everyone drank champagne.

A salute to the man!
A salute to the boy who became the man!

On 4 February 1910, after a difficult pregnancy and long labour, and without help from midwife or doctor, Rosalie Sybil Sinclair delivered her third child: a son, weighing ten pounds and seven ounces, who was named Hugh Macdonald Sinclair. The boy's father never let him forget that he owed his life to his mother's efforts. From this and other paternal instructions, the child would structure his future.

In the early 1900s pride in Empire and the wake of the enthusiasm that followed the Industrial Revolution still influenced fashion and attitudes. Hugh's parents were ideally suited to the time and to each other. Just after the turn of the century a chance meeting in the seaport town of Plymouth, England, had joined Rosalie Sybil Jackson, the intelligent and regal daughter of Sir John Jackson, and Colonel Hugh Montgomerie Sinclair, a tall and commanding older man. She had wealth. He had limited financial resources, but brought to the partnership a colourful military career and a name linked with aristocracy. Sinclair had served in Cyprus, India, and Russia, and had taken part in the expedition to subjugate the kingdom of Ashanti in West Africa. The fulsomely moustached Colonel traced his lineage from the Viking Monarch of Finland, Woldonius, and from the St Clair cousins of William the Conqueror. Closer to his own time, his family included the Right Honourable Sir John Sinclair, of Thurso Castle, Caithness, MP, founder and first President of the British Board of Agriculture. Sir John's second wife, the Honourable Diana Macdonald, was the only daughter of Alexander, Lord Macdonald (of the Isles). From this line, young Hugh received his second name and his strongest link with the Scottish aristocracy.

Rosalie's father, Sir John Jackson CVO, LLD, MP, received his knighthood and wealth through hard work, intelligence, and commercial success as a civil engineer. The Singapore naval base, and docks at Dover, Portsmouth, Swansea, Liverpool, Victoria (British Colombia), and Simonstown (South Africa) were among his most important commissions. Completion of the Manchester Ship Canal earned him his knighthood, but he is best remembered for building the piers of London's Tower Bridge. Money and contacts flowing from these efforts benefited his family far beyond his own years.

Those who knew Rosalie and the Colonel said it was she who looked the part of the aristocrat. She was straight-backed and imperious, while Sinclair was well seasoned. This unlikely couple might never have met but for an unfortunate incident involving the Colonel's spurs. The Colonel recounted the story in an autobiography he wrote for his family, calling it *Camp and society*. He had returned to England in 1902, after years of military duty in the Middle East and India, and was posted to Plymouth as Assistant Quartermaster General under the Command of Sir William Butler, whom he greatly admired. Soon after Sinclair's arrival, Sir William commanded that spurs be worn 'when necessary'. A few weeks later, Sir William and his wife hosted a grand ball. It never occurred to Sinclair that a formal ball might be a time to set aside field commands, and he arrived at Sir William's wearing his best uniform — and riding spurs. His error was obvious at once, but what could be done, other than join the party. Partnered with a charming young woman, Sinclair's blunder was soon ignored. Still adjusting to gentility after years of colonial military life, the Colonel's dancing was a bit stiff. Suffering a miss-step, he caught the jagged point of a spur in the hem of an elaborate gown worn by a woman swirling past with her partner nearby. Unnerved by the sound of tearing satin, the Colonel crossly complained to the girl on his arm that such long dresses should not be permitted at a ball. She giggled with delight, and before the next dance began, introduced the middle-aged Colonel to the wearer of the offending gown — her younger sister, Rosalie Jackson. By now this victim of the Colonel's disgrace had him well in her debt, and she said so. Soon laughter mixed with serious conversation, and a close friendship was almost instantaneous. Colonel Sinclair was 47 and she was half his age.

Rosalie plotted with her sister to find reasons the Colonel should be entertained in their home. Rosalie's father, only four years older than Sinclair, enjoyed his company from the start, because the Colonel was a sportsman, an inexhaustible raconteur, and (like Sir John) interested in engineering and railways. The two men shared common views on war, politics, and Empire. Perhaps helped by the fact that Sir John also approved of Sinclair's genealogy, he found him to be a good match for his daughter. Only a few weeks after they first met, the engagement of the couple was announced, and Rosalie's parents agreed they could wed at once.

Wedding plans were delayed, however, when reorganization within the Army took precedence. Plymouth ceased to be a Command and reverted to a coastal defence centre. Hugh was 'pitchforked upstairs', and posted as Assistant Quartermaster General to the Command Headquarters at Salisbury. He was given a few weeks grace before taking up his duties. Everyone agreed he and Rosalie should not marry until his new posting began. To fill the interval, Sir John invited Sinclair to join his family, servants, and staff, on a trip to the Continent, where he was to advise the Austrian Government on the naval docks at Pola. The days that followed sealed the friendship between the two men, and strengthened Sir John's resolve that Rosalie and her husband would want for nothing.

Sir John's party travelled to Paris, on to Vienna, and then to Venice, where Sir John inspected the docks while the Colonel enjoyed showing his young fiancée the architectural and artistic treasures he already knew. Soon after their return to England, early in 1904, the couple married in St Mary Abbots, Kensington [built by Sinclair's uncle] and moved to Salisbury where he took up his new posting. That posting was brief, as Sinclair was soon appointed Chief Engineer in Edinburgh. Largely through Sir John's influence, and by the judicious recounting of his own bloodline, the Colonel was accepted without hesitation by Edinburgh society. As for his residence, he rented Duddingston House from the Duke of Abercorn: a fine building set in a vast park that included an 18-hole golf course. Having no money himself, however, the payment of the rent fell to Rosalie's father, who was happy to oblige. The Colonel and his family remained at Duddingston House until the end of his Army career, enjoying the full fruits of his aristocratic ancestry and the excellent contacts of his wealthy father-in-law.

The Colonel's early family life had been modest. His father, William, went first to Cheam and then Winchester, where he saw the Church as his future career. However, William's father received an offer for his son to join the Indian Cavalry, and at the age of sixteen, the boy soon found himself assigned to the Madras Calvary, where he served for ten years. After his military career he returned to England and took half-pay for the rest of his life. This income supported his studies at St Mary's Hall, Oxford, after which he was ordained in the Church of England. Eventually, while in his early fifties, he took up a vacancy in the quiet English village of Pulborough, Sussex. Hugh's father's childhood in Pulborough held happy memories, but they were simple times, marked by neither wealth nor sophistication.

The Colonel and Rosalie's first child was born in 1905, and was wisely named 'John', after his maternal grandfather. Three years later, Rosalie

Helen (later known as Salie) was born. In 1910, the Colonel was presented with a second son, and to this child he gave his own name: Hugh.
Young Hugh's cousin, Sir Archibald Sinclair (later Viscount Thurso),
leader of the Liberal Party, was one of his god-fathers. Rosalie kept a written account of her young son's physical and social development: 'May 14,
1910, the baby's nails were cut; June 21, the baby crawled.' On 4 March
1912, The Duke and Duchess of Abercorn visited Duddingston House,
and Rosalie recorded that the Duke had danced with the children. He
showed interest in Hugh, and said he was the first baby to have been born
there. Rosalie also recorded glimpses of family life, such as 'Nurse Porter
left and Miss Ellis came as nursery governess'; 'Garden party'; and 'On
the yacht'. Decades later Hugh reconstructed the details of his childhood
from these accounts, paying special attention to his own nutrition:
'Breast-fed 6 months and 8 days', he recounted, 'December 11, egg in
milk; December 14, chicken broth'.

One short item recorded by Rosalie, in August 1912, held far-reaching
significance: 'Went to Barming House, our new home.' The Colonel had
retired from the Army, and Sir John placed Barming House, a fine estate
in Maidstone, Kent, in trust for Rosalie and her family. Its gardens, tennis
courts, stables, and servants' quarters were large enough to provide the
comforts Sir John felt his daughter and her family required. The Colonel,
now fifty-seven, settled down to the business of organizing and enjoying
family life. One wonders what play and family life were like for the
Sinclair children. The Colonel's own father was approximately half a
century old when he was born, so neither the Colonel Sinclair nor his
sons enjoyed the company or example of a father still physically strong
and finding his way in his own profession.

Of interest in this connection is the following extract from a letter sent
to Rosalie's mother by Lady Freemantle after a visit to Barming House.

14 September 1910
Dear Lady Jackson, Many thanks for your letter which I received while I was
with Rosalie. It was so nice being with her [and] Col. Sinclair, & seeing their
happy Ménage. The dear children are delightful and so well brought up.
They are such good parents & I think it quite touching to see that big man
on the floor for such a length of time playing with the children, so patient &
good to them, quite their … I have seldom seen so devoted a father — & so
unlike what I would have imagined he would be. The children are so good.

John has such nice manners, you would have liked to have seen him in his kilt yesterday but he looked too old quite like a boy of 6. Baba is such a sweet little girlie, but I think she has the strongest will of the three. The dear baby [HMS] was so good — but I felt his teeth were troubling him. I think you have cause to be proud of you granchildren … Rosalie is so happy too.

Three years after Hugh's birth, Catherine Julia Sinclair was born and the family was complete. Baby Catherine was Hugh's joy: here was a person smaller than he, someone to be shepherded and influenced. In a personal note penned late in his life he wrote: 'I can still remember being taken to see her and thinking how red she was.' Of all the members of the family, she would be his favourite. That same year, 1913, his elder brother, eight-year-old John Montgomerie Sinclair, left home to live at Stone House, a boarding school at Broadstairs, Kent, and Salie was soon sent to board at Brampton Down, in Folkestone. Their departures left the Colonel's younger son happy at home with his nursery governess and baby sister. But things would change. Following a visit by his parents to Stone House, Hugh received a postcard informing him that he, too, would one day leave Barming House to study there.

The Sinclairs planned their children's futures with precision. Good schools were essential and future memberships in the right clubs were a necessity. Among other efforts on her childrens' behalf, Rosalie Sinclair enlisted the help of her father and arranged that Hugh would become a lifetime member of the world famous Marylebone Cricket Club. The boy was three at the time. The Club was in need of funds to build a new stand, and lifetime membership required not only the signature of two prominent sponsors but a payment of £100 — a small fortune in 1913. Rosalie saw that the necessary sum was paid, and that the forms were completed and signed by friends of her father. When Hugh was older, all he would need to do was pay a small additional fee.

On the 28 June 1914, a Serbian nationalist assassinated Archduke Francis Ferdinand, the heir to the Austro-Hungarian throne, in Sarajevo, Bosnia. One month later, with Germany as her ally, Austria declared war on Serbia. Russia soon mobilized her troops and Germany declared war on both France and Russia. It was only a matter of time before Britian and the British Empire would enter the war on the side of Russia and France as part of the Triple Entente. During the last days of July, the Sinclair family travelled to Cowes, Isle of Wight, off the southern coast of England, to join Sir John and his party on his yacht, *Gelert*. As days passed the party became increasingly tense over the impending war. At last, on 4 August, a Bank Holiday, the waiting was over when Sir Edward Grey announced that Great Britain and all her Dominions were now at war. The conflict was to cost more than 10 million lives, leave twice that number of wounded, and eventually destroy Sir John Jackson's business enterprises and personal fortune.

Response to the declaration of war was predictable. Being a holiday, the shops were shut, but the wealthy found ways to fill their larders, buying up all the flour and other staple foods they could manage. The less well off endured the temptation to hoard until the next day. The country was a-buzz with rumours that every able-bodied man would be needed. The Colonel felt he had a duty, despite his young family, and immediately offered his services to the War Office. He waited for a response. Britain's opening volleys of the war were fired, but he still had no reply. The Colonel despaired; certainly they would need him despite the fact that he was almost sixty years of age. Filled with patriotic zeal and desperate to be part of the war, he began recruiting younger men to fight, speaking wherever he was invited. Eventually he was asked to address a mass meeting in Cheltenham. While there, he visited his brother John, Archdeacon of Cirencester. The Colonel informed the Archdeacon that he must set an example and instruct his son, a student at Oxford, to join up. Colonel Sinclair later wrote in his autobiography: 'The boy was reading for the Church, so it was a particularly terrible and bitter decision to have to make, but he gallantly rose to the crisis, and his son made good with eventually a double MC [Military Cross] and a wound in the last days of the war.' Sinclair then recounted how his five nephews all served through the war, and 'came through with three MCs, two wounds, and much credit. Few families,' he wrote, 'were so fortunate.' To be sure, he wanted a soldier-son.

At last the long awaited telegram from the War Office arrived, and the Colonel moved to Longmoor Camp, near Petersfield, Hampshire, England, where he was to instruct and drill four companies of railwaymen. Five years would pass before he again lived in Kent. Rosalie visited him when possible, bringing the baby and sometimes Hugh or one of the older children with her. But overall, letters were frequent substitutes for paternal affection. Hugh was only four years old when his first letter arrived from his father. Like the hundreds of other written messages from family and nannies that were to follow, it was carefully folded, returned to its envelope, and stored it away to be read again. These personal documents eventually became part of Hugh Sinclair's archive, and along with his childhood diaries describe a time of ponies, family plays, and music lessons which the Sinclair children enjoyed despite the war. They also reveal a lonely child suffering under the shadow of a bright and strong older brother who delighted his parents with success at sport and enthusiasm for the military.

Physically limited by age, when Colonel Sinclair saw his children he was unable to join in the rough and tumble games many fathers enjoy with their sons. Instead, he entertained them with stories of his exploits in the Army; one of these proved to be of particular importance to young Hugh. In his early military career, while in Cyprus, Sinclair had been 'put up' at Captain Kitchener's house, where he became seriously ill after eating contaminated food. The exact circumstances of his recuperation and the further friendship between these men are not detailed in *Camp and society*, but there is evidence that repeated stories based on these events made a great impression on his young son. During the last year of Hugh Macdonald Sinclair's own life, he prepared an autobiographical piece that suggests something unpleasant took place between his father and Kitchener. Much of this unpublished piece was prepared with the help of a friend, and is therefore written in the third person. It states:

> Hugh's father entered the Army after passing in and out of Woolwich Academy first, and as a young officer in Cyprus was a friend of Kitchener and his boy-friend Lord John Kennedy … At the time of the South African war, when he was the youngest colonel in the Army, Kitchener asked him [Hugh's father] to do something he regarded as dishonourable and he refused. Friends close to the Commander-in-Chief, Lord Wolseley, told Sinclair to apologize for disobedience, as he would never enjoy further promotion. The Colonel refused, despite the fact he knew a lack of apology would mean he would be overlooked for a Knighthood.

According to *Camp and society*, Colonel Sinclair had another disagreeable encounter with Lord Kitchener in January 1915. Orders arrived that four companies of Sinclair's railwaymen were to join a mass review by the French Minister of War. Sinclair's men were to march ten miles along rough and muddy by-roads to join up with the others. Their greatcoats were to be rolled and carried on their backs, and they were to be in place by 11.00 hours. The response would be arduous. At best, it would be a march of at least four hours. Bitter cold and falling snow made the going so hard that Colonel Sinclair ordered his men to unroll their coats and wear them. The snow melted when it hit the rough wool of the coats, and as several streams were forded along the way, Sinclair's men were soaked through and weighed down by wet garments before reaching the appointed location. They were on time, and told they were not to break ranks for any reason. They sang to keep up their spirits, but cold and exhausted, their waiting was hard. They remained standing in the snow for another two hours. Colonel Sinclair tells the story:

> At long last a cortège of motor cars arrived. Lord Kitchener, M. Miller and and their staff got out and started walking up the road in front of us as we presented arms. They passed in earnest conversation and neither of them turned their heads in our direction! A chill ran through the ranks, colder than anything caused by the snow. To have come so far in such discomfort. Not even to be looked at or our salute acknowledged, depressed the most enthusiastic and made us feel exceedingly small.

Hugh's father proudly and repeatedly told his sons these stories about Kitchener, encouraging them to stand up for what they felt was right whatever the consequences. If he gave Hugh nothing else, the Colonel endowed his younger son with a suspicion of authority.

Before long a number of Canadian railwaymen were stationed at Longmoor Camp for Army training, and the King, appreciative of their contribution to the war, decided to honour them by inspection. The Colonel had mixed reactions to the announcement, worried that such a visit could create jealousy unless all his men were to see the King. He replied that, unless King George could review the entire contingent of men, he would prefer that he didn't come at all. Sinclair got his way and the King agreed to inspect the men during training exercises. On the appointed day, the Royal party arrived and the King walked with Sinclair among the railwaymen, asking questions and observing training drills.

Seeing his father was always a memorable event for Hugh. Perhaps the most memorable occurred when he was seven. In 1917 the United States of America entered the war and an American railway regiment was sent to train under Sinclair's command. Just as had been the case with the Canadians, the King again wished to visit the camp. Rosalie and the children travelled to Longmoor for the occasion, and to be presented to the Royal party. She drilled the children in deportment while the Colonel patiently explained the importance of the event. On the day, four-year-old Catherine remained behind. The older children 'acquitted themselves passably,' wrote Colonel Sinclair in *Camp and society*, 'but when my small boy, Hugh, was brought up, he nearly bowed to the earth, his cap trailing on the ground. "That's a very fine bow", said the King, with a hearty laugh in which we all joined.' This is the only time Colonel Sinclair mentions any of his children by name in his autobiography. Perhaps for the last time, young Hugh had managed to impress his father.

The war years were difficult years for the boy. He was frequently ill, sometimes confined to bed for weeks with flu and a high temperature, and had symptoms of anaemia. He brightened, however, when air raids brought John home from school. To his delight, Hugh was asked to help his brother make plans to expand the laboratory he was building in Barming House. But their schemes were disrupted when Rosalie decided to move closer to her husband; she wanted all the children near both parents in this time of danger. Rosalie and the children remained at Longmoor until the greatest threat to Britain had passed.

In the early months of 1919, Colonel Sinclair's brother, John, then Archdeacon of Middlesex and Vicar of Kensington, died. This loss, combined with the end of his usefulness to the Army, left a great chasm in the Colonel's life and he never fully recovered. In May, he and Rosalie sent young Hugh away from the comforts of Barming House to Stone House School. This was the first time the boy had been separated from both his parents and nanny for more than a few days. The Head there called Hugh 'plucky', but the blow of leaving home was terrible.

The nation rejoiced when the Treaty of Versailles was signed on 18 June 1919, but there were problems ahead for the Sinclairs. The war had disrupted most of Sir John's construction projects and his wealth was draining away to pay debts. Then, with little warning, on 15 December, Sir John died. The Register of Death lists 'Syncope [fainting, or collapse] following diabetes and fatty heart' as the cause of death. Family financial problems soon reached crisis point, and Colonel Sinclair consulted Sir John's solicitor about the estate. The news was dire. The Sinclair family's future was at risk and the Colonel knew they needed to reduce their spending at once: his talk of moving to a more modest home frightened young Hugh.

A month after his grandfather's death Hugh contracted an illness that kept him in bed for weeks. His personal diaries record temperatures of up to 104° F, along with periods of delirium. On Hugh's return to Stone House, his worried father wrote frequently about the importance of hard work. In these same letters, the Colonel wrote of his own illness and his need to rest from time to time. Troubled and alone, ten-year-old Hugh's physical symptoms increased, and he complained of headaches and bad dreams. Unlike his brother John, who enjoyed preparatory school, Hugh loathed his days at Stone House. His masters reported that he could not read easily, his spelling was weak, he was poor at numbers and his written work was often spoilt by untidiness, although 'he learned easily by heart'. (In more modern times, one might have asked whether or not young Hugh suffered from dyslexia.) The Headmaster wrote to Hugh's parents, suggesting Hugh spend time during the summer holidays doing English, as he would require a 'push' to get him into a top college. In response, the ailing Colonel Sinclair tutored his younger son, and the boy rose to the challenge. Six months after this closer relationship, young Hugh was first in his form and received a school prize.

As his studies progressed at home, so did his loathing of Stone House School. In a private memoir written before his death, Hugh claimed that preparatory school 'taught him only to dislike masters, the headmaster being a sadist who was later imprisoned.' Whether this statement about imprisonment is true or not is unclear, but Hugh's diaries report beatings for 'technical matters', with blood drawn, beatings for untidy papers and beatings with a strap for ragging. Hugh knew that John was to become the soldier-son, and a feeling of competition took root. Not long after Hugh arrived at Stone House, John left the school to attend Winchester College, where he was enthusiastic about his studies. Being older and stronger, he did what he could to inspire Hugh's scholastic interests. 'I have begun chemistry here,' he wrote in a letter in the autumn of 1922, 'and am doing all sorts of thrilling things. I have written to suggest to Mummy that we should have some gas laid on and do some experiments on a proper scale, not like [sic] we have been doing. We could do some much more thrilling stunts, and, which is really more important, learn a little.'

The Colonel and his wife agreed that it was time they found the best possible school for young Hugh. Marlborough College was a fall-back option, as was Eton; but Colonel Sinclair wanted Hugh to attend Winchester College. The Colonel wrote to the school, asking that his son be admitted. It would be possible, he was told, if the average number of boys left and Hugh passed the required examinations. In April 1921 Winchester College conditionally offered Hugh a place during 1923.

Sport was important in the Sinclair family and Hugh was devastated when a badly damaged knee threatened to limit his participation. He had suffered a slipped cartilage while at school, and Colonel Sinclair wrote to the Headmaster complaining of inappropriate treatment following the injury. Possibly due to misdiagnosis of the injury, Hugh had been encouraged to play football at school, wearing only an elastic bandage for protection. The Sinclairs' doctor thought this very unwise and advised rest instead. As time passed it was obvious that, however the knee was cared for, the injury would prove to have permanent and painful consequences.

In February 1923, just before young Hugh's 13th birthday, he received a letter from his father:

> You are actually 13 on Sunday, getting quite big … and this is a most important year in your life. I am sure you will remember that Mummy and I are spending a great deal of money on giving you every advantage in training your mind and body so that you may make a successful career in life, and that it is for you to make the best use you can of these advantages.

That September Hugh entered Winchester College, where his attitude towards education was changed forever. There he found a personal freedom he had not enjoyed before and, to his delight, helpful masters. His work was uneven at first and he was prone to mischief, but after a time he took his studies seriously, and began excelling in science. While at Winchester, he developed a scathing wit and a taste for pranks that remained permanent parts of his character. Writing in *Sinclair*, a memoir published by the McCarrison Society in 1990, Hugh's life-long friend Maurice Partridge revealed something of Hugh's school-days:

> He deviated from its (Winchester's) traditions by combining a passion for experimental physics with an interest in the psychology of the authorities. This led to his first large-scale scientific experiment. A detailed mastery and adaptation of the wiring system enabled any one of the classrooms to be plunged into darkness at will.

Another letter from the Colonel marked his younger son's 14th birthday:

> I am thinking of the anxious time we had at Duddingston 14 years ago when

you made your appearance so suddenly and neither nurse nor doctor were at hand. Only your mother's wonderful nerve and self-control saved your life at that critical moment! Nothing you can do for her can make up for her heroism at that moment for your sake.' [The letter continues,] 'You'll be sorry to hear that I've had some trouble with my left eye. A cloud came over it some times for a few minutes, and I didn't see quite clearly with it. So I went to P. and he says I must have an operation to put it right.'

The Colonel wrote that he would be in hospital a fortnight before returning home.

Four days later Colonel Sinclair entered hospital. The following morning he wrote to Hugh and, along with a description of his hospital room and plans for his recuperation, he hinted of financial worries: '… we shall have no money to spare this year and must be careful … We've sold the cows — Joey and the cart — and the old car!'

Surgery on the Colonel's eye took place several days later, and he was soon sitting up and trying his hand at jigsaw puzzles. When well enough he returned home to Kent and the management of the family's financial affairs. Then, on 7 July 1924, a message to Hugh from the Colonel contained unwelcome news:

> My dearest Hughie, you will be interested — if not pleased — to hear that I have today signed the agreement to buy 'Gransden', Wateringbury. So in a week or two it will be ours and in the autumn after the holidays we hope to get it done up in order to move in — I explained to John why we have done this. It is really in the interest of our children to economize now. That we may have enough to give you all the advantages we want for you as you grow up — and as none of you will live here after our deaths. It's necessary, and its no good crying over spilt milk. Be thankful you've had so many happy years here … I had [Doctor] Tracers to see me on Sunday. I've had a lot of pain in my chest and get it when I walk. He wants me to see a specialist in London … your most affectionate father. Hugh M Sinclair.

The Colonel knew Hugh intensely disliked Gransden. It had no park, no tennis courts, no stabling, and no golf course, and, to his mind, only adequate rooms for the family and a few staff.

On 10 July 1924 Rosalie Sinclair sent a telegram to the Head of Winchester College, asking him to inform her sons that their father had died that morning. Hugh's diary for that day simply states, 'Father died

3.00'. With the exception of his servant, the Colonel died alone. Rosalie, in London with her mother, never forgave herself for being absent that night. As soon as possible after returning to Kent she wrote to each of her children. Several days later, standing at his father's grave and holding his young Catherine's hand, Hugh remembered his mother's letter to him. 'It was a dreadful blow and Daddy was the best of fathers and husbands — he was so proud of all of you and loved you so much … I do wish I had been here with him, I shall always regret that — he was always so good to me.' After the funeral, Rosalie wrote again, saying Hugh must try to be like his father. A dutiful son, Hugh would try to obey her wishes.

At the age of fourteen, Hugh had lost forever a parent who had been absent most of his life. Only since the end of World War I had the Colonel found time to give his younger son the attention he craved. Now it was all at an end; even the family home Hugh loved so much was gone, and the boy despaired. What was there to cherish, other than memories and promises made?

A seemingly unrelated event took place in the year the Colonel died that would deeply affect young Hugh's later life. Trustees of the Rockefeller Foundation, in the United States, offered the University of Oxford the sum of £75 000 to help establish a Laboratory for a Department of Biochemistry. It would be years before Hugh would hear about this gift, and even then he would not guess that this laboratory would be the place in which he would engage his most fierce adversaries, including the loneliness and self-doubt that took root during these early years.

CHAPTER 2

Manners makyth man

The Colonel's death cast a pall over Hugh's life. Some of his ghosts were laid to rest when, shortly after breakfast on an Out Day in March 1925, he borrowed a bicycle too small for his long body and headed away from School towards Longmoor Camp. He cycled hard over unpaved roads, crossing two fords and one flooded lane. Wet and uncomfortable, with his knees rising above the handlebars and bumping as he peddled, he lost his bearings repeatedly, making the journey longer than planned. At last a familiar house came into view. The outlines of the training camp where his father had drilled and readied the railwaymen for war began to unfold. Hugh had returned to the place where the King once walked with the Colonel among his men.

There was one family friend he particularly wanted to visit, a Miss Jackson, and he found his way to her house. Away for the morning, she was expected back after lunch, leaving time for Hugh time to continue his inspection of what remained of Longmoor. On reaching a certain landmark, called Apple-pie Hill, he found a shop and stopped for a drink and a rest. Asked by the shopkeeper if he knew the area, Hugh said he was the son of the Commandant there during World War I. The woman warmly recalled the Colonel, saying many living nearby missed him. She gave Hugh something he had come for: warm words from a stranger about his father.

After a quick lunch of sandwiches packed at school, Hugh explored further. Houses now covered ground once occupied by barracks and the railway yard. At the Commandant's house, fences were ramshackled, the kitchen garden was overgrown, and the pigsty was badly in need of repair. Seeing that the place of order and soldiering long held in his memories was gone, he returned to Miss Jackson's to share gossip over tea for what remained of the afternoon. Hours later, exhausted by travelling and remembering, Hugh cycled to the railway station and returned to Winchester. Before falling into bed, and without dinner, he reported everything in a letter to his mother. It was well past 10.00 p.m. when he finally wrote, 'Best love, Hughie.' Two days later he took to bed with flu and remained there for almost a week.

Rosalie Sinclair wrote to her children more frequently now, reporting the antics of family pets and various relatives. On black-bordered notepaper she expressed her concerns for their health and performance at school. It was all kindness. Unlike his father, Rosalie never worried young Hugh with monetary matters, although her financial situation was genuinely uncomfortable, and she was finding it necessary to sell jewellery and family heirlooms to pay for tuition. Looking for other sources of income, Rosalie decided to publish the Colonel's autobiography, *Camp and society*. The manuscript was a highly personal account written for the family, but, pressed for cash, she edited the stories of Empire and war for a broader audience. Then, as now, publishers were more sought after than seeking, and Rosalie failed in her attempts to find one interested in the Colonel's story. Still convinced there was money to be made, she paid to have the manuscript published and shamelessly wrote to hundreds of friends and contacts to personally announce the book's availability. Both the Queen Mother and the Director of the Imperial War Museum were asked if they wished to receive a copy. Free copies would, of course, be appreciated, she was told.

A copy of the red bound volume found a permanent place on Hugh's bookshelf after his birthday in 1926. By this time he ranked second out of 17 students in his division at Winchester College. Although some tutors complained that young Sinclair could be more 'concise' when expressing himself, his considerable intelligence and insight impressed others. An essay book from this time contains work on eugenics that suggests the boy's maturing ability and mindset. In part, Hugh wrote:[1]

> With regards to the application of biology to human life, the average prophet is silent, or at least contents himself with rather uninterested consideration of progress in medicine and surgery. A few go further, and predict improvements in domestic plants and animals, and a very few 'ride the steed of Inability sideways at the wall of Expression' and introduce the subject of eugenics ... if we follow J. B. S. Haldane across the purple sea of 150 years hence we find 'the eugenic official, a compound, it would appear, of the policeman, the priest, and the procurer ...' It is this that makes eugenics so interesting; it is essentially a science of the future. Biology is at present in its extreme infancy and the biologist is a little, round-shouldered,

spectacled, man, groping blindly amidst high-powered microscopes, engaging in bitter quarrels over 'gasticulation by invagination', entertaining perhaps severe unintelligible doubts, or waking one morning to find that someone whose name he has never heard had demolished by a few crucial experiments the work which he had hoped would render him immortal. But he is the man of the future; the biologist and physiologist will work together to perfect a race of perfect men.

Hugh's Confirmation, in November 1928, presented Rosalie with a difficult problem. The time had come to talk with him about things that need a father's guidance: Manhood and Personal Duty. In her clear, rounded script she wrote:

> My darling Hughie, this is the last chance of writing before the solemn occasion of Saturday so I want to send you a few lines about it such as I think Daddy would have written had he been here to do so … You are now getting on towards 'Man's estate' … You have one of the best guides in your school motto — 'Manners maketh the man' [*sic*] — but you cannot hope to live up to that standard without a firm and real belief in God … the future lies open before you and you must work for the profession you choose with firm determination and a good deal of self sacrifice … I think you have been warned of the serious danger that always lies before youths of your age in the temptation of evil minded companions to unclean talk and stories and to abuse of your body and its functions … You are now as vigorous and able in mind and body as any boy of your age, your rapid growth has not, I think, taken as much out of you as it might have done and you are broad and strong in proportion. God and your parents have given you a fine figure and a strong constitution … keep your mind sweet and clean and free from all impurity to take your part with vigour and success in the battle of life.

Both John and Hugh had been well educated as to the Respect of Manhood. Their formidable uncle, William Macdonald Sinclair DD (Archdeacon of London, Canon of St Paul's and Honorary Chaplain to the King) wrote an entire book on the subject. 'There are men', he stated, 'who would give hundreds of pounds if they had not seen some obscene picture, which has so photographed itself on the mind that it refuses to be obliterated, or has become animated and quickened into an almost ever-present thought or dominant passion.' How Hugh received William's advice is not known, but it is certain he was already very interested in the state of his own body and would keep detailed records of injury, illness, and self-experimentation until his death.

Hugh's academic work earned a succession of awards at Winchester, among them the Headmaster's Natural Science Prize in 1928, and the Senior Science Prize in 1929. He served on the committee of the debating society, and gave amusing talks on natural history, once daring to describe Stonehenge as a rival to Christianity. Affable and respected as a sportsman, he received various cups for house sports events, and his selection as Captain of the rowing team during his final year proved he was an 'all-rounder'. But above all, cricket — a sport for good friends on long summer days — was his passion. Since his mother had taken the first steps to secure his place as a Life Member of the Marylebone Cricket Club, the grand home of the sport, when he was only three years old, all that needed to be done now was for her to pay the final amount required for his admission. In December 1929, at the age of 19, Hugh Macdonald Sinclair became one of the youngest men ever to be received into membership.

Hugh's mother knew her money would not stretch to supporting the musings of a gifted dilettante: Hugh must accept a profession. But which one? The Church, perhaps? Having a powerful Archdeacon as an uncle could help his cause there, but he showed little interest in a Christian vocation. Certainly not the Army: John would carry his father's military banner. Engineering? Not a profession, in the gentleman's sense of the word. His mother and his masters fretted over the matter, but there was no need. Hugh had set his own course. Bored by classics, young Sinclair decided to spend his life working as a scientist: a scholar of the physical, as he put it. The menagerie of stickle-backs and guinea-pigs that held his interest during childhood still fascinated him, and he was a keen collector of butterflies: with a Winchester chum, Sinclair published his first paper on local species. But the human body now captured his imagination. He admired the work of the physiologist John Scott Haldane (1860–1936) who explored and defined the mechanisms of respiration, and he read the papers of Willem Einthoven (1860–1927), who received the Nobel Prize for electrocardiography in 1924, the year of the Colonel's death. In addition to the physical records Hugh had kept about himself, he kept notes on others. For example, he recorded in some detail the effects of a breakfast of wild mushrooms on the family. Surely he would become that 'spectacled, man, groping blindly amidst high-powered microscopes' he had described in his school essay on eugenics. And like that imagined man, would consequently engage 'in bitter quarrels' and entertain 'severe unintelligible doubts'. He would become 'the man of the future' when 'the biologist and physiologist would work together to perfect a race of perfect men.'

By the time Hugh knew he intended to become a scientist, new methods of chemical analysis were making it possible to investigate the links between the chemistry of the human body and illness. In the few years Sinclair attended Winchester College the anti-pellagra substance, riboflavin (vitamin B_2), was identified; Robert McCarrison showed that urinary calculi may be produced by feeding a diet deficient in vitamin A; and Carl Pieter Dam discovered vitamin K, the anti-haemorrhagic factor. Then, in 1929, Christiaan Eijkman, who had pioneered the idea that, what he called, 'vita mins' were necessary for health, and who had discovered the link between beriberi and thiamine [vitamin (B_1)], shared the Nobel Prize for Medicine with Frederick Hopkins. These men, in combination with the writings of the great agronomist Sir John Sinclair, Hugh's own great-grandfather, convinced the young student that food and nutritional substances were the underpinnings of human health. Aware that the foundation for future discoveries was contained in earlier research, young Hugh began collecting rare and historical books concerning medicine and illness. Heroes emerged.

Foremost was William Stark (1740–70), who is credited with being the first scientist to conduct human trials using diets restricted to one or several purified foods, such as honey, bread, and olive oil. As a young physician, Stark had placed himself on highly restricted diets, and carefully recorded the results: weight loss, skin lesions, feeling of depression, and so on. Repeated self-experiments using nutritionally incomplete food combinations finally resulted in his developing scurvy, and he died at the age of 29. (Many years later, when Hugh was lecturing to students at Oxford and Reading universities, he would point out that William Stark had received the distinction of having his post-mortem done by the great surgeon, Dr John Hunter (1728–93), who is credited with founding surgical and experimental pathology. This, for Hugh, was an appropriate reward for endangering one's life for science.)

When it was time to choose a university; the Sinclair family history dictated that it would be Oxford. But surviving Oxford in those days required more than an educated mind. J. G. Sinclair, a contemporary of Hugh's, not a relative, in 1931 wrote in his memoir, *Portrait* of *Oxford*:

> Class distinctions in Oxford are as numerous as the legs on a centipede. And
> as active. A mere shade in the colour of your spats sets up a subtle social
> standard. In the 'Queen City of Culture' it is not what a man is that matters.
> What matters is: What he appears to be. … The possession of money, or the
> appearance of possessing money, excuses everything …[2]

Among the other essentials for success were to possess one pair of plus fours; acquire a repertoire of pornographic stories; drive a Morris car; and develop a sneer that reflected an inexhaustible capacity for 'suppurating self-conceit'.

Tall, handsome, and regarded as gifted, Hugh already had a fine opinion of himself. He appreciated that social contacts were as important as a bank balance, and asked his mother to open the way for him. She obliged. By now he took pleasure in viewing authority with disdain, and had perfected a way of speaking in low tones that conveyed contempt without rudeness. In October 1929, he matriculated at Oxford University's Oriel College and began reading Animal Physiology as the first step in the pre-clinical studies that would lead to his Doctorate in Medicine. Rosalie Sinclair helped furnish his rooms with family pieces and some she bought at auction. She also provided a handsome allowance.

While reading physiology at Oriel College, Sinclair's tutor was Kenneth Franklin, came to admire this outspoken and brilliant young man. Franklin, known to his friends as K. J., was someone Sinclair could respect. Educated at Christ's Hospital, Franklin won an Open Scholarship in Classics to Hertford College, Oxford, but before going up, in 1915, he joined the Royal Artillery and served as an officer until the end of World War I. After that, the rigours of academic Oxford held only a modest challenge, and he gained first-class honours in two years. Franklin served as a demonstrator in Physiology under Professor Charles Sherrington, and the two men developed a friendship that lasted until Sherrington's death. During his clinical studies at St Bartholomew's Hospital, Franklin worked as a demonstrator in physiology under another distinguished scientist, Charles Lovett Evans. Through Kenneth Franklin, Hugh Sinclair met and claimed friendship with both Sherrington and Evans. Tall, fair, and only thirteen years older than Hugh Sinclair, K. J. was praised by his friends for his orderly mind, and was well known for his easily recognizable copperplate handwriting. It was later said that his physique and more than average good looks would have facilitated the transition to Harley Street.

Many of the traits later attributed to Hugh Sinclair were also characteristic of Kenneth Franklin. He was a respected historian of science and rebelled against many of the empirical methods of medical treatment practised during his early career. He also resisted the rigorous formality of the academic community. Somewhat shy, Franklin probably never communicated well with patients, but found tutoring students of ability vastly rewarding. Dullards were given little time. But books were not the only things of importance: sport and study were balanced in his life. Kenneth Franklin was elected a Fellow of the Royal Society in 1955; this was a goal to which young Hugh Sinclair now aspired. Until his death, in 1966, Franklin remained a source of inspiration, friendship, and stability in Sinclair's life.

In September 1930, Hugh began studies at Charing Cross Hospital, in London. Transportation became a problem, and in early October, he took ownership of a blue Alvis. Most weekends he broke away from the testing academic atmosphere of the university and returned to his family home in Kent. These were not always happy times, however. John was now serving in Egypt and his trips home were infrequent. Their mother enjoyed describing John's adventures to Hugh at length, while at the same time showing little appreciation of her younger son's trials in academic life, or of the subjects occupying her younger son's attention. Letters between brothers were few these days, mostly by Hugh's choice. The Guides now occupied much of both Rosalie's and Catherine's time, and they were avid followers of the theatre and 'the flicks'. Theirs was a world distant from that of young Hugh.

It was now time for Hugh to seek membership in suitable London clubs. As well as being quarters for personal relaxation and entertainment, Gentlemen's clubs defined a man's place in society. While standards of food were often modest, the availability of good wine salved disappointed palates and eased business conversation. But, more importantly for any ambitious young man, the proper club provided an excellent means of hearing the latest gossip concerning one's professional colleagues. The Naval and Military Club was a private shelter for serving officers and ex-military personnel, for example. Hugh decided the Athenaeum, in London's Pall Mall, was the place for him. Considered to be the formidable home of the intellectual elite, the Athenaeum was where Sherrington and Evans, and the other great men Hugh admired, dined. He again sought his mother's help.

Through influence and design, Rosalie had managed to arrange membership for Hugh in the Caledonian Club. She thought this would be a useful arrangement, as Hugh obviously enjoyed linking his family name with various members of the Scottish aristocracy. Instead, Hugh fretted and complained that some of his friends found his membership there amusing. Rosalie said she would do what she could about the Athenaeum, but in the meantime Hugh would have to be satisfied with the Caledonian Club. She knew gaining membership to the Athenaeum would take considerable effort. Members were selected after becoming established figures in their professions. Exceptions were those admitted without ballot, including the Speaker of the House of Commons, archbishops and bishops, cabinet ministers, high court judges, high commissioners and presidents of many of the Royal Societies, as they were assumed to have distinguished themselves by position. Others waited for as many as twenty years before the Membership Committee considered their applications. For Rosalie and Hugh Sinclair, such a delay was unacceptable.

Then they had a stroke of luck. It became known that the Athenaeum needed funds, and new members were being admitted at a faster rate. The time between one's nomination for membership and the actual election to membership had dropped from more than fifteen years to about four, although it was still expected that members would elect only men of established reputation in their profession. The entrance fee was 30 guineas, if paid in one sum, and the subscription was 15 guineas a year. Through some careful budgeting Rosalie could raise the money, but how could she expedite his nomination?

Introducing herself as Sir John Jackson's daughter, and highlighting some of his many achievements, Rosalie Sinclair wrote to the Secretary of the Athenaeum and asked for a list of members. The response was somewhat disappointing: circulation of the list was limited to the list of members. But, the Secretary stated he would be pleased to see her when she next came to London. Rosalie arranged to see the Secretary within days of receiving his response, and left their meeting with the membership list in hand. With almost military zeal, she and Hugh drew up and divided between them a list of the names of their most important and influential family, friends, and acquaintances, and checked these against the membership list. Just as Hugh was socially adroit and clever, his mother was influential and determined: together they launched an awesome campaign. Before the end of the 1931, when Hugh was still twenty-one years of age, a flood of letters to the Athenaeum supported his candidacy. They came from as far afield as King's House, in Jamaica, and Lanrick Castle, Doune, Perthshire.

Three other important events took place in 1931. The first was in February, when Hugh and thirteen of his friends celebrated his 21st birthday with a party at the Philistine Club. Some of the names of those in attendance were to appear on correspondence for decades to come. One name, however — Peter Meiklejohn — would have more professional significance than the others.

The second event took place a few months later, and involved one of Hugh's favourite pursuits: the purchase of a book. While taking a break from a meeting of the British Association, Hugh stopped in a bookshop in search of a way to invest a few shillings wisely. His attention was drawn to *Vital statistics: memorial volume of selections from the reports and writings of William Farr, MD, DCL, CB, FRS*, edited for the Sanitary Institute of Great Britain by Noel A. Humphreys, London, in 1885. Reading this text radically changed Hugh's view of medicine.

William Farr (1807–1883), a distinguished member of the Statistical Society with a considerable interest in the link between diet and death, published a paper in 1846 on 'Influence of scarcities and the price of wheat on the "mortality of the people of England"'. Farr's approach to population and census was similar to that used by Sinclair's great-grandfather: groups were studied by location and labour categories, and causes of death were examined. Farr reported that violent deaths, and all categories of 'sporadic Diseases', were higher in cities than in rural districts, but the opposite held true for deaths from 'age'. But there was something more. Recalling the incident years later, Hugh Sinclair noted that when Farr started his studies in 1841, he calculated the life expectancy of a man aged fifty as 20.03 years. Hugh recognized that those figures had hardly changed in ninety years, so that medicine in 1931 could do little more to keep a middle-age man alive than in its dark ages of the nineteenth century. In addition, Farr had noted that most middle-age men died of tuberculosis or pneumonia. The chronic degenerative diseases such as heart disease and cancer were rare.

The third interesting event during Hugh's 21st year was a serious effort in self-experimentation. Such practices were not unusual, and Hugh admired men, like Stark, who used themselves as scientific subjects. He would undertake self-experimentation several times during his life, but the first clearly documented work took place in 1931.

Based on correspondence at about this time, it appears that Hugh's own sexual nature held a bizarre curiosity for him. Just before Christmas, in 1931, a letter arrived at the Sinclair's Gransden home from a member of the Biochemistry Department, University of London, answering questions Hugh had put to him a fortnight earlier. Yes, 'oestrin' occurs in males, he wrote. Further, it had been found in both male urine and testes, and there was a large literature on its effects in male animals, including causing atrophy of the testis. Hugh appears to have described a self-experiment he planned to undertake.

We know little else of these events, but one of Hugh's relatives, Frank Hird, kept him informed of any strange sexual cases he heard of through his godson, Sydney, then a student at St Thomas's Hospital. Four months after the letter concerning oestrins, Frank sent Hugh a letter describing a hermaphrodite supposedly being transformed into a man, and a 'partially masculine' woman to whom doctors were restoring her original sex. He also mentioned that every year at least twelve alterations are made in the registry of Births of People whose sex has changed since the time of registration. 'Why not set your very clever brains to work?', he asked Hugh.

The year 1932 was a time of trouble. New damage to the knee injured at Stone House School almost eliminated Hugh's participation in sport. He was caught breaking a window in the Common Room at Oriel College, driving without a proper licence, and creating mischief by setting off fireworks near the centre of Oxford. More threatening to his university career, there was a College scandal. A drunken friend sought Hugh out at Oriel after curfew and, stumbling into the Provost's quarters, insulted him with lewd suggestions. Fearing his name would be blackened by the incident, Sinclair wrote to the Dean, who reassured him he was unaware of any specific charges against Hugh, and said he would 'do everything in my power to help you'. Within weeks, College authorities agreed the matter was closed.

Whether Hugh told his mother about all of this is unclear, but he kept her well informed on more positive matters. In May of 1932, Hugh was appointed Demonstrator in the Department of Biochemistry, where he would work under Rudolph A. Peters on the action of vitamin B_1 [thiamine] on the brain. He wrote to Rosalie:

5 June 1932

My Darling Mummy … I only sent in an application for the Demonstratorship in Biochemistry with very small hopes, because of the three other
applicants from here … But I got a letter from the Professor on Friday
saying I was elected at the Staff meeting. So I am very pleased. I will be able
to do research and will work for the degree of BSc (Bachelor of Science).
Also, it gives me 'Senior Status' or something so that I can have my car out
whenever I want it …

Concerning Maidstone Cricket Week; I think you said none of the spare
rooms were in use. So there seem to be five people who will probably be able
to come … Sir Charles Sherrington has sent my name with two others to the
President of the International Congress at Rome. They want to make it very
select, but as Sherrington is the greatest living physiologist, they are bound
to let us attend. One of the other two (son of Wilfred Trotter FRS, Sergeant-
Surgeon to the King and the man who operated on him when he was ill) is
probably going to motor with the other one (they are both taking Finals this
term). I have suggested that I should do the same. It seems to me that if I sell
my car (as I intend to) and get a 4-seater it will be cheaper and much nicer to
motor, as Peter Meiklejohn is going and would probably come with me and
Francis is very anxious to join us, as he knows some Italian (which would be
useful) and has got to learn the language next vacation. …

Best love — from your very affectionate son Hugh M. Sinclair

Being a Demonstrator in the Department of Biochemistry was an
achievement, but Hugh wanted something more — he determined eventually to achieve a Senior Demyship at Magdalen College. Along with the
Harmsworth scholarships of Merton, and the Senior scholarships at
other colleges, a Demyship at that time was given by Magdalen College
and provided significant financial support. Awards were made after a
competitive examination and were for two years.

Hugh's efforts were successful and he served as Senior Demy from
1932–34. His surroundings there are best described in a copy of the 1935
edition of the *Handbook to the University of Oxford*:

Of Magdalen one is tempted to say as Camden said of Britain, it is 'well
known to be the most flourishing and excellent, most renowned and
famous Isle of the whole world: so rich in commodities (from its garden on
the Marston Road), so beautiful in situation, so resplendent in all glory, that
if the most Omnipotent had fashioned the world like a ring, as he did like a

globe, it might have been most worthily the only gem therein.'

As Hugh's tutor, Kenneth Franklin had stimulated interest in his cause, enlisting Rudolph Peters, the Provost, and others to support him. On 2 July 1932, Franklin had written in Hugh's support to the President and Fellows of Magdalen College. He said Hugh had ability, initiative, and 'a legal clarity of thought in sifting out the truth from large masses of often-conflicting statements. In fact,' Franklin stated, 'I think he is in many respects the most able pupil I have had. ...' He also praised Sinclair for being both a pleasant and stimulating friend.

As his friend, however, Franklin was concerned about Hugh's apparent intention to focus his work on a single aspect of science too early in his career. He wrote and told Hugh that he should learn some German before going to work for Peters:

> A great deal of Peters' work involves a pretty good knowledge of German and you mustn't let him down in this respect ... Finally I want to say one thing. Don't let specialized research divert your attention entirely at this stage from physiology as a whole. Set apart a definite period of time each week to keep up general reading. Otherwise you will lose ground. I think Peters has got too much absorbed in one particular part of the field this last year, and hence this note of warning...

That summer Hugh motored to Italy with his friends. They walked all night on the beach, enjoyed the sun, and attended the international meeting of the Physiological Society. While there, he and others from Britain stayed in the rooms of the Rector of the Scots College in Rome. Shortly after his return to Kent the post at Gransden was heavy with letters congratulating him on receiving the Demyship. He was to be a Magdalen man. But before returning to classes, his damaged knee required surgery. As soon as he was able to get about with a cane, Rosalie packed up Hugh and Catherine and took them off to the Norfolk Broads for a family vacation: a trip Hugh recalled years later as one of the happiest times in his life.

CHAPTER 3

Robert McCarrison and other influences

Out of the earth are we and the plants and animals that feed us created, and to the earth we must return the things whereof we and they are made if it is to yield again foods of a quality suited to our needs.

Sir Robert McCarrison, The Cantor Lectures, 1936[1]

Complaining of its small size and lack of amenities, Hugh rarely visited Gransden, and entertained friends in his rooms at college rather than inviting them home. In more than miles, Kent had become distant from his life in Oxford. John was in Transjordan pursuing an increasingly successful military career: a career that his mother enjoyed describing at length to Hugh. Rosalie and Catherine were immersed in their own interests, and Hugh felt they spent too little time appreciating his academic work. Communication at home was increasingly conducted through letters, left on fireplaces or under doors, as this permitted each to have his or her full say without the fear of interruptions. As for Salie, she seemed to make a point of vexing Hugh; clever and good at sport, she had always outshone her younger brother more than he liked. During the last weeks of 1932, Hugh had written a letter to his mother about his own success and his efforts as a son. 'My darling Mummy,' he began:

I went in for the Francis Gotch Memorial Prize yesterday ... There were three of us competing for it (one a man who went down over 3 years ago and is now a qualified doctor). The examiners told me afterwards that 'the candidates were of such universally high standards that they have divided the Prize between the three'! So I'm very pleased about it. The examiners [included] Sir Charles Sherrington who had won this year's Nobel Prize of Medicine (the highest possible of awards); I think he was very unfair as he asked me a lot of clinical questions which of course I didn't know ... I am trying for a Senior Scholarship at Magdalen College. There is a poverty qualification attached to it — one's ... income must not exceed £250. So

since I get £159 from the University, and £105 or so from the war loan, I
would like to sell out and transfer temporarily to you or someone, a little of
the latter … I don't think there is any chance of my getting it, as they are
almost certainly giving it to someone doing non-biological work this year…

When Hugh wrote this letter he had, in fact, already received his Senior
Demyship, but he was anxious to obtain further financial support, and
needed Rosalie's help in demonstrating his need for 'pecuniary assistance'. There is no evidence, apart from this letter, that the Gotch prize
was shared with others this year.

Rosalie's second son continued by explaining how he had worked to
keep the family happy:

> … I have tried to do my best for you and Catherine; you were the only
> mother present at the Oriel Ball … you haven't missed an Eight's week or a
> Rugger night. And I am extremely glad that it is so. Parties on Rugger night
> are nearly always 'all-male' but I asked you and Catherine — just as I always
> have — because I wanted you to come and because I hoped you would like
> it. If Daddy were here, I know he would have arranged things for us … I
> realise fully and most sincerely that things are extremely difficult for you …
> I, for one, want to do my best for 'the greatest happiness of the greatest
> number' in our family … H. M. Sinclair.

There was no sweet salutation to the letter, as was his practice on other
occasions.

In fact, Hugh was awarded the Gotch Memorial Prize in 1933, after
obtaining a first-class degree honours in Animal Physiology from
Oriel College, and being appointed Departmental Demonstrator in
Biochemistry and Senior Demy, Magdalen College, in 1932. His friend
and mentor from Winchester College, Kenneth Franklin, suggested he go
on for clinical training at University College Hospital (UCH). He wrote
to Hugh saying he would give him letters of introduction to several
prominent scientists working there. Knowing Hugh's preference for
male-only environments, Franklin pointed out that UCH had the drawback of having both male and female students. Ignoring the complication
of women in the classroom, Hugh followed Franklin's advice and sought
admission to UCH and the position of Demonstrator in Physiology
at University College. Both were quickly his. Franklin was pleased, and
suggested a broader strategy for Hugh's career: he should qualify in medi-

cine, get a Radcliffe Travelling Fellowship to visit people in the United States, and then return to Oxford, where he could step into an academic slot. To support the plan, Franklin personally encouraged Sir Charles Sherrington to take an interest in the younger man, and sent letters introducing Sinclair to S. R. Douglas, a member of the Medical Research Council, and to one of University College Hospital's professors of biochemistry, Sir Jack Drummond. These contacts eased the way, and before long Hugh's intelligence, hard work, and well-placed friends soon gave him access to the middle circle in the expanding field of biochemistry. Rosalie may not have been able to help her son find his way into his chosen field of achievement, but others were there to take her place.

A year after becoming Senior Demy, Hugh's name appeared for the first time on a paper in a major scientific journal. This paper, on vitamin B$_1$, was the first in a long line of publications.[2] Reading this begs two questions. First, had Kenneth Franklin been right to suggest that Rudolph Peters' research work would be too restrictive for Sinclair? And second, was it possible for the subject of nutrition to become other than a stepchild to the more exact science of human biochemistry? The answers are partially found in the on going research that influenced Sinclair's view of science during his early years at Oxford.

In the early 1930s, Rudolph Peters and his team of gifted young scientists in the Department of Biochemistry were occupied by the meticulous evaluation of the action of thiamine (vitamin B$_1$) on the central nervous system of pigeons. Before Hugh joined the Department and became its Demonstrator, Peters established that pigeons fed purified diets deficient in vitamin B$_1$ develop a neurological condition in which the head is retracted in a very distinctive manner. Peters further established that minute amounts of both catatorulin (a substance Peters identified and named) and vitamin B$_1$ eliminated this neurological symptom. The task at hand next was to identify whether or not catatorulin and vitamin B$_1$ were the same substance. Using the best technology of the day (monstrously slow and cumbersome by modern standards) Peters' team was breaking new ground in the development of biochemical methodology. Bioassay systems, in which measurable changes in metabolic activity are used to determine the presence and strength of specific substances, and mathematical predictions, or statistics, were the primary means of testing theory. In what was then a world of glass tubes, incubators, and slide rules, Hugh was given the task of helping to assess whether or not catatorulin and vitamin B_1 had the same ability to stimulate oxygen uptake in a preparation of pigeon brain. Comparability proved to be so strong as to suggest they were identical. Further, the authors suggested that the methods reported promised 'to be of value as a new means of estimating the activity of potent vitamin B_1 preparations'.

Molecular biology (the study of the structure and function of molecules in living forms) was the vital focus in the most advanced scientific and medical laboratories of the time. For Hugh Sinclair, that made the study of nutrition a part of this growing field. After all, was an understanding of nutrition anything more than an understanding of the exact substances that interact in the metabolism of living tissues? What could be more exciting than working in an environment where the secrets of molecules, such as vitamins, could be examined in detail? Sinclair was well aware of the rich history of his subject. A man he greatly admired, Sir Frederick Gowland Hopkins (1861–1947), initiated the era of biochemical nutrition in 1901, when he found that the human body could not manufacture the amino-acid tryptophan, a necessary part of human protein. It had to come from the diet. Hopkins established the existence of other essential amino-acids, and then concluded that a number of essential 'accessory food factors' exist, which are vital for normal development and health. These he referred to as 'food hormones', 'growth determinants', and 'vitamines'. (In 1920, Sir Jack Drummond suggested that the 'e' was dropped from the word and, in line with the scientific nomenclature of the day, the term was defined as 'substances essential for life'.) Hopkins's work suggested that a deficiency in these essential substances could lead to a breakdown in normal cellular function, resulting in disease. Scientific investigators soon began isolating vitamins and identifying links between deficiency and acute illnesses, such as beriberi and scurvy.

In retrospect, Hopkins's revelations were not that surprising. For centuries it was known that certain foods could cure identifiable, and terrible, human conditions. The earliest scientific work was by Dr James Lind (1716–94), in his paper: *A treatise on the scurvy*, published in 1753. Here was strong evidence that the madness and bleeding of scurvy suffered by sailors on long voyages began when the fresh vegetables and fruits ran out, and could be treated by reintroducing these foods, especially lemons. Lind had been influenced not only by his own observations, but also by reports from others including the French explorer, Jacques Cartier, and Sir James Lancaster, of the Dutch East India

Company, who both claimed that certain foods cure scurvy. In his paper, Lind reported what is thought to be the first controlled study of nutrition. On board the ship HMS *Salisbury*, he took twelve men affected by about the same degree of scurvy, and treated groups of two with different substances: cider, a weak solution of sulfuric acid, vinegar, sea-water, or oranges and lemons. Only those receiving citrus fruit improved. But, as convincing as these observations and experimental evidence were, there were no technical means by which to isolate and identify the specific substance in food that cured scurvy, nor explain how this substance could reverse symptoms. The precise study of nutrition could make only limited progress until the early part of the twentieth century, when the scientific means of separating and testing individual constituents of food were developed. Progress was swift when it came, however, and by the time Rudolph Peters and his contemporaries took on the vestments of science, and began the meticulous investigation of molecular interactions, the foundation for the study of biochemical nutrition had been laid.

While working in Peters' laboratory Hugh contributed to a series of highly technical papers on vitamin B_1. But there were other means of scientific investigation he wished to explore. Kenneth Franklin had introduced Sinclair to men from a previous generation of scientists: men who discovered as much through observation and contemplation as measurement and prescription. One was Sir Charles Sherrington, the great neurophysiologist, with whom Hugh studied at Oriel College. Over years of research and teaching Sherrington drew together the established work on the brain and spinal cord, combined this with his own observations, and set about the task of developing an explanation of function within the nervous system. His genius was to see beyond that which had been previously tested, measured, and reported on by others; his genius was to hypothesize how these facts and observations were integrated in a functioning system. He concluded that sensory receptors in the human body provide information, or stimuli, from three locations: outside the body (exteroception), inside the body (interoception), and information from the positioning of muscles and joints (proprioception). All this information was fed into, and integrated, within the central nervous system, which then signalled, or stimulated, the appropriate body response. Responses covered all forms of activity: hormone release, muscle contraction, excretion, and so on. In 1906, Sherrington published *The integrative action of the nervous system*, which has formed the basis of our understanding of the subject ever since. Later, Sherrington's thinking combined the notions of discrete nerve impulses and the 'all-or-nothing' rule of nerve response, resulting in the conclusion that stimuli to a specific nerve must reach a critical level, or threshold, before that nerve will generate an impulse. In recognition of these achievements, in 1932 Charles Sherrington, together with Edgar Adrian (1889–1977), was awarded the Nobel Prize in Medicine. That same year he recommended young Hugh Sinclair for the Senior Demyship at Magdalen College.

Sherrington contributed significantly to Sinclair's concept of scientific research, and in later years the younger man was always pleased to mention Sherrington's name as a great source of inspiration. Like scientific thinkers before him, Charles Sherrington took time to observe, and then to think. He drew together the research findings of others, and then conceptually placed them in a cohesive framework, or explanation, of a specific phenomenon. He also drew inspiration from subjects other than his own, thus adding to the dimensions he brought to his final conclusions. Specifically, one can see the influence of Max Planck's quantum theory, published in 1900, on Sherrington's ideas about the definitive nature of nerve impulses. Broadly, quantum theory states that energy is not discharged in a range of values, but in indivisible amounts, called 'quanta'. Revolutionary in its time, Planck's work was the foundation for the study of quantum physics, and led to Einstein's work on electromagnetic radiation. But, why is this important to Hugh Sinclair's story?

Kenneth Franklin, who was a close friend of Charles Sherrington, encouraged a friendship between Sir Charles and young Sinclair to stimulate discussion of the nature of scientific investigation. Sherrington's approach differed from that of Peters and other members of the Department of Biochemistry, who were committed to the scientific method of formulating a hypothesis, and then testing ideas through repeated experiments based on precise measurement. Franklin wanted his young protégé to have a wider grasp of the nature of research, and his influence would be seen in years to come, when Hugh devised his personal plans for the study of human nutrition.

Sir Robert McCarrison was another scientist who influenced Sinclair. The two men met several years after Hugh began work with Peters, when he was well into his clinical studies at UCH. Major-General Sir Robert McCarrison (1878–1960), formerly Director of Research on Nutrition in India, retired from military service and returned to England, where he and his family took a house near Oxford. Although he stayed somewhat apart from the workings of the university itself, he became an important member of its community. Exactly how the men met is not known, but through his father's Army experience the younger man could relate to McCarrison's period in India and his military career. Their friendship lasted for decades, and the mutual admiration between them was obvious when, many years later, Sir Robert donated his extensive library and collection of scientific objects to Hugh in support of his plans for the study of human nutrition.

Some of Robert McCarrison's most important scientific work began with the simple act of observation. While in India he was interested in the distribution of cases of goitre, obvious due to the enlargement of the thyroid gland at the front of the neck. This condition is most commonly caused by a deficiency of iodine in the diet. More specifically, his attention was drawn to the fact that in a group of nine villages at the foot of the Himalayan mountains only eight had goitre. In a few villages people were not only afflicted by hugely enlarged thyroid glands, but also by cretinism, a condition in which mental and physical development are dwarfed. According to an account written by Sinclair,[3] who must have heard the story from McCarrison many times, the water drunk by the village showing no goitre was fresh from a local mountain spring, whereas that drunk in the other eight villages was polluted with sewage, coming from one village into the next. The further down the chain of effluence, the higher the number of cases of goitre. The water from all the villages was tested and found to contain like amounts of iodine. Certainly something was at play that went beyond the presence or absence of iodine. McCarrison had water from goitre and non-goitre villages put through an evaporation process to concentrate their contents, and fed amounts of the two residues to himself and volunteers. Those drinking the water from goitre villages, including McCarrison, developed goitres.

It was obvious that effluent in the water from the goitre villages contained high levels of substances that caused the condition. Based on these observations, he concluded that there are two causes of goitre: a lack of the essential mineral, iodine, and the ingestion of certain substances that are goitrogenic. Today we know these chemicals are found in many edible plants, and especially those of the *Brassica* family, such as kale and cabbage. In high concentrations these plant chemicals, for example the glucosinolates, reduce the thyroid gland's ability to absorb iodine, and therefore inhibit its capacity to produce the hormone thyroxin, which controls the rate of metabolism. Here was a case where observation, followed by preliminary measurement and controlled investigation, successfully led to the identification of a significant public health problem. McCarrison believed that 'the greatest single factor in the acquisition and maintenance of good health is perfectly constituted food'. In his quest for understanding, he observed the effects of diet on populations, conducted animal experiments, and carried out dietary studies on human volunteers. In 1927, he published a paper in the *Transactions of the Far Eastern Association of Tropical Medicine* entitled: Disease of faulty nutrition, in which he quotes Hippocrates as follows:

> … It appears to me necessary to every physician to be skilled in nature, and to strive to know, if he would wish to perform his duties, what man is in relation to the articles of food and drink, and to his other occupations, and what are the effects of each of them to every one.
>
> Whoever does not know the effects these things produce upon a man, cannot know the consequences which result from them.

Above all others, Robert McCarrison stimulated Hugh Sinclair's interest in human nutrition and influenced the course of his career. Sinclair claimed McCarrison's work was not given the recognition it deserved because it was considered by many to be passé. Work based on whole diets was not in vogue beyond the start of the twentieth century, and by the 1930s, most of the vitamins had been identified and isolated, making it possible to conduct animal experiments using purified, and highly quantified, mixes of nutrients. This, of course, supported the more systematic experimental approach preferred by Rudolph Peters. Nonetheless, it was McCarrison's view of nutrition that appealed more, and under his influence Sinclair became convinced of the need for a comprehensive approach to the study of relationships between diet and human health.

In 1936, Robert McCarrison was invited to present the highly prestigious Cantor Lectures, which were delivered before The Royal Society of Arts. The ease of style and the comprehensive insights contained in these three lectures caused them to be quoted by scholars for years. In 1953, Faber & Faber published these lectures and two earlier McCarrison essays on food. The book concludes with a chapter by Hugh Sinclair, entitled 'Nutrition and health: recent advances'. In that chapter, Sinclair not only comments on recent findings, but dares to expand his mentor's definition of nutrition itself to include both the absorption of substances dissolved in oil and absorbed into the body when rubbed on the skin, and the physical incorporation of nutrients administered intravenously. Sinclair calls this expanded view of nutrition 'nutriture'. Further, Sinclair wrote of the need to consider the social conditions of populations being fed, which includes their level of stress.

The concept of 'nutriture', and the methods Sinclair used to study this subject in later years, are indelibly printed with McCarrison's influence. For that reason it is useful to consider the contents of the Cantor Lectures. In the first of the three, McCarrison elegantly generalizes the contents of cells, the structure of the human body, the function of food, the processes involving 'the function of nutrition', and the 'pathological changes brought about in the organs and tissues of the body by derangements of nutrition consequent to faulty food.' 'Nutrition', he stated, 'embraces in its compass the many aspects of biology including biochemistry, biophysics, morphology, physiology, pathology, and medicine. In fact, knowledge of nutrition provides a necessary bridge between physiology and pathology, and is 'an essential foundation of rational medicine'.

McCarrison defined food as anything taken into the body that provided material for the normal growth, structure and maintenance of healthy tissues. As published in 1953, his lecture lists twenty-nine required nutrients in addition to water and oxygen: six vitamins (A, B_1, B_2, C, D, and E), ten amino-acids, the carbohydrate glucose, and eleven inorganic elements (calcium, chloride, copper, iodine, iron, magnesium, manganese, phosphorous, potassium, sodium, and sulfur). And he notes that one other, linoleic acid, is derived from fat. (This reflects the knowledge of essential nutrients current at the time.) In the second lecture, McCarrison laid out the role (as then identified) of each of these essential food elements. He drew attention to the fact that vitamins do far more than prevent specific illness (anti-scurvy, anti-beriberi, and so on); they have specific metabolic roles that are important to the structure and function of the body as a whole. He further stated:

Since the year 1921 I have used every occasion to emphasize that it is the lesser degrees of such deficiency that are of importance in Western countries. A recognition of this fact is, I believe, essential to the prevention and cure of many of the commoner sicknesses of mankind — sickness to which we cannot always attach a diagnostic label.[4]

Later on McCarrison wrote that maintaining nutritional harmony is the 'Melody of Life'.

The third, and last, of the Cantor Lectures addressed the problem of National Health and Nutrition. Recalling his early work in India, McCarrison reviewed the links between diet and the incidence of disease, and compared disease rates in different populations, attributing differing rates of illness to differences in diet. The diseases mentioned are not limited to those clearly attributable to nutrient deficiencies (such as beriberi, night-blindness, and goitre) but also infection and long-standing poor health. He warned that the link between poor nutrition and poor health is enough for any country to see the link between food and national defence. Speaking of conditions in 1934, McCarrison quoted a Report of the Adjutant General for the year 1934:

> What was disconcerting to any citizen with a care for the good of his country was that over 52 per cent of the men who went to the recruiting office did not come up to the physical standard laid down. In the big industrial areas of the north, the percentage of rejections rose to sixty-eight.[5]

This shocking fact reflected the state of the nation's nutrition following World War I.

The reason given for this crisis was malnutrition during childhood. And, if the men presenting themselves were rejected in such high numbers because of nutrition-related conditions, what could one expect of the population who failed to attend the recruiting offices? Throughout the remainder of his third lecture, Sir Robert stressed the importance of improving the nutrition of the British public as a matter of urgency. Children would achieve more at school, work performance would improve, and the cost of providing health services would decline. He also observed that a falling death rate in a population does not indicate a higher standard of health. Saving a life with drugs or surgery does not mean that the person saved is, in the end, necessarily healthy. McCarrison suggested that: if, during the course of a year, there were no deaths among the inhabitants of a home for the incurable, that institution's death-rate would be nil, and could be considered as 'the healthiest place in England', but it would not contain a single healthy resident. Englanders, McCarrison claimed, are most likely to be deficient in high quality protein, calcium, iron, and vitamins A and D. Based on his own opinions, he suggested that vitamin B_1 deficiency may also exist.

In the end, the matter of nutrition comes down to one of education. McCarrison advised that to build an 'A1' nation, medical students must learn more about the importance of food. A subject so central to preventive medicine should be given more emphasis in the medical curriculum, and students should be given more opportunities to see at first hand the damage to organs and tissues caused by poor nutrition. Second, the elements of nutrition must be taught to schoolchildren; they need to know how to sensibly feed their own bodies. Only by informing children, and changing their eating habits, will the bad food habits of the population be changed for the better.

This was heady stuff for young Hugh. It made championing nutrition the vocation that he needed to serve not only his intellectual curiosity, but also his sense of patriotic duty.

Although Hugh's work with Peters tied him to investigations with vitamin B_1, new work on essential dietary components excited him. Essential fats probably caught his interest at about this time. Even at this early date, the importance of dietary fat was a highly contentious area of investigation.

The scientific foundation for the study of fat had been laid down more than a century earlier, in 1823, by the French chemist Michel Eugène Chevreul, but the true nature and biological necessity of substances classified as fats remained unclear. It was well established that proteins, carbohydrates, and fats were the three primary constituents of food, but there was no reason to believe that animals need fat in their food to survive. In 1877, J. B. Lawes and J. H. Gilbert[6] found that pigs fed only protein and carbohydrate accumulated fat in their bodies. If this was the case, then how could fat be essential, as it could be manufactured by the body? By the time of World War I, it was known that fat-soluble vitamins A and D were necessary for normal growth and health, but it was difficult to show that any other vital substances exist in fats. Research published in the early part of the century suggested that healthy young men require no fat in their diet if adequate quantities of fruits and vegetables are eaten to supply their basic vitamin requirements. Investigators, including the English scientist Sir Jack Drummond, reported that laboratory rats flourished on diets devoid of fat, supporting this theory. However, there was an unforeseen mistake in their methods. Starch had been used as the source of purified carbohydrate in the experimental diets investigated, and both rice starch and corn (maize) starch contain fatty acids, including linoleic acid. There were those who believed that this simple molecule (a chain of eighteen carbon atoms containing two double bonds) may have profound influence on human health.

In the United States, H. M. Evans, who had discovered vitamin E in 1922, repeated these dietary studies, replacing starch with sucrose, a sugar, as the source of purified carbohydrate. The results were very different. A newly graduated scientist, George Burr, joined Evans' laboratory and together they showed that rats fed a purified diet containing casein (protein), purified sucrose (carbohydrate), a salt mixture, yeast, wheat germ, and a small quantity of cod-liver oil failed to grow, and young females failed to ovulate.

Burr married one of Evans' laboratory technicians, and together he and Mildred pursued the fatty acid story. In 1929 and 1930, while Hugh Sinclair was studying physiology at Oriel College, the Burrs published two definitive papers that established that linoleic acid is an essential nutrient. Its deficiency in the diet of the rat resulted in scaly skin, kidney impairment, failure to grow, impaired fertility, and increased water consumption without increased urine output. Soon there was further evidence that another fatty acid, linolenic acid, was essential for normal growth. Unfortunately, at the time there were few at Oxford interested in essential fatty acids, although McCarrison and others acknowledged their importance. It would be some years before Hugh Sinclair cast clear light on the subject by bringing to Oxford an international group of experts on essential fats.

Meanwhile, a letter from a relative, dated 3 June 1933, brought news of events in Germany:

> Have you heard that Hitler has closed the medical department that was making researches about glands in Germany, on the same lines as those you told me were being made at Oxford and in America? I have it on a good authority … [from a man] who has just come from the Finnish Legation in Berlin. Further, he says, Hitler has ordered all the medical records to be destroyed. He is uncertain as to whether the head of the department is a Jew or not, but the impression in Berlin is that the closing of the department and the destruction of the records are part of the 'moral' purge that Hitler is applying to Germany.

This was a time of moral dilemma for Sinclair, and probably the time when he began collecting books about sex and sexuality, among which was

T. Clifton Longworth's book *The Devil a monk would be: a survey of sex and celibacy in religion*. It was also a time when he subjected himself to rigorous self-criticism. Our best information about his personal struggle comes from a diary kept during the summer of 1933, in which he jotted down intimate thoughts and references to the Bible. Over the course of the pervious months, he seemed to have driven himself too far by overwork, until there appears to have been a period of emotional disintegration. The following jottings to himself are in order, but not inclusive:

> If you are doing it in order to be seen of men — don't do it.
> You are rambling and haphazard in your papers.
> You must get up earlier, you are not putting in sufficient time.
> The joy of combating one's sins is far greater and more satisfying than the joy of indulging in them, but they can only be beaten by keeping in close contact with God.
> … Avoid censoriousness, why should I judge others when there is so much in myself to condemn.
> Write to mother and tell her she must come up.

(The following text was written in large and bold letters. Was this Sinclair speaking to himself?)

> **'There is one person for you to start sharing with.**
> **Ask God who it is.**
> (My Father.)

Is the word **Father** used in the biblical sense, as in 'the Lord'? Or, is this the recognition of Hugh's longing for the Colonel, and the desire to follow his example? As the question calls for the name of a 'person', the answer must refer to Hugh Montgomerie Sinclair.

Hugh's note to himself about his mother in this list of thought is particularly important, because it is about this time he saw a property near Oxford which he felt would suit his personal and social requirements. The work schedule Hugh set for himself was exhausting, and despite the enjoyment of driving his Alvis between Oxford, Kent, and London, he wanted to consolidate his activities more around the university. Also, as his circle of influential friends grew, he increasingly felt the need to entertain them in suitable style. It was acceptable to invite school friends, such as Peter Meiklejohn and Kenneth Franklin, to stay at Gransden for the weekend, but Hugh felt the house was inappropriate for entertaining some of the older men he now claimed among his circle of friends. He needed to entertain in a more important setting. Turning to his strongest family ally, he asked Catherine to sound out their mother on the possibility of moving from Gransden. But Kent suited Rosalie and her daughters well, and she resisted the idea, saying they would miss their friends. In particular, they would miss contact with Robert Stephenson-Smyth, 1st Baron Baden-Powell, the General and founder of the Scout movement, and his sister, Agnes, who helped him found the Girl Guides. Guides activities took up much of Rosalie's and Catherine's time, and they were becoming well known in the Guide movement.

Hugh continued to fret. He complained that Gransden was cramped and embarrassingly ill-equipped for proper entertaining. Space was so limited that he and John were forced to share a room, he complained. (That lack of privacy would become an issue between himself and his mother in years to come.) Then a newspaper advertisement appeared which Hugh immediately clipped and sent to his mother.

> Sale on Wednesday next (21 June 1933) Very low reserve. 'Lady Place', Sutton
> Courtenay. Only 10 miles from Oxford. On outskirts of the village,
> approached by drive. Perfectly equipped. Music and 4 reception rooms,
> 12 bedrooms, 2 dressing rooms, 3 bathrooms. Main electric light and power.
> Central Heating. Constant Hot Water. Stabling, Garage and 2 cottages. Very
> picturesque Boathouse. Finely timbered. Pleasure Grounds including 2 hard
> tennis courts, rose and rock gardens, walled fruit and vegetable garden. In all
> about 14 acres. For sale by Auction at the London Auction Mart …

Certainly this was what Hugh was looking for. The Tudor style house had begun as a private residence, was later converted into a hotel, and then returned to private use. As a result it combined the features of a large family home with the facilities of a country-house hotel. The property had more attractions than detailed in the advertisement, including two grass tennis courts, a glasshouse in the walled garden, woodland walks, an orchard, and a 'Tea Room and Verandah' within the confines of a boat house. Within the house itself were a domestic office, a butler's pantry, a well-lit kitchen with wood-block floors and tiled dado walls, a scullery with its own independent boiler, a wine cellar, and assorted store rooms. From the moment Hugh saw Lady Place, he knew this was the home he needed to satisfy his yearnings for Barming house. Somehow this must be his.

CHAPTER 4

Lady Place: progress, and a blunder

Hugh had no money of his own, and was forced to convince his mother to finance the purchase of Lady Place. She would need to sell Gransden, but they could continue to count on her rental income from Barming House. Surely she would enjoy this place: Sutton Courtenay was a picturesque English village. Anglican and Catholic churches were close at hand, as was a fast train to London and a well-known golf course, Frilford Links. The market town of Abingdon, about 3 miles from Lady Place, provided reasonable shopping, and the open fields of the property were convenient for meets of both the Old Berkshire and South Oxfordshire Foxhounds. On a pragmatic level (always of vital importance to his mother), if the two cottages on the property were rented, and the gardens and pasture were put to good use, Lady Place could generate enough cash to help maintain it.

Although Catherine warned Hugh that their mother was not interested in moving, Hugh put the matter to her. Rosalie resisted even seeing the property. As good fortune had it, the auction of Lady Place was rescheduled from mid-June to 19 July, giving Hugh a few more weeks to make his case. At last she capitulated. She drove to Oxford in her Austin, and John and Catherine travelled over in the family Hillman. When Rosalie Sinclair arrived she understood Hugh's enthusiasm. The sale was only days away. Rosalie moved quickly and through her solicitors argued the price down from £5500 to £5000. She put down £500, and the auction was cancelled. On the evening of the day she bought Lady Place, Rosalie wrote to her younger son saying she would tell none of the other children that night about her purchase. She wished Hugh well with his examinations, and said she had just placed £50 in his personal account.

Hugh had just finished his examinations when his mother's letter arrived and he responded at once, heading his letter: Not for Family Circulation. First, he congratulated her on her purchase:

My darling Mummy, … I am full of admiration and congratulations really. I am convinced it was the right thing to do — If we are all decided to settle down permanently near Oxford — and we have all greeted the suggestion at one time or another — then it was obviously worthwhile waiting until somewhere really suitable turned up. … As you say, it will cost more than Gransden to keep up, but Gransden has always had the enormous disadvantage that it has so little ground attached; what ground there is cut up by oast houses etc.; and there are houses, roads, people etc. all clustered around. One does want a spot of garden to be able to walk in more or less privacy, 'far from the madding crowd'. Now we have it.

Shortly before Christmas 1933, Hugh was elected a fellow of the Royal Society of Medicine. During the weeks that followed he worked feverishly, tutoring students and preparing to qualify for the BSc. Hard work was put aside, however, on the day before his birthday, in February 1934, when Hugh and his mother made plans for their move to Lady Place. She moved in soon after, but it was not until 27 April that Hugh spent his first night under its roof. Rosalie had prepared as best she could for his arrival and the ensuing onslaught of his friends. Now that he had a place to entertain, Hugh lost no time in putting her to the test. On 13 and 20 May there were parties of tennis, lunch, supper, and bathing; on 18 and 28 June there were bathing parties. Further parties followed, and then on 2 August, the Sinclairs hosted their first large garden party, from 3.30 to 6.30 p.m., during which old friends from Kent mingled with new neighbours, and some of Oxford's most prestigious academics.

Lady Place was large enough to accommodate the needs not just of Hugh, but all Rosalie's children. Each had their own private sitting-room, each equipped with a grand piano. Hugh had a second grand piano in his rooms at Magdalen College, although he spent little time playing either. Hugh Macdonald Sinclair had found his home.

In 1935 Hugh began a race against time to complete all the tasks he had laid out for himself. But somehow, between tutoring, clinical studies, and running experiments for Rudolph Peters, he still found time to entertain and travel. Blessings were mixed. In August, he attended the meeting of the Physiological Congress in Moscow, and saw many of the places his father had visited with his grandfather. Impressed with Russia, Hugh wrote vividly to his mother about the All-Union Institute of Experimental Medicine, the Brain Institute, and the Mother and Child Institute. He enjoyed the Palace of the Tsars and, along with the others members of the Congress, was entertained grandly by the Leningrad Soviet in Catherine Palace, where they dined on *hors d' oeures, froid* and *chaud*, followed by soup, salmon, turkey, and ice cream. All this was washed down with seven different wines. The next day they travelled to the Kremlin. On the way, Hugh and his colleagues, most of them much older than he, were taken to Lenin's tomb, which had been recently closed but was opened in their honour. Once in Moscow, there was little time for sightseeing, but delegates to the Congress enjoyed a banquet in the Hall of St George. The highlight of that evening was a speech by Vyacheslav Miklailovich Molotov, then Chairman of the People's Council of Commissars. It was a grand time, not unlike the trips Hugh's father had enjoyed with Sir John Jackson, and Hugh felt well suited to it.

About this time a new name appeared, underlined, in Sinclair's notes of meetings and parties: <u>Howard Florey</u>. Previous pages were filled with names of the great scientists he knew, such as Sherrington and Haldane, but none had been underlined. Soon, a letter from Howard Florey, on his letterhead from the Sir William Dunn School of Pathology, University of Oxford, arrived stating: 'Sinclair — I would be very glad if you could come to see me the next time you are in Oxford.' Thus began a new influence on Hugh's career.

Howard Walter Florey and John Farquhar Fulton arrived in Oxford, during 1921, as Rhodes Scholars and students of Charles Sherrington. Both would have a significant influence on Sinclair's life. Fulton was an American, educated at the Harvard Medical School; Florey was from the University of Adelaide, in South Australia. Following their period as Rhodes Scholars, Howard Florey stayed on in Oxford, but Fulton briefly returned to Boston to resume his studies under Harvey Cushing, following which he returned to Oxford with a fellowship in natural science. He remained there until he was appointed the Sterling Professor of Physiology at Yale University, in the United States. The two men remained lifelong friends. Fulton was impressed by the importance Charles Sherrington placed on understanding the writings of past scientists, and (like Sherrington) became an avid collector of rare medical books and manuscripts. This shared interest sustained a long correspondence between these men and stimulated the exchange of numerous manuscripts. In particular, the work of the French scientist, Jean François Fernel (1497–1558) interested Sherrington, and thereafter Fulton. Fernel, the first man to write a book entirely devoted to the subject of physiology, and the first person to call the subject by that name, was Charles Sherrington's scientific hero. Fernel inspected and wrote in detail about both normal and abnormal conditions of the body, throwing medical thinkers of his day into turmoil. He saw and described acute illnesses, including appendicitis and endocarditis, and in the second part of his great treatise, *Medicina*, published in 1554, defined and named the subject of pathology for the first time. In his book *The endeavour of Jean Fernel*,[1] Sherrington reminds his readers that Fernel advised physicians to reject tradition and study the human body for themselves. All this eventually influenced Hugh Sinclair, who also adopted Fernel as a scientific hero.

According to Sherrington's biographers, John C. Eccles and William C. Gibson, Florey's manner of working was deeply influenced by Sherrington's, and he never forgot that his mentor was first a pathologist, and then a physiologist. In 1923, Sherrington, then Head of the Department of Physiology, invited Florey to become Demonstrator for the mammalian class. Florey later attributed his success to the good fortune of being taken into the Sherrington fold. As Florey's term as Demonstrator came to a close, Sherrington convinced him that good pathology is the same as abnormal physiology. By the time Hugh Sinclair became tutor in physiology at University College, Howard Florey was elected to the Chair of the Sir William Dunn School of Pathology, in Oxford. Florey would eventually try to convince Sinclair that nutritional biochemistry was only another aspect of pathology, and that Hugh should join his department, rather than risk working with Rudolph Peters in his relatively new Department of Biochemistry.

Howard Florey's election to his Chair involved a number of key individuals who would eventually shape Hugh Sinclair's academic fate, and so it is worth considering the circumstances of the event. Florey's appointment was strongly supported by Sherrington, but Sir Charles had given up his place on the Election Board in favour of Sir Farquhar Buzzard, the Regius Professor of Medicine in Oxford. The outcome of the debate by the Board was finely balanced between Florey and another strong candidate. However, only a few weeks before the final decision, Sir Edward Mellanby, of the Medical Research Council, and a close friend of Sherrington's, was asked by Lincoln College to act on its behalf on the Electoral Board. Sherrington took heart, believing that Mellanby was likely to support his choice of candidate. On the day, Mellanby's train from London was late, and he almost missed the final vote: the decision was going against Florey when Mellanby arrived. As a highly regarded member of the group, Mellanby's views were immediately heard, and his support of Florey swayed the decision away from the other candidate. Florey never forgot Mellanby's influence and efforts on his behalf.

Sir Charles Sherrington's influence affected Hugh Sinclair's life both in and outside Oxford. In October 1935 Hugh was informed that he was to be put up for election to membership at the prestigious Athenaeum during the following year. His mother had managed to get his name listed among the nominees for membership some years before, and now the time had come for the final step in their process. He was asked to give details of his degree and any special 'attainments' — although he was assured that he need not be too alarmed by the latter requirement. Again, Hugh and his mother solicited help for every appropriate quarter. In the end, however, only one voice counted. During January 1936, a month before Sinclair's 26th birthday, Sir Frederick Kenyon wrote to inform him that he had been elected to the Athenaeum under the new system, being accepted into membership by a Committee, rather than the General Membership. He wrote: 'Your qualifications were described, with Sir Charles Sherrington's letter in your support, and you were unanimously elected.' Hugh penned a note to his mother at Lady Place saying:

> My darling Mother,
> Just a line to say that two days ago, greatly to my surprise, I was elected a member of the Athenaeum. Bob Trotter had told me that his father — who is the King's surgeon, an FRS and a great physiologist — got in 4 years ago after waiting 20 years; and I heard that Rudyard Kipling created a record by getting in at the age of 27. So I did not hope to be elected for at least 50 years …

Rosalie's budget remained tight, and she turned to her mother for the sum needed to secure Hugh's membership. From her home at 24 Palace Court, Lady Jackson sent her daughter a letter of congratulation on her son's success, and a cheque for £66.3.0 to cover Hugh's membership fee, annual dues, and other costs.

Not everyone was as impressed with Hugh's potential as Sir Charles. The previous year Hugh had managed to draw a cloud over his reputation through what he probably considered to be a clever bit of humorous writing. Early in 1935 Hugh added the position of editor of the University College Hospital magazine to his exhausting list of responsibilities, and in the January–February issue published an editorial entitled 'Turks, Infidels and Jew CH'. It read:

> The editorial Pen was worn out; the Editorial Wrist was aching. Coffee, we thought — even Refectory coffee — might stimulate the Editorial Brain. We sat at a table. On our left two men, one elderly, were chatting noisily with two women, one matronly, about the treatment of some filthy surgical condition. On our right a lot of canvas hoarding and mess indicated that our Refectory was about to be improved. A sudden burst of hammering told us that the process was active. As we walked away, deafened, we wondered why it takes weeks to knock a wall down.
>
> We thought that at 11 a.m. the Lounge would be empty; keen students would be pressing round dying patients on Visits and in OPD (the bigger the lump, the keener the interest and the more hopeless the prognosis). We were correct. We sank into a tattered chair and turned to thoughts Editorial. Two men came in and sat near us. They began to converse in a language not usually heard in Europe. 'What a wonderful Medical School we must have,' we thought, 'since men from all parts of the Empire — from all over the world — flock here. We get the cream of the would-be doctors of all nations; what matter if the cream be sour?' The Editorial Breast filled with pride at this glorious realisation, but quickly emptied again because an exotic smell entered with the pride. Perhaps we were wrong. Perhaps we get the dregs of the world's students because we can get nothing else? That might account for the lack of interest in the Medical Society meetings, in the Medical Society Ball, in athletics, in everything except work. Are there no Medical Schools in Palestine, in Poland, in India, Africa, or Baluchistan? A breath of fresh air blew in from somewhere, and we became less bitter. We would welcome most foreign students if only they would regard themselves as guests at a British Medical School and not always consider solely their own interests. The only tolerable foreigners are the unselfish ones, and they are few.
>
> Our musings were interrupted by some workmen who wanted to mend the chair we were sitting on and put new carpet beneath us. We went into the Silence Room. It wasn't. It was filled with polychromatic faces, broken English and strange odours. We could not get near the Ping-Pong table (except on a Saturday when the Synagogues are open).
>
> We went to a place usually labelled 'gentlemen.' Here at last we found no one.

There were murmurs among the medical community about the piece, although some of Hugh's college chums supported and congratulated him. Not surprisingly, the *Jewish Chronicle* found Hugh's sentiments offensive. In the 22 March issue, under the caption: 'An irresponsible effusion', Hugh's article was summarized with the following comment:

> The sad point about this sorry stuff is that it was published in the magazine of a Hospital whose Chairman and Treasurer, Sir Herbert Samuel, is a Jew by race. The Hospital itself is an adjunct of a College which was built as a gesture to the new but now apparently obsolescent spirit of tolerance, and in its creation Jews bore an honourable part. Little more need be said to stigmatise the effusion we have referred to; but we are glad to record that an extraordinary general meeting of the Medical Society of the College was quickly convened, and that this meeting, attended by considerably more than one hundred members, carried, practically unanimously, a resolution proposed by non-Jewish members dissociating the Society from the article, 'of which it strongly disapproved.' The issue of the magazine in which it appeared is, further, to be withdrawn from circulation, and an apology is to be inserted in the next number. This prompt action does credit to a great institution and is an exemplary vindication of the finest British traditions. We hope that it will be noted by any other editorial institution here, and especially abroad, in which the miserable spirit of intolerance might, in these difficult times, show itself. The writer of the article was perhaps more bent on showing the cleverness of the Editorial Fist than on airing any considered views. We regard the whole incident as happily and honourably closed.

In his own defence, Sinclair penned a statement, which he apparently read at an Extraordinary General Meeting of the UCH Medical Society on 18 March. Sinclair stressed that he had not expected the article to be taken seriously: the phrase 'Jews, Turks and Infidels' was a well-known one, he stated. And, as 'not being a Christian myself', he claimed rank among the Infidels. What is more, Sinclair argued, the epithet, Jew CH, which he chose to use, had been applied to the institution before. He thought it 'absurd' that his article was taken as a piece of 'religious or political propaganda'.

> Everyone [he said] is entitled to his own views ... and I believe as strongly as anyone — that science knows no confines of race or creed ... I am justified in taking the view which I did take upon the interpretation which would be

placed upon the Editorial, as shown by the fact that a member of the Staff told me that he and other members of the Staff read it, laughed, and laid it aside. And I know that some foreigners did the same.

His more formal apology appeared in the next issue of the UCH magazine.

> We regret to find that the editorial in the last number caused offence to many people. We intended it to be neither religious nor political propaganda. We therefore apologise to those who believe that it was a direct and serious attack upon race and creed as such, for not foreseeing that that was a possible interpretation; and we assure them again that no such attack was intended.

Sinclair's career would continue to progress at Oxford, but the arrogance of his plural-self was not easy to forget, and may have helped form opinions of some of his work during the world war now brewing in Hitler's Germany.

The rumblings of that were beginning to roll across the Channel. In the early months of 1935, Hugh Sinclair obtained two issues of *Paris-Médical* containing the results of a symposium, edited by Médicin-Commandant Moynier, on chemical, or 'gas', warfare. In the May–June 1935 issue of the UCH magazine, Sinclair published 'Poison gas for the masses', an editorial defining the types of gas available, their modes of action, and how doctors and nurses will be taking part in the teaching of 'anti-gas manoeuvres'. He implored medical students not to attend to the Government's call for them to help convince the public that a defence against chemical weapons is plausible. He wrote:

> To do this they require more than politicians. They require people whose profession is so honourable that all their statements are taken as truthful and authoritative by the public — they require the doctors, nurses and students to disparage the protective system as infallible…
>
> It is upon the attitude of the doctors of this and other nations that Governments must rely to carry out their plans, our refusal will mean that they will be forced to seek Peace in some other way and we shall have retained our honour and be worthy of the trust the public place in us.

Sinclair offers his readers a bibliography, supplied to him by the Inter-Hospitals Socialist Society. He suggests that a comparison of that list of publications with the war propaganda and press notices provided by the Government would make interesting reading.

Soon after Hugh's debacle with the Jewish establishment, Florey asked Ernest W. Ainley Walker to contact Hugh and sound him out about joining his department. In late May 1935, a letter arrived stating:

> *My dear Sinclair,*
> I am writing this letter at Professor Florey's request. It will need some thought; but the matter must be dealt with within the next few weeks or so … Vollum is going to be in charge of the teaching of bacteriology in our Department and Florey wants (either now or presently) a younger man — a researcher, and perhaps a biochemist by nature — who would be willing to come in as a demonstrator in bacteriology. Florey has knowledge of a suitable man, who is available now if a decision were reached shortly to take him. But he would prefer to have an Oxford man, and would wait some little time (i.e. two years or more) if he had the right man in view. Florey has somewhere picked up the idea that you look as if you were likely to succeed me at University presently, and are intending to pursue a career in research on the biochemical side. He has heard a good bit about you up and down. He wants to know whether his job would attract you at all; and if so to arrange to have a talk with you pretty soon to see whether he can come to terms with you.

Ainley-Walker further said that the position with Florey would allow 'abundant leisure for any line of biochemistry', and that he thought the Departments of Biochemistry was beginning to fill up with staff, many of whom were quite senior. The appointment would take place in October 1937, and although Florey would want Hugh to take a Diploma in Bacteriology, his requirements would fit around Hugh's plans for a Radcliffe Travelling Fellowship in February 1937.

About this same time, Hugh approached his good friend Kenneth Franklin for advice as to how he might pursue the honour of becoming a Fellow of the Royal Society. Franklin said that working with 'the person in question' (Florey) was a swift way for a physiologist to be considered for Royal Society. Franklin also defines Hugh's true nature:

> There will come a time, *experts crede*, when tuition is for you also a weariness of the flesh … If you have a definite bent for research, give it the best opportunity for developing. Discovering something new is the really joyous thing … If you renounce E.W.A.W's job, Peter Meiklejohn might perhaps act as substitute. All best wishes K. J.

Probably to his regret, Hugh did not follow Franklin's advice. Years later, when Hugh talked about Florey's interest in him, he ended with the footnote that he had told Florey he intended staying with Peters and studying nutrition. Florey is quoted as saying that nutrition is an uninteresting subject, as most major discoveries had already been made. Hugh would add, after allowing his audience time to smile, that Florey later apologized to him; and that Florey's biographer, Gwyn Macfarlane, had told him this was the only time Florey apologized on such a matter.

As time approached for Sinclair to take his medical examinations he followed the example of others before him, and qualified for the medical degree at the Worshipful Society of the Apothecaries. It was not unusual for university students at the time to qualify through the Apothecaries, earning the LMSSA, before being examined within the university, as this allowed them to practise answering complex clinical questions. It also allowed them to respond in the affirmative if a university Professor inquired as to whether or not they had already qualified in medicine. It was then possible to qualify to practise medicine by two systems: through an apprenticeship with the Apothecaries, or through university examination. Many a young man, unable to afford or to obtain a place at university, would do his apprenticeship, receive his LMSSA, and set up his medical practice the next day. (Although Sinclair remained a faithful member of the Apothecaries, and rose to its highest ranks, he rarely mentioned his LMSSA.)

On 18 June 1936 Hugh Sinclair received his MA, BM, and BCh, an achievement celebrated with a party at Lady Place.

One of the most significant events in Sinclair's life took place during the summer of 1937. Having received a Radcliffe Travelling Fellowship, he journeyed to the United States and spent three months visiting friends and making new contacts for both inspiration and pleasure. The stay could have been longer, but Hugh had been elected a Fellow and received a Tutor-ship at Magdalen College, and was needed back in Oxford by the start of the October term. He had been planning the trip for some time, collecting names of those he wished to see. John Fulton suggested he visit George R. Minot (1885–1950), Director of the Thorndike Memorial Laboratory, in Boston, Massachusetts. If possible, Fulton suggested, Hugh should spend an entire year working with Minot, as he knew of no other laboratory (perhaps other than that of Sir Thomas Lewis, in London) where someone could gain a richer opportunity in scientific medicine. Minot was best known for devising a diet of raw liver to treat pernicious anaemia, for which he had shared a Nobel Prize with William Parry Murphy. But Minot and his co-workers were also interested in vitamin B_1, and therefore anxious to hear more about Sinclair's work with Rudolph Peters. Unfortunately, during the time Hugh planned to be in Boston, the Thorndike was unable to offer much support, as Minot was away. Hugh was welcome to spend time in the laboratory, but there was no money to pay a stipend, or for animals and supplies he would require. Minot wrote to Fulton about Sinclair: 'Experience has taught me', he wrote, 'that men coming for a short period of time should look upon their experience as a cultural one, rather than that they would necessarily have time to accomplish a thoroughly worth while clinical study.' When Hugh Sinclair finally arrived in Boston, he spent only a short time there.

Hugh planned his Fellowship to study 'methods of research in scientific medicine in the USA and Canada'. His road was made easier by letters of introduction written by his friends. John Fulton wrote more than a dozen to important medical scientists, including Herbert M. Evans, at the University of California, Berkeley; William J. Mayo, of the Mayo Clinic; and Walter B. Cannon, of the Harvard Medical School Department of Physiology. Each letter described Sinclair in a particular way to enhance him in the eye of its recipient: in one case, he was interested in vitamin B_1; in another, he was a friend of Sherrington's; and in several, Hugh was identified as a collector of rare medical books. Fulton even wrote to his Aunt Sue, who already had some knowledge of Lady Place, and was acquainted with Sir Charles Sherrington.

Soon after arriving in New York, Hugh visited the Rockefeller Institute, the Cornell Medical Center, and the Presbyterian Medical Center. He then travelled to New Haven, Connecticut, and Yale University, where John Fulton introduced him to Professors Harvey Cushing, Chairman of the Department of Neurology, and J. P. Peters, Chairman of the Department of Medicine. Charming, a fine example of a British gentleman, highly intelligent and well informed, it was easy for those he met to recall Sinclair's name in the years that followed.

Hugh did spend some time with Professor Castles doing Ward Rounds at the Thorndike, and nine days at the Woods Hole Marine Biology Station, in Massachusetts. He travelled to Canada, where he took time to visit some relatives between visits to McGill University and the Banting

Institute at Toronto. From there, he went on to Michigan, Chicago, and Madison, Wisconsin, where he stopped for a short time at the Sprague Institute. Then he went to the Mayo Clinic, where he met most of the senior staff. He left the Mayo Clinic greatly impressed.

Moving on to San Francisco, he visited Professor H. M. Evans, at Berkeley, who was working on vitamin E and essential fatty acids. (In writing about the trip later, he mentions research on vitamin B_1 being carried out in Evans's laboratory, but made no mention of either vitamin E or vitamin F, which is what Evans called essential fatty acids.) Turning back towards England, Sinclair stopped at St Louis University, and Johns Hopkins University, in Baltimore, and then travelled on to Washington DC, Philadelphia, and Boston, where he met Professor Cannon and his associates at the Harvard Medical School Department of Physiology. After a final trip to Yale to visit John Fulton, and to meet again with Harvey Cushing, Sinclair returned to New York. His final meeting was with the Head of the Rockefeller Foundation. On his arrival back in England, at the beginning of October, he returned to Oxford and began teaching at once. He did not find time to write a report to his benefactors about his trip until February of the following year.

Although Hugh was slow preparing the trip report required of him, almost as soon as he returned to England he laid out a scheme that he called the *General plan of research: diseases of the nervous system*. (See Appendix B.) The scope and organization of this plan was influenced by close friends, like McCarrison, and by the experts and laboratories Sinclair visited while in North America. Unlike work he would plan for the Department of Biochemistry, this new idea involved several disciplines, and looked at the relationships between identifiable illnesses in a new and comprehensive manner.

He split the subject of diseases of the nervous system into two categories: neuritis; and psychoses and psychoneuroses. According to Sinclair, conditions in this last category were to be those caused by 'deranged metabolism' of brain cells, and those caused by toxic substances or 'anatomical derangements', under which he included liver disorders, Pick's disease, and Alzheimer's disease.

Neuritis was to be studied as a condition caused by the 'deranged metabolism of the neuron, and was to be investigated' in four ways: by subjecting brain cells to neuritic agents *in vitro*; by inducing neuritis in animals, and then studying the respiration of nerve cells *in vitro*; by comparing the clinical manifestations of different sorts of neuritis and then attempting to explain the differences; and by 'investigating clinical aspects of B_1 deficiency'. This last point would justify beginning work on this research model in Peters' department.

By laying out his plan of how medical research should be pursued, Sinclair achieved two things. First, he demonstrated his ability to define large multifaceted, multidisciplinary investigations, which combined clinical and laboratory research. Second, and of great political significance, he cut out for himself a research approach that opposed the more narrow and rigorous step-by-step strategy followed by Peters and most other senior scientists of the day. At the time Sinclair devised his plan it had little relevance to his work, but soon it would become the foundation for decades of scientific activity.

Hugh was not alone during his period in America. Archibald Malloch MD, Librarian at The New York Academy of Medicine, sent a letter to the Librarian at Woods Hole, Massachusetts, telling her that two Radcliffe Travelling Fellows from Oxford wished to see the library there; the second Fellow was Dr A. P. Meiklejohn, who subsequently took a position with the Rockefeller Foundation, in New York.

Catherine accompanied Hugh on much of his trip, having left Liverpool on 20 August and arriving in Boston on 29 August. Rosalie was also with him in spirit. In her firm, clear hand, she wrote frequently, telling Hugh she was pleased to hear he was having a nice time, giving instructions about making suitable arrangements for Catherine, informing him of John's travels, her taking tea with friends, and the antics of canaries nesting in the aviary. Her only request was that Hugh should buy her a book on canaries, to be signed by the author, if possible.

Rosalie's interests were too far removed from her younger son's to see that problems were mounting. She would not know that a warning shot came in a letter from Peters that briefly mentioned a conversation he had had with another investigator whose experimental results differed from Sinclair's. It was Peters' habit to communicate frequently with Hugh by note, and on occasions he had expressed oblique concern over the accuracy of Hugh's data: numbers in calculations pertaining to research results had obviously been corrected, had Hugh checked the final figures? Peters was being pressed into war work on the development of anti-Lewisite, (an antidote to Lewisite, a persistent form of poisoned gas) causing him to be away from Oxford for days — or even longer. Sinclair's attention to detail was critical because Hugh was now responsible for many of Peters' experiments.

As already seen, Hugh had expressed strong feeling about the role of medical scientists in war, and was now a member of the Medical Peace Campaign, which was being steered by its President, Professor John Ryle, of Cambridge University. (These early contacts with John Ryle set the scene for later events in Sinclair's career, when Ryle moved from Cambridge to Oxford University.) The Dons at Oxford saw the university as having a part to play in war work, and Hugh would certainly be pressed into some form of related activity. When his challenge came, however, it would involve work far beyond the walls of Oxford University.

Human nutrition, a subject viewed by many as a sub-set of issues within the more important field of biochemistry, suddenly gained considerable national importance. During World War I the British were unprepared, and the food rations to the population were so low that there was concern that the population would not have been physically strong enough to withstand a much longer conflict. This time, things would be different. The Government would seek scientific evidence on the state of the nation's health. It would also maintain a high level of good health though public education and popular involvement in the production of food.

From Sinclair's point of view, these objectives had major significance. It was now possible that he could have access to large populations of people with vitamin B_1 deficiency, because there was evidence that a significant portion of the British population did not consume enough of this vital nutrient. For example, an article published in 1938, in the *Journal of Hygiene*, stated that 70 per cent of the population in Britain consumed less vitamin B_1 than was necessary for good health.[2] Action was necessary, and who better to take it than the experts at the Oxford Department of Biochemistry? With a little good fortune, Sinclair would have a means by which he could begin collecting information to fit into his master research plan for the study of disease of the nervous system. Certainly this would be the subject that would secure his name in the history of Medicine.

CHAPTER 5

War

On 1 September 1939, the British Prime Minister, Neville Chamberlain, demanded that Germany cease its invasion of Poland or face a declaration of war: the ultimatum was repeated on the following day. These efforts proved futile, and on 3 September, at 11.30 a.m., the people of the United Kingdom were told hostilities had begun. British ships in the Western approaches and elsewhere were set upon by U-boats, and at 9.30 that night the passenger liner *Athenia* was torpedoed with the loss of 112 lives.

On hearing the news, Hugh unburdened himself in a letter to his mother. He thanked her for the care and kindness she had shown him, recalled the good times they had enjoyed together, and said he loved her. He admitted that he had been 'dreadfully tired' at times, and confessed to two bad breakdowns through overwork. If he had been unkind to her, he wrote, it was due to exhaustion, and never the lack of love. After placing the finished letter and a copy of his will in a small brown envelope, Hugh sealed shut the flap with red wax.

Hugh's will reveals more than the letter to his mother. Hugh estimates his worth to be approximately £7000, most of which was in stocks and shares purchased with money given to him by Rosalie. The sum also included his collection of rare books, which at that time he estimated would fetch about £2400 if sold at auction. With the exception of leaving a sum of money and the gift of his father's gold pocket watch to John, and providing a sum of money for Salie, the remainder of his wealth was to go to Catherine. Hugh left Rosalie nothing, but said she should have the use of any of his furnishings until her death: then, all was to revert to his younger sister. In the event that Rosalie outlived Catherine, everything would go to his mother for her lifetime, after which it should be put to use for scientific research, 'to which I am devoted', he wrote. His further stipulations were that if:

> … Catherine die unmarried after my mother has died, she (Catherine) will leave anything that remains of my property to the Royal Society in order

that, should they wish to do so, they may found a scholarship for research in medical science to be attached to the Department of Biochemistry, University Museum, Oxford. It is my wish that, after the death of my mother and my sister Catherine, whatever is left of my money shall be devoted to the advancement of science in the way I have indicated above.

War left Rosalie at Lady Place with the family servants, and John's dog, Blood. As the Colonel would have wished, she undertook voluntary work with the Red Cross, assumed a leadership position in the local community, and worried about the safety of her children. John had been posted to the Middle East: his letters were infrequent and censored now. For her part in the war, Catherine had joined the FANY: The Women's Transport Service First Aid Nursing Yeomanry. Rosalie approved of Catherine's choice of war activities, despite knowing something of the danger she might encounter.

Founded by a cavalry sergeant in 1907 as a small unit of nurses on horseback to link fighting units at the Front with field hospitals, the FANY now drove ambulances and set up troop canteens. The majority of recruits were well-educated women from the best British families. But intellectual skills were not enough — they were also expected to act with bravery in the cause of their country. Unlike other women's services, the FANYs could carry and use small arms, and many became active agents of the SOE (Special Operations Executive). More than 2000 women worked in ciphers and signals, administered Special Training Schools, and worked as agent-conducting officers. Many were dropped into occupied France, and took active part in espionage. Thirteen died in concentration camps. Three were awarded the George Cross, two posthumously.

Hugh stayed close to home, living sometimes at Lady Place and sometimes in his rooms at Magdalen College, but his life was far from unaffected by the circumstances of war. Some of his activities are well documented; others leave only shadows. For example, shortly before he died, Sinclair told friends that he had had three fiancées during his life, one of whom killed herself with cyanide when caught behind enemy lines. Catherine's post in the FANY could have made a tragic liaison between her brother and one of her colleagues possible, but no evidence of such a romance exists among Hugh's papers. And it must be said that, considering the exhausting schedule of research and teaching he set for himself, one wonders when he would have found time for an affair.

In the ancient refinement of Oxford and its colleges, preparations for war were well underway. Air-raid shelters were built, including one on a strip of land between the Department of Biochemistry and the University Museum. College windows was blacked out, and everything valuable (including stained-glass windows) was removed to storage. Along with Professor Rudolph Peters' work on anti-Lewisite, Oxford scientists turned their hands to other areas of great military importance, including the testing and synthesis of penicillin, radar, and the atomic bomb.

The twelfth volume of *The history of the University of Oxford*, edited by Brian Harrison, recounts some of the peculiar circumstances created by the war.[1] The Oxford Union remained open, in sharp contrast with the period of World War I, when it was closed for the duration. Many college and university buildings were requisitioned by the Government: the Examination Schools and St Hugh's were converted into hospitals; most of Balliol College was occupied by the Political Intelligence Department of the Foreign Office; and the Controllers of Fish and Potatoes at the Ministry of Food were lodged at St John's. This last group was often referred to as 'the biggest fish and chip shop the world has ever seen'.

Debate about the war was allowed, but censored, and any direct criticism of its conduct forbidden. Shorter and more practical war degrees replaced standard academic programmes. Refugees arrived, and those with skills were given work to replace staff specialists who had entered government service.

Hugh now combined departmental demonstrating and research in the biochemistry laboratory with long hours of tedious experimentation for the Ministry of Supply, assisting Rudolph Peters in his secret development of anti-Lewisite. Some of this anti-poison-gas work led to the publication of three research papers after the war, on arsenic compounds and pyruvate oxidation; but from the standpoint of Hugh's academic career, that was small reward. The war work slowed what had become Sinclair's prodigious flow of research publications. By the end of 1939, at the age of 29, his work on vitamin B_1 was becoming well known among the international community of biochemists. He was sole author or co-author of 15 scientific papers, and had published long discussions of vitamins in several major works. Now that flood of publications had almost ceased. Worse still, research on anti-Lewisite restricted Hugh's time to pursue the comprehensive research plan he had developed after returning from his three-month trip to America for the study of diseases of the nervous system. But the realities of war were to change that.

Sinclair's plan required the simultaneous investigation of brain disorders by biochemical analysis and clinical assessment. Biochemical data presented no problem: he was steeped in it. Thus far, however, he had had few opportunities to collect the clinical data he needed. But, when the British Government announced it wanted to know the frequency of vitamin B_1 deficiency among the British population, Hugh saw his chance. His interests in human nutrition now took on a more pragmatic and patriotic focus. By applying himself to the problem of B_1 deficiency within the nation's population he could create opportunities to collect the clinical data he sought. Equally important, he would have a visible means of making a vital contribution to the war effort, which was certain to please his mother.

The British Government's concern about the nutritional state of the population had a strong foundation. With war came the possibility of malnutrition, even famine, and it was thought that civilian populations suffering nutritional stress were a risk to the defence of the nation. The inadequate supply of food during World War I, followed by the consequences of the deep depression of the 1930s, had left many British civilians in a poor state of health. The great British nutritionist, Doctor John Boyd-Orr, estimated that one-third of the British population lived on inadequate diets owing to ignorance or lack of funds to purchase healthy food: estimates by other experts were even higher. In response, the Ministries of Food and of Health, in cooperation with the Medical Research Council, set nutritional standards for the nation, and undertook steps continually to assess the state of dietary well-being by the use of various surveys.

Particular attention was now being paid to those groups of workers who would be needed in the event of another all-out military conflict: miners, farmers, railwaymen, and munitions workers. The nutritional levels of the families of these men were also studied: the data were depressing, and experts worried that — owing to the nutritional decline of the country — it might be difficult to find adequate numbers of healthy recruits to fight for their homeland. Ultimately, if these islands were threatened directly, and the civilian population was in a poor state of health owing to inadequate food, they might not be strong enough to defend their homes and resist invasion. Part of the problem resulted from poor food habits; part had to do with the availability of things as basic as fresh fruit and vegetables.

During the years between the wars, efforts to increase a general understanding of what constitutes a healthy diet had enjoyed some success; but by the time war with Hitler became inevitable, the problem of availability had become paramount. Food was produced at home, of course, but a large proportion of what the British had been eating during peace-time was brought to their shores by ship. Between the wars, approximately 40 per cent of food was imported from the Channel Islands, Holland, New Zealand, the Caribbean, Canada, and elsewhere. As war approached, new food supplies were desperately sought, especially from the United States, but experts questioned whether or not sea routes for civilian shipping could be kept open. Deeply concerned about the possibility of enemy attacks within the shipping lanes, members of the Government agreed that every effort must be made to monitor the nutritional state of the public, and to help people provide their own food supplies. Part of the answer lay in increasing the land available for agriculture.

An article in the 15 December 1939 issue of *The Times* announced the Ministry of Agriculture's plans for a large increase in the home production of food. If the British people could produce more from their land, the savings in foreign exchange and shipping could be allocated for the purchase and import of raw materials and munitions, it reasoned. Therefore, the Government planned to secure the ploughing up of at least 1 500 000 acres in England and Wales and 2 000 000 acres in the whole United Kingdom before the following summer. Great Britain was asking men and women to go to war with spades and ploughs as well as guns and bombs. By the early part of 1940, it was estimated that about 900 000 allotments dotted the country, and another 500 000 were needed. Along with the millions of private gardens already in place, it was hoped these personal efforts could supply the tons of cauliflower, broccoli, carrots, and potatoes needed.

Some of the flat and fertile land that could have provided space for crops was being pressed into other wartime uses. For example, the expanding use of airpower removed large areas of arable land from production. During the course of World War II, hundreds of new airfields for the Royal Air Force and the United States Air Force were built, mainly in agricultural areas. At the peak of construction, in 1942, the RAF alone was creating a new airfield every three days, each covering many acres.

By the end of 1939, a new subject excited British health experts: synthetic vitamins. Could these mass-produced chemicals help nullify the clinical significance of restricted diets? Britain was not the only nation interested in this question. According to reports from Rotterdam and elsewhere on the Continent, an increasing number of stories about night-blindness were appearing in the German press. With blackouts, the condition had become more noticeable, and limited food supplies were being blamed. During the World War I, German prisoners held in French camps were suddenly attacked by night-blindness, but found that the condition reversed itself when they were moved to locations where there was more food. The curative factor was thought to be an increase in the quantity of food containing carotene and vitamin A, such as eggs, butter, and fresh fruit and vegetables. By 1940, the realities of war, and the preferential supply of food to the German armed forces, had reduced the availability of these foods for German civilians, and synthetic substitutes were being considered for distribution. The idea of synthetic vitamins was becoming increasingly popular with British health experts as well, and nutritionists around the country were consulted on the matter.

Hugh was willingly drawn into meetings with various governmental groups to discuss a range of subjects pertaining to the nutritional status of the British population, undertaking the preparation of reports and position papers when asked to do so. He also decided to help provide accurate nutritional information to the public by writing letters and articles in newspapers, and giving talks to community groups. This gave him influence, which he enjoyed, and he was pleased when his mother took notice of something in the paper that quoted him. All this was done in addition to his long-standing teaching and research obligations. He seemed never to tire, but he became irritable and, with good reason, Rosalie worried that he was again exhausting himself.

With the New Year, 1940, came a succession of new horrors. Himmler established the first Nazi extermination camp. There were dogfights over the Channel. In May, the Germans outflanked the Maginot Line. Rommel and the 7th Panzer Division crossed the Franco-Belgian frontier to face only the French 2nd and 9th Armies, without anti-aircraft artillery or anti-tank guns for their defence. Mothers wept as letters arrived stating in cold and simple language the essential circumstances of the wounding or death of a son.

But more than weapons kill during war: infection is the great silent slayer. During the World War I, and likewise during the American Civil War, disease exterminated more men than did munitions. But infection need not kill to win: an outbreak of dysentery in a fighting unit can disable it, for example. For centuries, commanders knew that disease often plays the winning hand in a military campaign. To defeat Germany and her allies, finding new antibiotics was a priority. Important progress had been made since World War I. Sulfonamides became available in 1936, and would save many lives in the years of combat to come, but they had unpleasant side-effects on the kidneys and blood. Alternatives were necessary. In 1928, Alexander Fleming found that certain of the bacterial cultures in his laboratory were killed by the growth of a mould called *Penicillium notatum*. The active agent, which he called penicillin, was effective against a wide range of pathogenic organisms, including members of the highly infectious bacterial genus *Streptococcus*. Of equal importance, experimental evidence showed penicillin did not damage human cells. These observations were important, but Fleming was unable to isolate and purify the active substance, and therefore conducted no clinical tests. However, scientists at the University of Oxford saw the potential usefulness of Fleming's antibiotic, especially during war, and pursued the matter.

In 1940, a team led by Howard Florey published proof that bacterial infections could be cured by treatment with a substance extracted from Fleming's mould. Under primitive conditions, Florey and his team had found a way to isolate the antibiotic and produce small quantities for clinical tests conducted on critically ill patients with no hope of survival. The results were astounding. Temperatures broke, tissues began to heal and, unlike the sulfonamides, there were no damaging side-effects to vital organs. Here was the lifesaver war surgeons had prayed for. But its availability was so limited that it was difficult to conduct proper clinical trails, let alone cure the battle wounded. A vast infusion of money was required to develop the methods needed for mass production. Already financially stretched by the cost of the war, the British Government was unable to provide the necessary sums. As one of the many ironies in Sinclair's life, he played an accidental, but fundamental, role in finding the support needed. But that story will be told in due course.

It is also ironic that Howard Florey's partner in this work was Ernst Boris Chain, a Jewish refugee, who had filled the position Florey had offered Sinclair several years earlier. In 1945, along with Fleming and Florey, Chain would receive the Nobel Prize for medicine for his work on penicillin. If Hugh had chosen to work with Florey, rather than pursue his interests in human nutrition, the Nobel Prize may have been his. Certainly he would have enjoyed membership of the Royal Society, which he openly sought.

During the first few days of May 1940, Rudolph Peters told Sinclair that the Accessory Food Factors Committee of the Lister Institute and the Medical Research Council were anxious to investigate whether or not B_1 deficiency existed in Britain — and if so, to what extent. The original idea was to measure levels of the vitamin found in the blood. On 7 May 1940, Sinclair wrote a memorandum defining how he would undertake such an investigation. To estimate mild deficiency, he suggested studying 200 individuals by both clinical and biochemical methods. Vitamin B_1 blood levels could then be correlated with any clinical findings. He further suggested that he and Dr Robert H. S. Thompson (who also worked with Peters on anti-Lewisite) begin such a survey as soon as they were officially asked to do so.

There was no response to Sinclair's suggestion from any quarter. He fussed, complained to Peters, and continued his teaching and research.

Great Britain was at war and the prospects of possible devastating consequences were terrifying, but at the same time everyday life took on a surreal quality for many. For Sinclair, at Magdalen College, the early days of the war must have been an inspiring and even attractive break from personal pressures. In his *History of the University of Oxford*, Brian Harrison records that Trinity term 1940 was a tranquil, sunlit interlude for undergraduates. The world may have been falling down around them, but while men were returning from Dunkirk many college students sat on the lawns of Oxford, listening to Handel's *Water Music*.

In June, while clearing the cellars at Lady Place for use as air raid shelters, Rosalie Sinclair received a letter from Hugh. France had surrendered to the Germans, and he felt now was the time for her to store food. He wrote:

> Several countries in Europe are even now on the borders of starvation … In France the position is critical. If the Germans get the French Navy … it should not be difficult for them to make the food situation extremely difficult in this country. Therefore it is obviously everyone's duty to store as

much food as possible <u>now</u> while it can still be imported fairly easily. Unfortunately, the Government can't advise this officially as it could lead to panic. It is difficult to realize what starvation means … I do think you would be well-advised to store whatever you can — food that will keep indefinitely … It cannot be wasteful even if we win the war quickly (which unfortunately seems improbable!) because we can easily eat them later. Canned fruit or vegetables, honey, rice, oatmeal (for porridge), condensed milk, olive oil (I think I should get a lot of this — it provides far more calories than any other food. I'll buy what you don't use from you after the war) — indeed anything you can think of that will keep. Soap is a thing that tends to go short also, and <u>seeds</u> of vegetables (potatoes, lettuce and cabbage etc.) may become available and will keep. Post going. Great haste. Hughy.

Bombs now rained down on Britain. Cowley, east of Oxford, and an industrial area about 12 miles away from Lady Place, had become an important centre for the production of munitions. From the time the Blitz began in September 1940, this area was targeted and repeatedly bombed; but the nearby golden-spired academic citadel of Oxford University, admired by Hitler for its beauty, books, and history, remained untouched.

Lady Place and the little village of Sutton Courtenay in which it stood also remained intact, and life remained strangely normal. During that summer, Hugh went to hospital for surgery on his knee, and Rosalie enjoyed a few weeks by the sea. Mother and son exchanged letters frequently, but as the weeks went by their communications became increasingly acrimonious. What had once been an occasional skirmish between them soon became a series of battles. For some weeks after Rosalie's return from holiday, Hugh spent almost all his nights at Lady Place, but mother and son continued to communicate by letters left on a table, or on a mantelpiece, or nudged quietly under a door. Most contained petulant and trivial arguments. He wrote to scold her for the use of the word 'serviette', when he felt table napkin was the correct term. His mother reminded him that it was difficult to remember the newer phrases when speaking, although she did use the term 'table napkin' on her laundry lists. She defended herself in the way that cut him most: 'Daddy preferred serviettes,' she said.

Hugh argued that he wanted more say over how their land was used. Specifically, he wanted fields replaced by fruit gardens; Rosalie was against this. It would be too great an expense for her, she claimed. Money was now a problem that ate away at their relationship. Then the real storm between them broke. Rosalie wrote:

> … When we took Barming House, we thought and grandfather told me the week before he died, we should have at least £10 000 a year from his Estate. We get less than £1600 i.e. about £1530 and my marriage settlement is nearly halved too. Daddy took Gransden as he realized we could not afford to go on at Barming House and now we have really more than we can afford at Lady Place, but I did it for your sake and I hope it has helped you. Health, wealth and intelligence will not give happiness unless there is love, unselfishness and sympathy. Also we have not wealth — and I think courtesy or consideration for age should be included … I gave up everything when I married Daddy; I mean youth, friends etc… I was younger than most. It was not easy so I lost touch with friends and life has been very lonely since Daddy died …

War between the Sinclairs raged for months. Rosalie claimed Hugh showed little interest in the alterations she had made at Lady Place, complained that he ignored her dinner guests, and responded to her in monosyllables when she asked him questions. He even provided the servants with more information about his comings and goings than he gave her. From his side, he felt she made no effort to understand the importance of his work. Finally, Hugh had had enough, and announced that he wanted to buy a property of his own.

At first, Rosalie thought he intended to marry, and was pleased by the prospect. Then she realized his real intent was to purchase Lady Place from her, taking over the role as decision-maker in the family home. He had already paved the way for such a move. Months earlier, such a proposal had been discussed with both Catherine and John. Rosalie resisted selling Lady Place to Hugh, and received some support from Catherine, but Hugh persisted. At last, Rosalie capitulated, and wrote a note explaining that if John, the older son, did not want Lady Place when she ceased to live there, Hugh could purchase the property by paying each of the others one-quarter (£1250) of its then current value. Or, she said, she could add a codicil to her will, saying Hugh could buy Lady Place after her death under terms agreed with the others. If he agreed with this last suggestion, Hugh could get on now with planting the meadow and other improvements he desired; however, he was not to increase her annual expenditure on upkeep.

Hugh was not satisfied with either proposal, and protested that he wanted a house of his own now, whether he married or not. If Catherine did not marry, she would probably live with him, he claimed. Because of the war, large houses were now cheap and he wanted to purchase one as an investment. Ideally, he said, he sought a place he could improve by planting trees. 'I am not only very fond of trees', he wrote, 'but also realize that the great anxiety of the Forestry Commissioners about our trees is justified: the Government during the last war cut down trees wholesale, and did not plant.'

Rosalie did not soften her position: Hugh was not to take full control of Lady Place from her now. Polite words exchanged were rife with bitterness. Matters were made worse when Catherine and John came home on leave from time to time; Rosalie gave them full run of the house with little thought for Hugh. At last, overworked, and again at physical and psychological breaking point, Hugh sat in his room at Lady Place and wrote to his mother.

> You asked me … if there was anything that you could [do] to help my work … I try to arrange my work now so that I write up the day's experiments between dinner and 9 p.m. and also work out the next day's programme. Then I listen to the headings of the news (because I hardly have time to read the papers at present) and then do my main work. Last night I had hoped to finish a long and intricate memorandum for one of the Ministries and after the news I got the whole thing thought out and had started to write it. Your two interruptions completely defeated me. It may seem very important to you that you should say to me: 'Call Blood [John Sinclair's dog], because I can't take him with me', but from my point of view it is very cruel … I am dreadfully busy at present because we do not show many signs of winning the war … Unless it can be clearly understood that I am not to be disturbed unless absolutely necessary (and I do not include Blood's alleged comfort in that) then I shall have to live in Magdalen …

Whether out of concern for her overtly stressed son, or as a way to stem her guilt over any lack of attention to him, Rosalie capitulated on the purchase of Lady Place. To be as fair as possible to all her children, she began arranging the settlement of her estate among them.

Hugh's work with Peters continued, but other interests also evolved. He had remained in contact with many of the American friends he had made during his time as a Radcliffe Travelling Fellow. Among these was the biochemist Herbert Evans, at the University of California. Hugh and Evans exchanged information mostly pertaining to vitamin B_1, although among the research papers Hugh received was one discussing essential fatty acids. He had little interest in the subject at that time, and filed it away.

Peters now spent most of his time in Cornwall, at Porthgwarra, working for the Ministry of Supply on anti-Lewisite. He frequently communicated with Sinclair by notes about the vitamin B_1 work in the department, but also warmly encouraged Hugh to pursue the survey plans previously sent to the Medical Research Council. Hugh delayed following Peters's advice. Finally, in August, Peters suggested that Hugh speak directly with Sir Edward Mellanby, Secretary of the Medical Research Council, who was a friend of Peters and the leading authority on nutritional surveys in Britain at that time. After much thought and procrastination, Hugh found a fresh idea with which to approach Sir Edward.

A man by the name of Davis Steven was coming from the United States to work in Peters's laboratory. He had some experience with nutritional surveys, particularly as they pertain to vitamin A deficiency. Hugh proposed to Sir Edward that his original plans for the vitamin B_1 survey be expanded to include vitamin A. This was timely in the light of growing concern regarding night-blindness. Sinclair pointed out that Steven had spent time working with Dr George Wald, who was conducting groundbreaking experimental work on the molecular basis of vision. While with Wald, Steven devised a simple and portable apparatus to measure the clinical (visual) symptoms of vitamin A deficiency. Would the Medical Research Council support Steven's work with a maintenance grant, if he were to join in the survey Sinclair first proposed, Hugh asked?

The possibility of working with Steven would achieve two things: first, it would broaden the appeal of his proposal for a nutritional survey by increasing the number of vital nutrients investigated. Second, it would give him a new contact, and possible ally in Peters's Oxford laboratory. Over the past few months, Sinclair had suffered a major falling-out with Robert Thompson, who was highly respected by Peters. The conflict, from Sinclair's point of view, is reflected in the following extracts from a letter to Thompson. It provides insight into the kinds of problems Sinclair suffered when working with men of approximately his own age and at a similar stage in their scientific career:

Private — My dear Robert, it seems that sooner or later there is bound to be a crisis because we both seem to be pursuing the same sort of fundamental research. I am not writing about war-work. Let me put the position as I see it as directly as possible. When I returned here in 1936 I decided to work on neuritis, studying both the peripheral form and diseases of the brain that I thought were probably metabolic (Alzheimer's, Pick's, Wilson's, and even psychoses) as opportunity presented itself. I started off to see what part B_1 played ... I was determined to choose a line I could work on alone, and I was certain this one was all right — the Professor knew of none of the diseases I have mentioned ... But in the summer of 1938 I discussed my programme with Peter [Meiklejohn] and made sure that it in no way ran across his (work). I therefore set about collecting some of the chemicals ... causing neuritis ... The idea was to study these in vitro and in vivo ... But there was a difficulty in that the Professor did not want me to go outside clinical aspects of B_1 ... He did give me permission to have some rats to produce arsenical neuritis. I started some work and wrote to Peter [Meiklejohn] about it in late 1938. Goodhart's arrival inhibited further work because of the pressing need to study clinical aspects of cocarboxylase ... In these years I got together a great deal of neuritis literature and discussed it with neurologists when I met them (e.g. Denny-Brown sent me some unpublished observations about diabetic neuritis). When Goodhart left I went to my second International Neurological Congress with my plan of research before me (cf. the concluding remarks of my paper). Then the war broke, but in the past year I have not lost opportunities of trying the compounds I am interested in ... (I suspected a year ago that you were becoming interested in rather similar lines when you asked me for some references ... But I thought you were [using lead] ... and I kept quiet because it seems to be generally believed now that 'lead neuritis' is not a neuritis but a myopathy. I did take the precaution of asking David Fisher what peacetime plan of research you had. But I suddenly realized today that our paths had crossed. As I am spending a good deal of time writing up results, and am only engaged 30% for the Ministry, I thought Waters and I could finish this month by further investigating the neuritic chemicals; then I could get down to the B_1 work in October. I am fully aware that some of them are relevant to the war ... and for that reason as well I have tried them with pyruvate and succinate. You saw some of the chemicals lying on my bench on Friday and asked what I was doing. I paid no particular attention, but could not fail to notice today how you appeared to hide from me the fact that you were using [copper] ... as I had done on Friday. I noticed it when I came over to show you the report, and you quickly tried to cover your notebook — too quickly to be successful and so quickly that it attracted my attention and I therefore read the ill-concealed heading on your experiment. I have not the slightest wish to cash in on anyone's research, and I hate all questions of priority. My one desire is to choose a line that is promising and neglected and work upon it steadily and alone … I have done so … This letter simply defines approximately my research path so that there can be no possible misunderstanding. I hope and expect that I have misunderstood entirely the position and therefore this letter has been wholly unnecessary, if so, forgive me.

After that salvo across a departmental colleague's bow, the possibility of conducting a nutritional survey outside the university became a priority for Sinclair. It would allow him to move into another aspect of vitamin B_1 investigation, which now held Rudolph Peters' interest, and it offered an opportunity for involvement with Government experts. Of even greater importance, the survey work he prescribed combined clinical observation with biochemical determinations. If such a survey developed, Hugh Sinclair was determined to be its director.

In October 1940, Sinclair enlarged the scope of the proposed survey to now include vitamins A, D, B_1, and calcium. He discussed his ideas with Dr Jack Drummond, an old acquaintance from University College and now a Scientific Advisor to the Ministry of Food. Drummond encouraged Sinclair, and on 28 October 1940, at a meeting at the Ministry of Health, it was agreed Sinclair should proceed with what was called 'The Oxford Physiological study to determine the nutritional state by means of efficiency, clinical and blood tests'. It was further agreed that this work 'should be done in the fullest possible cooperation with the precise dietary studies of the Ministries of Food and Health.' Senior officials of both gave their blessings to the project.

Records do not suggest the extent to which Sinclair discussed the expansion of his survey plans with Rudolph Peters and others at Oxford, but the work was soon regarded as a 'war project' within the scope of Peters' department, with Peters as its primary investigator. Sinclair had Peters' support, but was distressed when he was told Thompson would be a member of the survey team. Nonetheless, he carried on and wrote a research protocol in which he stated: 'A comprehensive nutrition survey of a group of persons calls for an investigation of their food consumption (by a <u>dietary survey</u>), and an enquiry into the state of nutrition of the individual person (by a <u>nutritional survey</u>).' This statement clarified and expanded his original mission, placing it in sharp contrast with previous and current surveys undertaken by others. Sinclair explained:

> Almost all the surveys that have been made in this country have been *dietary* surveys and have either related food expenditure with the calculated minimum cost of a diet, or analysed the food consumed and hence estimated the quantities of the various constituents that are eaten; both methods are liable to very large errors. <u>Nutritional</u> surveys have until recently been used mainly for obtaining anthropological data. At an international conference at the end of 1932, the Health Committee of the League of Nations considered the possible ways of making such surveys. A further meeting of experts at Geneva in 1936 recommended three types of enquiry: first, somatometric data; secondly, additional somatometric data combined with a medical examination and laboratory tests; thirdly, the same as the second type with the addition of a complete medical and psychiatric examination. For the purposes of the present survey, the first is inadequate and the third impracticable. It is suggested that the second type could be put into practice somewhat as follows …

Sinclair then described a sampling technique to include various groups, including schoolchildren, agricultural workers, recently captured German prisoners, surgical patients, recruits for the Services, and men and women who have 'been upon the Service diet for several weeks', Information to be collected included:

Somatometric data — name, age, sex, address and occupation, general appearance, height, weight, skin thickness, vital capacity, and sitting height.

Clinical measures — pulse-rate, blood pressure, routine examination of the heart, ankle-pitting, ankle-jerks, calf-tenderness, vibration sense in legs, appetite and digestion, teeth and mouth.

Physiological — blood estimates (20 ml sample — haemoglobin, clotting time, red and white cell counts, phosphate levels, total vitamin B_1, cocarboxylase, vitamin C, riboflavin, nicotinic acid, possibly vitamin B_6.

Urinary estimates — comparison of two samples taken four hours apart comparing nitrogen, vitamin C, and vitamin B_1 levels.

Assessment of visual dark adaptation using the methods Stevens and Wald had developed in the United States.

After expenses are estimated for ancillary staff, travel, and so on, Sinclair suggests that the cost of his survey would be between £750 and

£1000. (While this was a considerable amount of money at the time, Sinclair ran the risk of promising more than his proposed budget would allow.)

The surveys were designed with one purpose in mind: to answer questions within the British Government about the state of the nation's health. This task was necessary and appropriate, but it introduced certain limitations into Sinclair's work that would later haunt him. As outlined, his work was meant to collect data from various populations for comparison against previously set national nutritional standards. That was an adequate objective. However, from a research standpoint, the surveys had to be looked upon as exploratory in nature, offering opportunities to compare and contrast results after the fact. No stated hypotheses were to be tested. No attempt was made to set up any controlled investigations.

Support was forthcoming from the Ministry of Health, and on 9 November 1940, Dr E. Bransby, a member of its scientific staff, wrote to Sinclair. He made a request that was politically impossible to refuse, but it added another complication to the work Sinclair proposed to undertake.

> I think a most useful study would result from combining a dietary investigation with your vitamin investigation from blood samples. Do you think it could be arranged to undertake the food study in co-operation with the work you are doing? We will, we hope, be able to supply well-trained investigators, and the organization of our side of the fieldwork need not throw any strain upon you … if you think that this joint work is possible it would be a good plan to have another talk to settle more details.

Enjoying academic credibility as part of Rudolph Peters' Department of Biochemistry, and supported by two powerful ministries, Hugh's primary difficulty now was finding the money to undertake the work. Peters had financed the early developmental work within his department, but much more funding was needed. Sinclair learned that Sir Farquhar Buzzard, the Regius Professor of Medicine, and the surgeon Professor George Ernest Gask had considered the desirability of conducting a nutrition survey in Berkshire, Buckinghamshire, and Oxfordshire under the Nuffield Regionalization Scheme. On 29 November 1940, Sinclair outlined a proposal for a survey meeting the organizational requirements of this scheme, and sent it to the Regional Medical Advisory Committee. A subcommittee was appointed to consider the plan: Professor Joshua H. Burn, Professor of Pharmacology at the University of Oxford, was appointed Chairman.

Some members of the sub-committee disagreed with the concept of the survey from the beginning, but it had Burn's support.

The sub-committee procrastinated. The war moved on. Sinclair fretted. Frustrated by delays, he began the work on his own. In later years, when the history of the surveys became important to his career, he claimed Burn had advised him to ignore the sub-committee and move ahead.

Talk of the Oxford Study or, as it was becoming known, the Oxford Nutrition Survey, excited considerable interest among staff working in the Ministry of Health. Dr. E. Bransby now was ready to move, and set his sights on influencing the design of the work. A few days later he wrote to Sinclair. Describing the work he wished to have done:

> … covering approximately 2000 working-class families four times a year. Two hundred and fifty families are being studied in London, Birmingham, Sheffield, Glasgow, Liverpool, Reading and Rural Districts. The technique of investigation is not so exact as normally employed in such work but we think that the results will serve a useful purpose … We would if possible like to survey the families studied twice a year to form an idea of the changes in food consumption. Your plans may not be to continue the investigation in this way but if we continue the food study you may, at some time in the future, consider continuing your work. If you decide to extend the laboratory study to Berkshire, Buckinghamshire and Oxford I am sure we could arrange the food study accordingly.

The introduction of Bransby's ideas into the Oxford Survey was popular with the Ministry of Health, who were willing to finance parts of the survey that interested them, but it introduced several uncontrollable factors. As we will see later, the most important of these is the matter of detailing the nutrient composition of food.

In early December, Dr Bransby talked with Dr Magee, his superior in the Ministry of Health, who had discussed the survey with the Chief Medical Officer, and cooperation with Sinclair was given warmest blessing. Bransby wrote to Sinclair and said he had tried, but failed, also to involve the Army and Air Force in the work. Sinclair replied that he was very glad about the blessings of the Ministry of Health, but was disappointed with the news from the armed services. Further, Sinclair stated:

> The Regional Medical Council of the 2 counties around here have set up a sub-committee to draw up details. Professor Burn is chairman, and other

members of its are Sir Robert McCarrison, Professor Peters, Keith Murray and several members of this department who are taking part in the survey.

It was only days before complications arose in the loosely formed cooperation between the governmental groups and the Department of Biochemistry. On 16 December, Bransby wrote to Sinclair:

> … there will be, I am afraid, a slight alteration from our original proposals. In the first place we asked for £4000 to carry out this work but have only obtained £3000 and from this it will be necessary to undertake a number of *ad hoc* studies as they arise. … It was only found possible to allocate one investigator to Oxford. I do not think it will be possible for one investigator to cover all the families that you will be contacting, but it should be possible to relate nutritional conditions from the blood samples and the diets of the families studied. We will be very glad to hear from you when your plans are made. …

Many more experienced investigators would have demurred at this point; problems of insufficient funds and the demand for a number of *ad hoc* studies would have raised warning flags. But Sinclair persisted.

Burn and his group met on 27 December and considered Hugh's statement defining the need for the survey, the plan for its execution, and the costs in terms of personnel and expenses. The proposal was approved and forwarded to the Regional Medical Advisory Committee, which met on 8 January 1941.

In the United States, another man was anxious to play a part in bringing about the end of the war. Dr Hugh H. Smith, then working for the Rockefeller Foundation in Columbia, South America, requested relocation so he could 'serve in some relation to the present war in Europe'. He wrote to his superior, Dr W. A. Sawyer, Director of the International Health Division. Sawyer had been planning to visit England himself around the end of the year. At the time, cooperation with Great Britain was limited to influenza vaccine; but Sawyer had written to Sir Wilson Jameson, of the Ministry of Health, and asked for suggestions about expanding their joint efforts. Sawyer could see advantages to having a Foundation staff member in London. After a long wait, Hugh Smith received Sawyer's reply: he was going to London as a member of the Rockefeller Foundation Health Commission to Europe. There would be danger in the posting, but the Foundation saw that, through Smith's experience and skills, they could make a wider contribution to the war effort.

Smith gladly accepted the challenge.

CHAPTER 6

Hugh Smith and the Oxford Nutrition Survey

As the Nazi shadow moved over the Continent, Hugh Smith prepared to leave for Europe. The British were anxious to have a team of experts survey the nutritional status of factory workers, and one of his first tasks would be to facilitate this work. To Smith's surprise, he would not make the journey alone. A young British research scientist, interested in nutrition, and currently on fellowship at Harvard University, wanted to return home for war service; the nutritionist was Dr Peter Meiklejohn, Hugh Sinclair's long-time friend. They later bitterly quarreled.

It was April 1941 when Smith and Meiklejohn sailed for Lisbon on the American Export Liner, *Excambion*. The United States was still neutral, although the passengers included US Army Air Force pilots in civilian clothes, heading for Britain and training in war-time tactics. After leaving the ship arrived in Portugal, Smith and Meiklejohn were forced to wait for two weeks before boarding a DC3 with blacked-out windows and heading for Bristol, England. Arriving at night, they were driven through blackness between piles of rubble and bomb craters to a small hotel for a few hours sleep. The next morning, 13 May, the pair took an early train to London. The Luftwaffe had heavily bombed the capital three days earlier and as they approached the city fires could still be seen raging in gutted buildings.

Departing from the train, the two men were greeted by an old friend of Hugh Smith's, Dr Patrick O'Brien, and the first matter of business was to make the newcomers comfortable at the Savoy. While they lunched in the hotel dining room a few hours later, they were seated only yards away from Prime Minister and Mrs Winston Churchill. Smith watched as Churchill settled down after his meal to enjoy a brandy, a cigar, and the privacy London then offered its public figures.

That evening, Smith and Meiklejohn dined at the Athenaeum with a Scotsman, Sir Wilson Jameson, Chief Medical Officer of the Ministry of Health. The Athenaeum and its library impressed Smith; and so he was delighted when he learned that his position warranted honorary membership there while he was in London. Sir Wilson was well known by the Foundation, having received many Rockefeller fellows and recipients of Rockefeller travel grants while he was Dean of the London School of Hygiene, in London. Both Smith and Meiklejohn were to receive all their instructions from Jameson. Meiklejohn was to begin dietetic and nutritional surveys at once in the industrialized areas; as the more experienced member of the pair, Hugh Smith was given time to look around and see what was going on before he decided how to apply himself.

More bombing raids on London were expected. After a few days of luxury at the Savoy, Smith and Meiklejohn sought lodgings in a building with a steel frame, as it would be more likely to withstand pounding and blasts during a raid. Along with other members of the Rockefeller staff, flats in an elegant building at 52 Curzon Street, in Mayfair, became home.

Smith began the task of finding his way around the British hierarchy. Among the first men he called upon was Sir Edward Mellanby. Smith found him somewhat abrasive, but respected both the powerful position he held and Mellanby's earlier work on vitamin A. Allies in key positions were vital to Smith's work, and Mellanby had his hand on almost all the nutritional survey work being conducted in Great Britain.

Through Meiklejohn, Hugh Sinclair soon learned of Hugh Smith's arrival, and travelled to London to meet him. Their mutual interests in nutrition were enthusiastically shared, and Sinclair invited both Smith and Meiklejohn to see his work in Oxford, on 1 June, some few days hence. The two men were also invited to stay at Sutton Courtenay while visiting the University, and they gladly accepted. After issuing the invitation, Sinclair then needed to enlist his mother's help in entertaining such important houseguests. Fortunately, Rosalie warmly remembered Peter Meiklejohn from her son's schooldays, and was pleased to have him visit Lady Place again. She issued a string of commands: have Hugh's scout at Magdalen collect Peter and Dr Smith's two days' ration tickets and bring them to her; bring home sausages, suet, a boiling fowl, and some scones. Hugh was drafted in to do more dog's work than he had anticipated, because Catherine's car had been hit broadside by a skidding lorry. She was able to get off the road without injury to herself, but her car was badly damaged. Repairs seemed to take forever.

Hugh Smith enjoyed his stay at Lady Place immensely, and while he remained in England, this old rambling house, Rosalie and Catherine were to become pleasant relief in his hectic life. He soon began corresponding with Rosalie, and several weeks after his first experience of her hospitality wrote that he favoured the early entry of the United States in the war, which appeared more certain than ever.

The trip to Oxford and Lady Place went well for everyone. After discussions involving both Sinclair and Meiklejohn about the planned survey work, Smith wrote a letter to Sir Wilson Jameson that helped settle any questions about its inevitability. Saying he would help with its clinical investigations, he wrote that the Oxford Survey was ready to start in earnest, and that all the initial difficulties of the organization had been overcome except for two problems: inadequate financial assistance and more clinical support. Unfortunately, this message of enthusiastic praise was overly optimistic. Staff illness and administrative problems had already slowed early work on the Ministry of Health dietary survey in Oxfordshire, and a dietary investigation undertaken by Sinclair at Magdalen had been rushed. Within the Department of Biochemistry, financing the survey was beginning to drain resources that could be used for other work.

The Rockefeller Foundation, an American philanthropic organization based in New York, had a longstanding interest in the development of the Oxford Department of Biochemistry. Immediately after World War I, Sir Walter Morley Fletcher, then Secretary of the Medical Research Council, took decisive steps to improve the teaching of medicine at the universities of Oxford and Cambridge. Biochemistry was a subject of growing importance to medicine, and to compete internationally as leaders in this field, both universities needed strong departments. In 1919, the Board of the Faculty of Medicine at Oxford urgently appealed for a Chair of Biochemistry. Fletcher approached the Foundation among others, for financial support. The Rockefeller Foundation warmed to the idea, especially if their support would facilitate a more research-based approach to medical education.

Fletcher did not have the full support of the Oxford faculty, however, opportunities for change were not as warmly received at Oxford as they were by professors at Cambridge where, under the leadership of Professor Fredrick Gowland Hopkins, biochemistry was a growing area of scientific investigation. Despite this, Fletcher persuaded the Rockefeller Foundation to fund a new department and, in 1927, his efforts resulted in a grant of £75 000. As a result, the study of 'Physiological Chemistry', then part of

Charles Sherrington's Department of Physiology, was split off into the separate field of Biochemistry, and Professor Rudolph A. Peters became the first Head of the new department. For many at Oxford, these developments, as well as organizational changes affecting the departments of Pathology and Pharmacology, were adequate for the time. However, The Rockefeller Foundation had other ideas.

Based on an extensive study of medical education in the United States and Europe, Abraham Flexner, Director of the Institute for Advanced Study, at Princeton University, in the United States, recommended broad changes in the way medical training and research were integrated. His ideas had a major impact on universities in the United States, with the result that clinical research and practice were being raised to previously unattainable levels of advancement. The Rockefeller Foundation saw one of its roles as being the stimulation and support of work that would encourage these same developments on both sides of the Atlantic.

In 1927–28, a delegation from the Foundation visited London in the hope of finding a responsible body within the University of Oxford with which it could discuss, with confidentiality, plans to enhance medical research and postgraduate education further. There was little enthusiasm. Soon after those meetings concluded, however, the influence of the new Regius Professor of Medicine, Sir Farquhar Buzzard, was felt. The Rockefeller staff tried again, and this time they found a welcoming and innovative partner in stimulating change. Buzzard remained Regius until 1943, and his influence remains imprinted on the teaching of medicine and clinical practice to this day.

At first, Buzzard was committed to the development of a postgraduate programme in medicine, and was satisfied with this as his ultimate goal for Oxford. However, generous gifts from Lord Nuffield stimulated interest in expanding the scope of medical education to include an undergraduate school of medicine. This concept was supported by younger enthusiasts, including Howard Florey, who had found his intellectual home in the Department of Pathology, and Hugh William B. Cairns, a neurosurgeon of considerable influence in medical teaching at Oxford. As time passed, tension would develop between those who wanted only moderate change in medical teaching, and those who had a vision of a significant research role for Oxford within the international medical community. Both sides saw opportunities to advance their arguments during the period of chaos forced on them by war. Hugh Smith's arrival from the United States was to be one influence on the outcome of the debate.

Smith was pleased by what he witnessed in the Department of Biochemistry, and was anxious to support new work there, especially if it aided the war effort. Sinclair impressed him, and Smith believed the expanded proposal for nutritional studies, as placed before the Ministries of Health and Food a few months earlier, would serve a useful purpose. Therefore, as one of the department's war-time activities, Smith encouraged Meiklejohn and Sinclair to combine efforts and begin a few nutritional surveys as quickly as possible in certain industrial towns, and among schoolchildren. The department's problem was finding funds to continue and expand the work already under way. Smith said he would encourage the Foundation to contribute towards these costs. Without funds from the Rockefeller Foundation, Sinclair's fledgling Oxford Nutrition Survey would not have survived.

The Rockefeller Trustees had set aside one million dollars for humanitarian and health work in war areas, and Smith felt sure the surveys were an appropriate activity to be sponsored from this source of money. He assured Sinclair that funds would be available for a proposal of this kind *if* he gained support and cooperation from three key organizations: the faculty and administration of Oxford University, the Ministry of Health, and the powerful Medical Research Council. Sinclair felt certain Rudolph Peters would support him within the University, but there remained Sir Wilson Jameson, Chief Medical Officer of the Ministry of Health, and Sir Edward Mellanby, Head of the Medical Research Council, to impress. Hugh Smith would eventually intercede with both men, and gain the needed support. From that time until his return to the United States, Hugh Smith remained Hugh Sinclair's greatest political ally. Smith would remain convinced of the Survey's importance, even to excess. In his autobiography, written in 1978, Smith states that the Survey '… produced valuable reports of nutrition in Britain during the war and led to the founding of a permanent Institute of Human Nutrition in Oxford after the war.' Unfortunately, these words do not reflect the true course of events.

Smith's role in London was administrative. For Sinclair, one of Smith's most important contributions was obtaining equipment and other resources that were impossible to obtain in England, due either to government regulations at the time, or to cost. For others, however, Smith's greatest contribution was his ability to see potential in existing research. That returns us to the matter of Howard Florey, and his work with penicillin.

During Hugh Smith's first visit to Oxford University, made at Sinclair's invitation, a meeting took place that is arguably one of the most important in the medical history of the war. While visiting the Department of Bio-chemistry, Sinclair suggested that Smith should also visit Howard Florey. Florey and his colleague, E. B. Chain showed Smith how they could pro-duce small amounts of pure penicillin by using a complicated and crude machine they had designed. They had already demonstrated that this pure substance killed pathogens *in vivo*, but they had hardly enough of the drug for limited clinical testing. Only five cases, all terminally ill with acute infections that had failed to respond to sulfonamide, had been treated by the time Smith visited Florey. The clinical results were impressive; howev-er, enough of the drug could not be made to fully cure such stricken men. Knowing British resources were too limited to pursue this valuable line of research, Smith suggested that Florey contact the Rockefeller Foundation for support. Through the efforts of Smith, and his Foundation colleague, Pat O'Brien, arrangements were made for Florey to visit Washington DC and present his findings to experts in the United States. As a result, the United States Government provided $15 million to underwrite develop-ing the means to mass-produce this new miracle antibiotic. Major phar-maceutical companies, universities, and United States government agencies, together representing thousands of scientists, engaged the prob-lem. By 1942, means had been developed to produce enough penicillin to permit large-scale tests. After the usefulness of the antibiotic was estab-lished, the British Government agreed to manufacture mass quantities. By that success, the treatment of infection had changed forever.

While bombs fell on London and the industrial cities of Britain, Oxford became a town of refugees and evacuation agencies. Pavements were crowded, theatres were packed, and the mix of languages gave the place a certain continental atmosphere unknown before. Airmen and raw recruits passed through the town on their way to training. The unifying effect of the war was felt everywhere: all had the same purpose — to beat Hitler. The mundane problems of careerism and plans for tomorrow were pressed to the back of the mind. Now, winning was all that mattered.

In this atmosphere of camaraderie and militarism, Hugh Sinclair began building the Oxford Nutrition Survey (ONS), while Rosalie determined to fight the war in her own way. Mindful of how her dead husband might have responded to current conditions, she reminded Hugh of how active and successful both John and Catherine were in distinguishing themselves in the war effort. Hugh now felt it necessary to prove himself. Although he was a tutor at Magdalen College, and a Demonstrator in the Department of Biology, he became Director of the ONS without cutting back on either responsibility. To meet the growing demands of the survey, he began building a staff. A young woman, Angela Lousley, joined him as his secretary: she would be loyal to him throughout the war. One of the first scientists who came looking for work was to become a lifelong friend; his name was Brian B. Lloyd. Like Sinclair, Lloyd had distinguished himself as a scholar at Winchester College. Recently a student at Balliol College, where he had completed enough academic work to earn him a Special Certificate in Chemistry, thus enabling him eventually to get a War Degree, Lloyd was highly regarded by his tutors. In the eyes of some, however, he had one flaw: he morally objected to all war. After enduring the painful scrutiny required to establish his authenticity as a conscientious objector to military service, and registering with the Ministry of Labour, Lloyd was faced with several choices of work: land-work, forestry, or agriculture — or some job in the medical field. One of his Balliol tutors, who knew about Sinclair's plans for the Survey, suggested that Lloyd should go to Magdalen for an interview. Sinclair could see the younger man was keen and skilled, and he respected his commitment to his own ideals. He also knew that, because Lloyd was a conscientious objector, he need not be paid more than the salary of a land-worker. Sinclair hired Lloyd as soon as possible to work on vitamin C, at a weekly salary of two pounds sterling.

Although preliminary survey work was underway, Hugh had not yet obtained approval by the three organizations Hugh Smith felt vital to the survey's success: the University, the Ministry of Health, and the Medical Research Council. The sub-committee asked to review Hugh's plans disagreed on its value. Some members argued that the extent of nutritional deficiency was already known from other sources, so the ONS was redundant. One critic of the plan questioned whether or not those employed by the Survey could not be used to better advantage in war work. Some warned that results would be needed quickly to be useful: could Sinclair's complex blend of nutritional and clinical investigation achieve this? In the end, the Region sub-committee, chaired by Burn, and mentioned earlier, agreed. However, the higher body of the Medical Research Council, the Medical Advisory Board, was not as impressed.

Sinclair wrote to the Regius Professor, Sir Farquhar Buzzard, asking for support. Sir Farquhar responded, saying he would do his best to promote the survey, and that he was not impressed by the arguments against it. If Mellanby and Jameson favoured it, then the work would go ahead. A second letter from Sir Farquhar arrived a few days later. The contents were disappointing. Sir Farquhar had talked with Sir Edward Mellanby, and found that he did not support the Survey. However, he had invited Hugh and Sir Farquhar to lunch at the Athenaeum some weeks later to discuss his concerns. This meant that the matter could not move ahead quickly, but all was not lost. Whether Sir Edward had a specific objection to the Oxford proposal, or whether he simply thought that there were too many surveys already taking place, is unclear. But it was worrisome that so many people around the country were making a job of checking what people ate. Several controlled nutritional studies on human subjects were already underway. As examples, just as the Accessory Food Factors Committee of the Medical Research Council had approached Rudolph Peters about undertaking studies of vitamin B_1 in the British population, they also approached Sheffield University about studies involving vitamin A. This work was to be conducted on volunteers, and carried out under the exacting eye of the highly respected scientist, Dr Hans Krebs. Also, at the University of Cambridge, Drs R. A. McCance and Elsie M. Widdowson were conducting dietary studies on human subjects to determine the digestibility of high extract wheatmeal and its effect on calcium absorption in the gut.

Supporters of Hugh Sinclair argued that his approach to the study of nutritional deficiency differed from those of the Sheffield and Cambridge groups. Where the others used volunteers kept for long periods on carefully controlled deficiency diets, Sinclair planned a comprehensive investigation of groups of people in the community, living on normal diets. His detractors said that the controlled diet studies could determine a direct causal effect between the development of observable pathology and specific diet deficiency. In Sinclair's proposed investigations, observed pathology could only be correlated with what were thought to be dietary deficiencies, as calculated from reports and measures of food consumed. In the end, it was agreed that Sinclair's work was scientifically crude, compared with controlled diet experiments; but it had the advantage of allowing the identification of pathology caused by dietary deficiency in target communities necessary for national defence, such as miners.

Sir Edward Mellanby did not write to Sinclair himself concerning the ONS, but communicated through a lower member of the Medical Research Council hierarchy, Dr Platt. Platt wrote to Sinclair, saying that he wished to visit Oxford, and hoped that they could then decide upon a definite plan of action for the survey: the Medical Research Council was determined to impose its stamp on the structure and purpose of the ONS. The men met and negotiated the content of the ONS. Based on their agreements, a proposal was prepared and submitted to the Oxford Registrar, Douglas Veale, with a request for his support. The following are key points in its content:

First, field experiments were needed 'to discover the existence and extent of malnutrition in the population'. To this end, Sinclair and Platt would prepare an illustrated handbook of the signs and symptoms of nutritional disease. This clinical field research, which he called the *Standard Nutritional Survey*, would be run in conjunction with a Dietary Survey — to identify what samples of the population ate — and a Socio-economic Survey — to discover, within practical limits, the nutritional habits, including cooking methods used, within the population. It was thought this work would be conducted with the Ministry of Health. (These particular words would soon stimulate a perceived link between Sinclair's work and that of the emerging study of Social Medicine, which members of the Oxford medical community and others were supporting at the time.)

Second, the ONS would undertake laboratory experiments, including work on volunteers much like that being conducted at Sheffield; these investigations would provide information about methods that could influence directions for new research.

The day following the submission of the proposal, Rudolph Peters championed Sinclair's proposal by writing to Veale, stating that the survey work proposed by Sinclair was part of his Department of Biochemistry. Peters requested £800 at once to support it: £400 for the nutrition survey itself, £200 for an ultracentrifuge, which could have wider applications within the study of biochemistry; and £200 for 'the study of antiseptic in relation to burns'. Peters' letter concluded:

> Sinclair is badly in need of money to start work upon the nutrition survey which he is organising. This is a pioneer research in what will be ultimately a big scheme, and I have little doubt that finance for its continuance will

come from Government sources; but these are slow in mobilising and much delay will be saved if he can start at once.

With mixed support, Hugh began the survey work Jameson wanted to see undertaken within British industry. On of his first approaches was made to the Pressed Steel Co. Ltd. To help get the work started, the Managing Director of Pressed Steel inserted a notice about the survey in all the pay envelopes, asking employees to fill in a form if they were interested in participating in the study: an impressive 160 forms were returned. At Morris Motors a less useful means of seeking volunteers was employed. A senior staff member decided the information about the Survey should be publicly posted on a bulletin board — only two volunteers came forward. Despite this patchy start, Hugh had some subjects for the industrial survey, and he began work at once. True to past form when starting a new part of the ONS, Sinclair immediately extended the scope of the investigations. This time, he wanted to include a study of teeth. Sinclair wrote to Platt for help: could he assist in finding a dentist to look at teeth? On a daily basis Air Force Officers were examining recruits from a large area around Oxford; would it be possible to include some of these men in the nutrition survey?

Hugh also turned to Bransby for assistance. Vitamin supplements could be obtained from the United States, and as the British Government wanted to know if vitamin treatment would help protect factory workers from illness, it seemed reasonable to include questions about supplements in the developing survey. To get support, he wrote to Platt, who responded in 13 May 1940:

> I have discussed both with Dr Magee and Dr Henry, of the Ministry of Labour, the question of an experiment in a factory to examine the effect of vitamin treatment. There seems no reason at all why such a piece of work should not be done provided the ground has not been too much over-run by the distribution of Adexalin, which has already taken place within the Nuffield factories. Do you think you could make a few enquiries as to whether it will be possible to undertake the work? I am informed by the Ministry of Labour that there is an increasing distribution of vitamin tablets and it would be a most excellent thing to run a controlled experiment to see whether they have any beneficial effects or not. It is probably that there will be a number of difficulties to overcome in measuring the change in health or in efficiency due to vitamin treatment but these should not be insuperable …

Not shy about discussing his work with anyone who would listen, Hugh made his efforts widely known, and as a result was frequently asked to address important public meetings. From time to time he invited Rosalie to attend with him, so she could observe first-hand his contribution to the war-effort. One such invitation, in mid-May, received a typical written response from Rosalie. At the time, John was home on leave and wanted to get the boats into the water, she said, and Catherine would be home at the weekend, so — although she would like to hear him, it was not possible.

As the weeks went by, Sinclair's proposed survey design was becoming more closely tied to the goals of government ministries and the Medical Research Council than to his work with the Department of Biochemistry. Platt and Sinclair prepared and signed a memorandum on 'Nutritional Surveys'. In this, 'mal-nutrition' was defined as 'a state of imperfect physical development or ill-health which could have been prevented by a good diet'. They agreed on paper that standard survey methods — including analytical techniques — were needed, as the surveys would be conducted in many different places and at different times. Other key points of the agreement were:

1. Clinical condition of the subjects — the examination must be exhaustive and must be designed to detect or to exclude any available evidence of existing past malnutrition.
2. The habitual diet of the same subjects, including not only the amounts of the various foods consumed, but information concerning the treatment the foods received before they were eaten.
3. The environment and social habits of the subjects, in so far as these might have some bearing on their food habits and state of nutrition.

All this pushed the design of the survey further away from Sinclair's original research scheme. To make matters worse, a bureaucracy was growing to ensure that all the necessary parties and agencies had a voice in the content and direction of the work. The potential for disagreement and delay was obvious in these lines from the agreement with Platt:

Immediate action should be to establish a planning committee to decide details of the survey, and should be empowered to co-opt outside experts. Field trials, at first limited to clinical examinations, should be carried out under the auspices of the Committee before procedures and techniques are put into widespread application.

Meanwhile, Hugh's problems at home increased. Lady Jackson, Hugh's maternal grandmother, was dying. Rosalie was distraught and said she needed to have Hugh spend more time at home now. Her request was underlined by a letter from John to his younger brother, in which he emphasized the importance of Hugh's family. As a result, Hugh left his work in Oxford and returned to Lady Place for a matter of weeks, during which he spent a great deal of time improving the garden. Later, after Hugh had returned to his rooms at Magdalen, he received a letter from his mother. She ignored the sacrifice he had made to be with her, but complained that she could not get creosote off the trousers he wore while working on an out-building. Did he not know, she demanded, that he had damaged a little table that she particularly liked by placing the open can of creosote on it?

She had no idea what those weeks away from the development of the ONS may have cost her son and his control over his work.

Hugh Smith was convinced Sinclair's surveys would be important for the war effort, and wrote enthusiastically to New York with a request for financial support. Influenced by Smith's positive attitude, in mid-July 1940 Sinclair wrote a confidential letter to Sir Farquhar Buzzard, saying that the Survey was now financially viable, as the Rockefeller Foundation was prepared to spend £3000 a year for a minimum of two years on the project. Everyone connected with the Survey, Sinclair said, felt that some central premises should be found for the work, preferably near the Radcliffe Infirmary. It was his hope that they might be able to find a temporary hut that would suit the work.

Now the three major issues that would determine the future of the Oxford Nutrition Survey had been engaged: these being control, finance, and accommodation. Unfortunately, things were not going to be as easy as Hugh Smith's positive attitude suggested. Conflicts involving control of the ONS had been developing for months. Within Oxford, it is noticeable that Thompson's name no longer appears on survey documents, suggesting the animosity that Sinclair expressed towards him had somehow been resolved by his absence. Far more important, however, were the serious problems growing between Sinclair and Sir Edward Mellanby. Sinclair directly involved the Regius in the conflict when in July, he wrote to Sir Farquhar:

Following your advice, we have explored with Professor Gunn the possibility of using rooms in the Nuffield Institute for the clinical examination under the

Ministry of Health's Nutrition Survey in Oxford ... I had a talk with Sir
Edward Mellanby and Sir Wilson Jameson three days ago. There was a
difference of opinion between Sir Edward and myself as to why he had held up
the Survey on the Medical Advisory Council, and what criticisms he
produced when you and I lunched with him at the Athenaeum. I understood
from him that he wished the Survey to be only clinical and not laboratory,
whereas, in my original memorandum, I suggested it should be a combination
of both. Sir Wilson was very enthusiastic about it, and said that it must go
ahead whether Sir Edward wished it or not. He told me he would talk to Sir
Edward about it yesterday, and would then formally apply to the International
Health Division of the Rockefeller Foundation for financial support ...

Smith intervened with Mellanby, and then wrote to Sinclair. He reported that he had earlier talked with Sir Wilson, who suggested Smith
request funds from New York to support the Survey on a temporary basis.
Before doing so, however, Smith decided to bring the matter to Sir
Edward Mellanby's attention. Their discussion of Hugh's proposal was
lengthy. In the end, Sir Edward agreed that the survey should go ahead,
but on a more simplified basis than stated in the memorandum Hugh
had agreed with Platt. If simplification could be achieved, Mellanby
would give the plan his blessing. He also objected to the £3000 request
from the Rockefeller Foundation for the project. Instead, he suggested
that the Rockefeller Foundation and the Medical Research Council each
put up £800 towards the survey's first year's budget. With another £400
from the Nuffield Trust, Sir Edward calculated that Sinclair would have
£2000 to get started. If the surveys appeared valuable, additional money
should be obtained later. Smith sent a cable to New York, asking for £800,
knowing this was far less than Sinclair expected. He was not aware that
Sinclair had already informed Sir Farquhar of a much larger sum.

Smith was in a difficult position with Mellanby, and others knew it. In a
letter to the Foundation's Head Office, in New York, the Head of its
London Office reported that Smith was doing a 'solid job'. He had had
the wisdom to bring Meiklejohn into 'friendly relations' not only with
Jameson, but also Edward Mellanby, who, it was said, controlled the
future of all nutrition work in England.

Sinclair felt he could no longer determine events, and chose to respond
to the disappointing information in Smith's letter by writing a long,
emotional message to Platt:

... A possible interpretation of Sir Edward's inexplicable attitude is that he objects to me. If that be so, I very much hope that you or he will say so as I have no intention of hindering the Survey on personal grounds, and have more than once thought that my most useful contribution would be to follow [others into] the RAF ... What was the objection to Sir Wilson accepting the £3000 if the [RF] was prepared to give it ... I am sorry to inflict all this upon you, and I hope you will not think this letter childish or ill-mannered. As [a] scientist I believe in plain speaking, and I feel very strongly that it is impossible for the Survey to continue at present knowing that Sir Edward has criticisms of its programme and has vetoed the finance it hoped to get ...

Rudolph Peters was concerned about the lack of support outside the University, and to emphasize the importance he placed on Sinclair's survey work, Peters wrote again to the Registrar:

At present there is developing in this Department a new enterprise known as the Oxford Nutrition Survey, which is under the direction of one of my University Demonstrators, Dr H. M. Sinclair. Two other University Demonstrators are also concerned; Mr C. W. Carter carried out certain blood examinations concerned with the survey, and Dr E. Walker is acting as Administrator (and Bursar) ... The group concerned is attempting to correlate in a comprehensive manner biochemical data on blood etc. with the first development of clinical symptoms and signs that may appear as the result of dietary deficiency. This is of importance in the war effort, as it is hoped to be able to obtain warning of the possible approach of dietary insufficiencies before the deficiency has become serious enough to cause illness. It also carries implications for the future of medicine. Linked with the work centralised in Oxford, it is projected to have mobile units capable of carrying out ad hoc surveys in industrial and other districts ... Now that the enterprise shows signs of vigorous growth, I think that it should be put upon an official basis so far as the University and my department are concerned.

At Lady Place, the war continued to cause misery for Rosalie. Both Catherine and John had returned to their posts, leaving only Hugh to provide family contact. On 7 December 1941 news arrived of the Japanese bombing of Pearl Harbor; America was in the war at last. Three days later, John wrote to his brother:

My dear Professor, How is our [Christmas] present to Mummy progressing? Are you producing a circular saw, or shall I have to get something else? Please let me know soon. What yellow-bellied blighters the Japs are. The Prince of Wales and Repulse are serious losses, and the Yanks seem to have been caught pants down, but we must see them off soon. Elsewhere the war is going well. … Life is very hectic. I wish we could have a crack at somebody, for we could fight really well.

Lord Woolton, Head of the Ministry of Food, told the nation that no immediate changes in rations were contemplated. However, the outbreak of war in the Pacific would place pressure on shipping and the sea-lanes, possibly causing some cuts. He explained that his aim was to give people enough food to keep them healthy and energetic. If successful, this would have effects lasting long after the war. It would keep the workers strong and future generations even stronger. The Ministry of Food would soon need information resulting from Sinclair's ONS.

CHAPTER 7

The foundations of Social Medicine

Intended to lower British morale, the Luftwaffe 'Baedeker' raids struck at the cultural heart of the country. Using a guidebook by Karl Baedeker describing the finest architectural sites in the country, the bombers dropped their loads at night on treasured cathedrals, towns, and cities. In Oxford, in the darkness, one could hear German bombers throb overhead on their way to the Midlands and beyond; but still, the City of Oxford was not targeted.

Largely due to the effective action of the Royal Air Force, by early 1942, Britain had withstood the worst of the Luftwaffe attacks, and government officials began looking ahead to rebuilding the nation. Two years earlier, Ernest Bevin, the Minister of Labour, had appointed William Henry Beveridge, an economist and social reformer, to conduct a rapid survey of peace-time manpower requirements. In 1941, however, Beveridge's mission was changed, and he was appointed Chairman of a committee on social insurance. The work was comprehensive and included a number of social surveys. Added to all the other surveys under way, members of the public complained that every time one ate, or coughed, or failed to attend work, there was another person measuring, timing, and counting what had occurred.

Beveridge published his report of the committee's work in 1942, *Social insurance and allied services*, a document that would reshape the structure of British society. Although the final document was not formally published until December of that year, rumours of its findings and intent boosted the feeling that the nation was on the verge of great change. The report identified five 'giants' of social misery: illness, ignorance, disease, squalor, and want. To counter these social evils, Beveridge believed there should be a national insurance system designed to protect the entire population comprehensively, 'from the cradle to the grave', by ensuring basic requirements for a reasonable standard of living. Recommendations

Sinclair in January 1942 when Head of the Oxford Nutrition Survey.

included a policy of full employment, family allowances, and social insurance. The concept of a National Health Service was added later. Beveridge's ideas would not be implemented until after the war, between 1945–50, when they formed the foundation for the Labour Government's social reform legislation. However, between the publication of the report and the implementation of its recommendations, many attitudes held by leading British institutions were to change.

For Hugh Sinclair, the Beveridge report was important because it pointed the way to Social Medicine: a concept that would eventually challenge his firmly held view that human nutrition should be investigated as an independent subject. Sir Wilson Jameson and Sir Farquhar Buzzard, two of Sinclair's strongest allies, were also staunch champions of Social Medicine, and saw it eventually having an important role in the developing medical school curriculum at the University of Oxford. Under the evolving concept of Social Medicine, interdisciplinary research on topics as diverse as housing, employment benefits, and rates of illness were to be developed and studied cooperatively by a range of experts. In the study of Social Medicine, it was not only important to know what people ate, and how many days they worked each month, but also to understand how one factor influenced the other. As months passed, Sinclair became aware that he would need to adapt the ONS programme in some way to fit this changing environment.

Early in 1942 Sir Wilson gave his support to Sinclair's search for suitable accommodation for the ONS outside the Department of Biochemistry. On the surface, one might have seen this as a sign that things were at last moving in Sinclair's direction. Administrative oversight of the Survey continued to be ambiguous at best. A Survey Advisory Committee had been formed consisting of Sir Wilson Jameson, Sir Edward Mellanby, Professor Peters, Professor Robert Macintosh (Nuffield Professor of Anaesthetics), and Dr Hugh Smith, as a representative of the organization providing most of its financial support. This Committee was to consider the progress of the work, and make recommendations to higher bodies both within and outside the University about the Survey's future. People who would have considerable influence over the future of the ONS, such as Professor Burn and the Registrar, would have their say through other University committees that, at times, would strongly disagree with the Advisory Committee.

After considerable discussion, on 29 June 1942, the Congregation of the University accepted responsibility for the Oxford Nutrition Survey and passed a Decree to that effect. Specifically, it was decreed that the purpose of the ONS was to investigate the economic, dietary, and clinical aspects of nutrition. A committee was named to oversee the Survey, which included: the Regius Professor of Medicine, until his place could be taken by a Professor of Social Medicine (who would serve as *ex officio* Chairman); the Whitley Professor of Biochemistry; the Waynflete Professor of Physiology (*ex officio*); Sir Robert McCarrison; Sir Wilson Jameson, Chief Medical Officer of the Ministry of Health; and H. M. Sinclair. It was specifically stated that if Sinclair should relinquish his membership of the Committee, the Hebdomadal Council of the University would appoint some other person.

Introducing the decree before the Congregation of the University, Sir Farquhar Buzzard thanked the Rockefeller Foundation for its support, and said that it must recognize that valuable research carried out by

Professor Rudolph Peters and his staff over a number of years laid the foundation for the nutrition surveys. About two years earlier, Sir Farquhar continued, Dr Hugh Sinclair and one or two of his colleagues began applying this knowledge in a serious effort to establish whether or not the population of Great Britain was consuming adequate food, both in terms of its quality and quantity. It was hoped that this work would assist the Ministry of Food in developing its rationing programme, and it was hoped that the work at Oxford would form the basis for a permanent plan for nutritional studies that would be maintained and enlarged during times of peace. Sir Farquhar further stated that the Oxford Nutrition Survey was not only conducting work of immediate value to the nation, but was also opening investigations into obscure clinical problems that would contribute to the advancement of social and preventive medicine.

Years later, Hugh Sinclair would protest that Oxford University had failed to see the importance of nutrition as a subject in its own right, and had thwarted his progress in establishing a Department of Nutrition there. Based on the evidence in the files, it would seem that the opposite was true: despite the uncertainty surrounding the Survey and its future, by midsummer 1942 it had found a home within the University. There was funding to support it, and the complicated matter of space, the availability of which was greatly limited by the war, was soon to be resolved. Marcon's Hall was about to be vacated, and a Director of the new Institute of Social Medicine would soon be appointed. Certainly the Survey would be an important part of this imminent venture.

Despite the fact the University had accepted the Oxford Nutrition Survey as one of its parts, Sir Wilson Jameson and the Ministry of Health were gradually assuming more authority over its design and execution than Rudolph Peters, the main academic sponsor of the project. Sir Wilson championed the Survey, adding to its credibility by mentioning it during press conferences and speeches. Ministry staff, including Bransby, saw the Survey as an activity to support their own work and did not hesitate to tell Sinclair how specific investigations were to be conducted. From a more positive viewpoint, however, Sir Wilson's enthusiasm meant he used his considerable authority to plead Sinclair's case when necessary with higher authorities both within and outside the University.

By this time, Peter Meiklejohn had undertaken a number of rapid community surveys using a mobile unit as his research base. Sinclair accepted that these should become part of the Survey, although he resented the degree of autonomy Meiklejohn enjoyed. Old rivalries resurfaced, as both men wanted to control and receive credit for the work. Finally, it was agreed Sinclair should direct the Survey, and Meiklejohn should conduct the rapid mobile investigations. Dr E. Walker was another key player in the Survey, and as its Administrator was primarily responsible for budget control and the search for new accommodations.

Within the University, voices raised against the Survey grew louder. James A. Gunn, the Professor of Pharmacology, felt that the Survey was of little value to the University, and he objected to housing such a large staff in what was becoming ever more overcrowded research space. Rudolph Peters was more supportive of the Survey than Gunn, but by now was less convinced of the importance than was Sir Farquhar, whose word as Regius Professor carried considerably more weight than either of the other men. Sinclair felt Gunn was influencing Peters, and complained to Hugh Smith about what he called 'a strict trade union among professors'.

However, the most dangerous problem looming over Sinclair was not the University's pedantic approach to the administration of his work, but rather the more practical matter of data control. There was a lack of any clear understanding of how the accumulating masses of survey data would be aggregated and stored. There was no consensus among the Survey staff, the Department of Biochemistry, and the Ministries regarding which information should be collected, or how data would be tabulated, analysed, and validated. Although Sinclair attempted to meet all the research expectations of the Ministries, Peters, and the Foundation, he personally insisted on defining the research protocols. Unfortunately, he was not always clear in his own mind how best to tackle a problem, and he frequently changed research methods and means of measurement while work was in progress. To complicate matters more, scientific standards that were acceptable six months earlier to experts within the government and elsewhere were now either being questioned or were in the process of being changed. Most worrying, new methods for estimating nutrients and human nutritional requirements frequently appeared in the scientific literature and in government documents. For example, on 23 April 1942 Bransby wrote to Sinclair mentioning that new food vitamin values had been prepared by the Medical Research Council. The impact of this on the Survey's work already completed, and on the procedures employed in Sinclair's laboratory, were considerable.

A few weeks later, in May, Bransby visited the Survey office to study Sinclair's data. He was concerned to see that when pregnant women were examined, their dietary data was collected only by questionnaire. While Bransby was in the office, one of Sinclair's staff asked Bransby which socio-economic data he thought should be collected, thus hinting at the existing level of uncertainty regarding the structure of the Survey. Bransby immediately brought this conversation to Sinclair's attention and demanded that data collection be stopped temporarily. Certainly, Bransby protested, this haphazard way of developing the research information base must come to an end, and a proper plan be put in place before new work was started. A committee was needed to design projects, he said, to strengthen the scientific approach to the work. As for the analysis of data, Bransby wrote to Sinclair:

> It is axiomatic that only trustworthy results should be presented in the report on any scientific study. Hence, if any of the data already obtained by the Survey are not reliable, they must be discarded. Presumably you know which of the biochemical data are reliable, and it should be comparatively simple to separate these from those which are of no value … I agree that no major change in technique should be made in the middle of the Survey unless it has been shown that an improvement will result. Consequently, any such suggestions at the introduction of the individual method for dietary studies, the examination of families by social class, and so on, are automatically shelved until the old methods have been shown to be unsatisfactory. I am not hopeful that, without a special effort, the material for such an examination of methods will not be available for some time … I wish therefore to suggest that, if need be, the work of the Survey be temporarily stopped to get the data up to date and that preparations be made for their analysis. Decisions should be made as to what relationships are to be determined and the order in which such determinations should be made … This is of course a personal note in my capacity as a member of the Survey Team and not as a member of the staff of this Ministry …

These last words were Sinclair's salvation, as it was an admission by Bransby that he had no right to influence the pace of Sinclair's work, either by adding to its workload with new ideas, or slowing it by objecting to the methods used.

Sinclair's first report to the Rockefeller Foundation concerned what he believed to be a riboflavin deficiency among naval recruits. It was informal and contained few statistical details. [This was typical of almost all Survey reports issued by Sinclair.] Smith wrote a letter of congratulation to Sinclair, suggesting that the findings would be of use to the Admiralty. Unfortunately, the lack of statistical validation of Sinclair's conclusions eventually led to criticism, and more murmuring about the value of the Survey. When Smith knew of the criticisms, and understood how they were affecting Sinclair's morale, he wrote reassuringly to him:

> … we want the studies to be organised so that they will have the greatest possible significance and the more favourable outlook for permanence. … If you do get a separate department of your own, additional funds will be required. I think there is a good prospect that the Foundation would make an appropriation for a term of perhaps three to five years and they might increase their contribution to some extent … I think that you should not feel discouraged, as we have really made a lot of progress during the past year and I am certain that with the backing of the Chancellor, Mr Veale, Sir Farquhar Buzzard, Sir Wilson and ourselves the project will be put on a reasonably satisfactory basis very soon. You can count on me to help you in any way possible. …

The Survey staff was small, and had had little time to coalesce into a well functioning unit. Nevertheless, Sinclair took on whatever new projects were requested of him, and then added a few new ones of his own. He was like a whirling apprentice to the twin sorcerers of power and insecurity. He enjoyed being described as the man who would guide the hand of the Ministry of Food as it developed its plans for rationing. He enjoyed the feeling of power when he lunched at the Athenaeum with men like Jameson, Buzzard, and Mellanby. How miserable the plodding work of the bench scientists must have appeared. On the other hand, he appears to have been unable to refuse any addition to the Survey, even if he knew they were beyond its capability. Was he afraid to turn down a request in case such a refusal would suggest a limit to his capacities, or worse still — incompetence?

Such criticism may be too harsh, however. Hugh Sinclair may have seen what had become a patchwork of Survey studies as a means of collecting the data he needed to exploit his original scheme of integrated nutritional research developed for the study of diseases of the nervous system. In a report to Sir Wilson, Sinclair described the following ambitious complex of projects under way or planned by the Oxford Nutrition Survey:

> … the complete nutrition survey (economic, dietary, clinical and laboratory) is being made upon a random sample of the Oxford population, to test and

develop methods of assessing the state of nutrition, and to train personnel in the use of these methods. In addition … a small area … will be selected and a complete study made … of all the households in that area. Where possible, all these subjects will be examined again at intervals of six months.

A clinical and laboratory examination will be made of RAF recruits coming to Oxford for their entrance interview. …

Possibly an examination will be made of men attending one of the two civilian recruiting boards for their first medical examination. Unlike the RAF recruits, these men have not passed through the filter of a previous medical examination.

Medical, and possibly other University, students will be examined.

A supplementary feeding test will be carried out upon volunteers from one or more factories [for example, Morris Motors Ltd or the Pressed Steel Co Ltd.] in Oxford.

A supplementary feeding test with women attending an antenatal clinic …

A supplementary feeding test has been in progress for a few months, giving yeast daily to half the boys in an Oxford orphanage.

The nutritional state of different economic groups (e.g. Radley College and an Oxford Secondary School) will be assessed by the Survey method.

With the collaboration of Dr A. E. Barclay, the incidence of rickets in infants and children will be assessed.

Mobile Survey …

We are about to place undergraduates upon a diet deficient in vitamin A, and frequent examinations will be made of skin, cornea (with the slit-lamp microscope), visual adaptation, and blood vitamin A.

Experiments with volunteers upon diets deficient in one factor have been started to define the earliest signs and symptoms of deficiency, and the most satisfactory methods (whether clinical or laboratory) of detecting them. [Twenty-four] students at Heythrop College recently lived upon a diet deficient in vitamin B_1 (Twelve being controls).

Two things should be noted here. First, Sinclair subjected himself to the protocols used for most deficiency experiments, even when they weakened him. Second, the work mentioned as being conducted at Heythrop College was of particular interest to Rudolph Peters, who was anxious to obtain its results for his own investigations into vitamin B_1. Unfortunately, Sinclair did not give the Heythrop project the priority it deserved. Despite repeatedly asking for information, more than a year was to go by before Peters received any of the Heythrop College data from Sinclair, and these would be limited.

This lack of attention by Sinclair to Peters's personal research interests, along with Sinclair's growing lack of attention to his work as a Demonstrator in Peters' department, undermined the more senior man's faith in him. Peters also felt that his own work was intruded upon by the constant flow of experimental subjects coming through his department to participate in Sinclair's studies of night-blindness. This annoyed him, as did the increasing amount of space Hugh's survey equipment and staff occupied within his laboratory. For a time, Peters merely muttered his disapproval, and sent brief notes; then, he confronted Sinclair face-to-face. He reminded Sinclair that his reservation from military duty was based on his being a 'Whole-time University teacher'. He complained that the Survey was drawing precious financial resources away from the department. And he accused Sinclair of not taking his teaching responsibilities seriously, pointing to a lecture Sinclair gave concerning the effect of sherry on dark adaptation as an example of frivolity. He asked Hugh to consider giving up demonstrating while working on the Survey, and suggested that a sum of £250 from the Survey budget be paid to the Department so Peters could hire a suitable replacement. Hugh objected strenuously, saying he would do more for the department in the future, and that finding space for the Survey, separate from the Department, would eliminate most of the problems. Wavering, Peters calmed his anger.

Fortunately, Sinclair continued to enjoy Hugh Smith's support. Smith knew that Sinclair's work was off to a slow start, and that he was under pressure from a number of outside sources, and so intervened repeatedly with the Ministry of Health, the Medical Research Council, and even with the Foundation. In addition to securing funds and equipment from the Foundation, he also recruited another clinician to help with the mobile surveys. Dr V. P. Sydenstricker, from the University of Georgia, in the United States, joined Peter Meiklejohn. Smith went so far as to have some Survey reports edited and retyped before they were submitted to the Ministry of Health. In Hugh Smith's mind, the Survey was now reaching its final form, and he wanted to see it succeed. He again wrote reassuringly to Sinclair:

> The whole project is gradually evolving in everyone's mind and should be on a satisfactory basis within another few weeks. The period of preparation, which took up nearly the whole of 1941, was unavoidable, I think. This year

will be one of steady progress and 1943 will see the methods generally
accepted by all who have their eyes and minds open. You should feel
encouraged to go straightforward and not mind the doubters and critics.
Do let me know whenever I can be of any service to the Survey …

Hugh Sinclair responded to his friend, claiming to be pleased that
others '… agreed completely with my previously unexpressed view that
most of the clinical work (and probably some of the biochemical) done
during this preliminary period has not been of much value; I believe
Peter [Meiklejohn] thinks otherwise.'

As the volume of data increased, Sinclair realized he needed access to the
fastest and most sophisticated means of analysis possible, and turned to
Bransby for technical support. Specifically, he wanted access to a Hollorith
machine, a large and cumbersome precursor of the modern computer, in
which data punched on special cards could be aggregated, counted, and
compiled. Hugh designed a format for the punched cards, covering all the
parameters, or the data set, that he felt should be included in the basic
Survey. As one might expect, details of what was included on those cards
became a cause for disagreement between Sinclair and Bransby.

They also disagreed about the interpretation of data. 'Have you no faith
in the lab's data?' Sinclair once asked in a letter. In fact, Bransby had many
questions about their validity.

Over the months to come, the process by which the Hollerith cards
were prepared and analysed frequently slowed the reports from the
Survey, and added to the seeds of Sinclair's eventual downfall within
Oxford. Young scientists today, who are blessed with the advantages of
speed and accuracy provided by modern computers, cannot imagine the
laborious process of data analysis faced by Sinclair and his team. Coding
data by hand, entering it on punch cards, separating and aggregating
variables by simple card-sorting devices, and performing with pen and
paper statistical investigations, such as multivariant analysis, was laborious. On a few data sets the work was possible; on masses of data as large as
those accumulated by the ONS, the task was gargantuan. Hugh was
proposing to identify causal relationships between diet and disease that
required analytical tools that had not yet been invented. It is a sad fact
that Hugh and close colleagues worked to analyse the data on these cards
for the rest of his life. Mining the information collected and stored during
the Survey would not be possible until after his death.

To help maintain his position in Peters's laboratory, Sinclair also continued animal experimentation within the department. To supply this work, Hugh converted a tennis court at Lady Place into a pen for breeding guinea-pigs. Rosalie had little say in the matter: under the agreement recently made in favour of her children, Hugh now formally owned Lady Place. Somehow, Rosalie fell heir to the onerous task of feeding and counting the rapidly multiplying guinea-pig population. The health and size of the herd became a frequent source of considerable irritation between mother and son as she struggled to maintain constant supplies of proper food for the beasts, and fussed to make sure Hugh came from time to time to remove as many animals as possible to his Oxford laboratory.

But Rosalie was not the only person irritated by Hugh's animal experimentation. Clive Staples Lewis and Sinclair occupied nearby rooms in the Madgalen College residential hall, known as New Buildings. Best known for his children's books chronicling the land of Narnia, C. S. Lewis also spoke and wrote on religious themes. Among his writings is a little book entitled *The problem of pain*. As Lewis was an outspoken anti-vivisectionist, in this book he extends his subject to life other than human:

> The problem of animal suffering is appalling; not because the animals are so numerous (for, as we have seen, no more pain is felt when a million suffer than when one suffers) but because the Christian explanation of human pain cannot be extended to animal pain. So far as we know beasts are incapable either of sin or virtue: therefore they can neither deserve pain nor be improved by it.[1]

Within the halls of New Building and over dinner at the College High Table, Sinclair and Lewis often and loudly debated the value of vivisection. Eventually, this argument would even be heard as far away as America.

By early March 1942, Peter Meiklejohn had completed his first rapid nutrition survey in the small industrial town of Accrington, with a population of approximately 4 000. He reported that about 20 per cent of adults, mostly women, showed some evidence of malnutrition and a slightly smaller proportion of children were malnourished. Vitamin A deficiency was the biggest worry, although laboratory investigations also demonstrated some anaemia in women of all economic groups, and low blood levels of vitamin C. When found, real malnutrition was due to economic factors, he said. Practically every case was in a family who spent less than ten shillings per person per week for food. As for the case of the women, when food was available they tended to feed their husbands and children before they fed themselves.

While Meiklejohn applied more rigour to his survey than that used in Oxford, he still left his work open to criticism for its lack of statistical methodology. Some months later, when American experts reviewed the report on Accrington, it fell short of expectations. Sampling techniques were less than ideal and not all members of every family studied participated in the survey. While these criticisms were valid, the experts appeared to overlook the complexity of conducting a mobile community survey, especially under war-time conditions in populated areas.

For now, however, Hugh Smith was pleased with both Meiklejohn's and Sinclair's progress, and took a big step into the future. He suggested in a letter to Dr Sawyer, Head of the International Health Division of the Rockefeller Foundation, that the Oxford Survey Group should be used in post-war nutrition studies in Belgium, Holland, or France. When Sinclair was informed of this, he was keen on the idea and, despite being overstretched both in his teaching and survey responsibilities, immediately began drawing up plans for the Survey's post-war future.

As mentioned earlier, support was growing within the Oxford community for a new Chair in Social Medicine, and an Institute of Social Medicine. Sir Farquhar Buzzard was among the strongest champions of the idea, and he hoped to get such a Chair funded and its incumbent in place before his retirement the following year. Peters suggested that the nutrition Survey should be part of this new endeavour. Sinclair probably saw himself as the man who should become responsible for both the Institute and Oxford's continuing work on human nutrition.

By this time, the Survey had a staff of twenty, including clinicians and technicians. With Dr Walker, the Survey administrator, Sinclair began drawing up extensive plans for the use of Marcon's Hall. Knowing Sinclair's view that human nutrition must be made a leading scientific discipline in its own right, it was not surprising that Social Medicine was allocated less space than Nutrition. Writing to the Registrar, Sinclair said that the building '… seems almost ideal for accommodation for the whole survey … I assume that the Survey would be able to use most of Marcon's Hall, as I think you said that the Department of Social Medicine would only need some accommodation for offices at the top of the building.' This was a vital underestimation of problems arising from scheduling and demands by the Institute of Social Medicine.

Sir Wilson Jameson began looking towards his retirement, and was now anxious to see both Social Medicine and the Oxford Nutrition Survey formalized as part of the Oxford medical community. He thought that the Nuffield Regional Hospitals Trust would help fund the Survey, and he saw the new Chair in Social Medicine financed by the same organization. By tying the two concepts together, it seemed reasonable that they should enjoy receiving simultaneous support from the same sources. Partly influenced by their close association with Sir Wilson, Hugh Smith and others at the Rockefeller Foundation agreed that Social Medicine was important in future developments at Oxford, and should have a dominant role over the Survey. The general thinking was that social and economic factors dominate food choices and eating patterns, and as factors in social medicine are more complex than those involving food, the Nutrition Survey would be of less importance than the overall study of social influences on health. Ultimately, everyone but Sinclair agreed that the Survey should become a sub-department of the Institute of Social Medicine.

About this time, Sinclair submitted an extensive proposal for funds to the Rockefeller Foundation. In it he defined the complex nature of the study of human nutrition. In the text he states that 'The science of human nutrition is no narrow subject, for it stretches from the farm through the market, the kitchen, the clinic and the laboratory, to the hospital ward. In the narrower and immediate practical field of judging the state of nutrition of an individual there is no easy method: the assessment rests upon a tripod, whose three limbs are the dietary, laboratory and clinical methods.' In these few words, Sinclair stated his unique vision of the scientific study of nutrition. His vision required cooperation between himself and experts in other medical disciplines, but not necessarily those to be found in Social Medicine.

Perhaps unsettled by reports from Dr Bransby, Sir Wilson was becoming restless, and asked to see hard facts generated from the Survey. Sinclair deflected this request by sending a copy of a recent proposal seeking funds for the Survey. He states in the covering letter:[2]

> As I have mentioned at the end of the report, our results are tentative, because it is difficult for us to produce valid results until we have got settled in our new premises and have the help we are about to obtain for this specific purpose.

Rumours were now circulating within the University concerning Hugh Sinclair's future. Growing tension between himself and Rudolph Peters was well known. In the end it was Peters who brought the matter to the surface:

> My dear Sinclair … I understand from the Registrar that you will be resigning your University Demonstratorship. Is this correct, because I want to get things tidied up for next term … Magee said that you had finished the Heythrop blood results. May I see them?

Hugh responded to Peters by letter, emphasizing that he had not told the Registrar he planned to resign. He goes over the contents of previous discussions with Peters on the matter, and concludes:

> I have to weigh the pros and cons, taking into account particularly what I can best do to help to win the war and trying to disregard particularly an obvious personal disinclination to terminate the career I should like to follow: research and teaching in biochemistry … [the] scales are tipped that way in my mind because I cannot at the present see how I can effectively Tutor in Magdalen, direct the Survey in Marcon's hall and hold a University Demonstratorship.

Peters responded: 'In view of all you say, would it not be best for you to give up your Demonstratorship temporarily while on war work. I am afraid that this would mean financial compensation to the department, but surely something should be got from the Survey funds. …'

Hugh let the matter drift. He was beginning to find that meeting his financial obligations to Lady Place was difficult, and his general financial situation would not readily allow him to abandon any source of income.

As if sent by some kind god to revive Sinclair's ebullience, Sir John Boyd-Orr, the world-renowned nutritionist, had heard about the Survey and was impressed with its growing reputation. In a letter, from his office in the Rowett Research Institute, Bucksburn, Aberdeenshire, Boyd-Orr wrote:

> Mr dear Sinclair,
> I would very much like to have an opportunity of discussing your work with you. It is full of interest. It is exactly along the lines of our Carnegie Survey, which we have never been able to write up because all the people who were engaged on it are in the Fighting Services or other war work. I would like to visit Oxford. Possibly we will get the chance of meeting and discussing it sometime when I am in the south. In the meantime I would like to say that in my opinion it is a first-class bit of research, though I doubt whether you

will get out anything quick enough to be of value during the war ... With kind regards and best wishes for the success of your work ...

Hugh was delighted. But such was his state of mind at the time he could not accept even the hint of criticism that his work would not have relevance until after the war, and so he responded:

> I have of course been fully aware that our main Survey is on precisely the lines of your Carnegie Survey, and therefore we have all been longing to have the opportunity of your advice and criticism ... We would not have undertaken the work had we not had Sir Wilson Jameson's strong support, since he is convinced that we can give him the results that will be of value in war-time. For that reason we have mobile surveys that go round the country ... and send in quick reports within a month of completing the work ...

Hugh's work was receiving quiet, but influential, support from another channel. Jack Drummond was passing encouraging information about the survey to the Information Division within the Ministry of Food, so that abstracts could be published in confidential summary documents and circulated among higher officials there. Of greater importance, Drummond had in his files preliminary plans for post-war nutritional surveys on the Continent, and Sinclair was listed as a participant.

CHAPTER 8

The Institute of Social Medicine

During the presentation of the budget on 30 June 1942, Ernest Brown, Head of the Ministry of Health, reported that the Ministry was cooperating with the Ministry of Food, the Board of Education, the Medical Research Council, and the University of Oxford to improve the health of the nation. Regional Medical Officers and the Emergency Hospital Service were supported by local nutritional surveys helped by funds and expertise from the International Health Division of the Rockefeller Foundation. Although Hugh Sinclair's name was not mentioned, reference to the University of Oxford clearly identified his work, thus adding to his personal credibility.

Now in demand as a well-known expert, Hugh involved himself in the initiation, or reorganization, of various professional nutrition groups: the powerful Nutrition Society was one such organization. A friend of Hugh's, Harriette Chick [later to become Dame Harriette Chick], of the Lister Institute, was a leading force in the Society. Hugh, Harriette, and others wished to shift the dominant membership of the group from a collection of food producers and manufacturers, to one controlled by scientists and nutrition-related professions, thus pleasing many within the medical community. Unfortunately, his work with the Society took precious time away from work on the Survey.

Hugh Smith had told Sir Wilson Jameson that he could expect monthly reports from Sinclair. When asked by Sir Wilson for these documents, Sinclair pleaded his 'deep preoccupation with making plans for the new building and the laborious work of getting out estimates.' Not surprisingly, this eventually annoyed Jameson. Instead of formal reports, from time to time Hugh sent brief statements of current activities, or delivered verbal reports over lunch at the Athenaeum.

The Rockefeller Foundation was interested in aspects of the Oxford Survey other than Sinclair's work. Meiklejohn was well into his second major mobile survey, in Merthyr Tydfil, a mining area in South Wales with a population then of about 30 000. Hugh Smith (who confused matters by referring to the Survey as a 'department' when writing to Dr Sawyer, in New York) was most anxious to support Meiklejohn's work. In Smith's view, the first mobile survey resulted in a well-written report, which was needed to build credibility for the Survey in America. Surprisingly, to many observers, Peter Meiklejohn's early findings indicated that, with the wide availability of war-rations and public education concerning nutrition, the nutritional status of Britain's poorer populations was better now than during the hard years of the Depression.

In mid-August 1942, Dr Walker, the Survey Administrator, died suddenly and unexpectedly. Mrs A. Fisher, MA, MSc, joined the Survey staff in his place. Her qualifications and previous work made her ideal for the position, as she had worked in both the Lister Institute and the National Institute for Research in Dairying, in Reading, near Oxford. Until the Survey was terminated some years later, her skills and training provided Sinclair with a source of wisdom, expertise, and quiet support, which he often under-rated.

Marcon's Hall, hereafter known as 10 Parks Road, required refurbishment before the scientific work of the Survey and the Institute of Social Medicine could move in. Dr Walker's duties had included securing and upgrading this space for the Survey; Sinclair now assumed full responsibility for this task, even helping with the physical work of construction and plumbing. By doing so, he further squandered time that should have been used for the analysis of Survey data.

By some, including Hugh Smith, the Oxford Nutrition Survey was now sometimes referred to as a separate university 'Department'. This terminology was inaccurate, as the Survey administratively remained little more than a cost centre within the Department of Biochemistry. Although Rudolph Peters, Sinclair, and others discussed the possibility of a Department of Nutrition after the war, probably with Sinclair as its Head, it was obvious that the Survey could not form the basis for its development; it lacked the teaching and basic scientific research demarcating a university department. This seemed not to bother Sinclair, who appeared indifferent to ongoing plans for medical education at Oxford. Although he was invited to help devise a new curriculum he made few efforts to participate, thus leaving others to decide how human nutrition should be taught to undergraduate medical students.

Sinclair's interests now lay beyond the study of basic nutrition; his appetite was whetted for the much broader challenge of Social Medicine. He hinted in a letter to his brother that he saw himself as Director of a new Institute, and even his descriptions of the Oxford Nutrition Survey now placed a firm emphasis on social and economic factors in nutritional deficiency. In late August Sinclair wrote to Sir Farquhar on the matter of the relationship of the Survey to the Institute, saying that both he and Hugh Smith believed the Survey should become a sub-department within the Institute. Would Sinclair have been willing to have the Survey buried in the Institute if he did not see himself as its Director?

Demand for space at the University was acute, and Sinclair soon learned that more than the Institute and Survey would be housed at 10 Parks Road. The Ministry of Home Security wanted to move the anatomist, Dr Solly Zuckerman, into rooms on the top floor of the building as soon as possible. As his part in the war effort, Zuckerman had abandoned his colony of baboons for more practical pursuits, including the preparation of highly secret plans for D-day. Also, the Land Army occupied rooms in the building, and there was little chance of their making a hasty move. Hugh protested, but to no avail. At times, Mrs Fisher found the comings and goings of people outside her jurisdiction disagreeable, and petty theft became an irritating problem.

The Ministry of Health continued its interest in studying the effects of nutrition and vitamin supplements on the health and efficiency of factory workers, and wanted the Oxford Nutrition Survey to conduct relevant industrial studies, thus further expanding its work-load. In mid-September, 1942, Sinclair wrote to the Factory Medical Officer at Morris Motors Ltd, saying that a nutrition study had been discussed and agreed with Lord Nuffield. The work had also been discussed with the Ministry of Labour and the Industrial Health Research Board. Three hundred male volunteers would be examined for early signs of nutritional deficiency and tested for blood vitamin levels. Half of the men would then receive vitamin capsules provided for this purpose by the USA to the Ministry of Health, and the other half would be given placebos. The men would be followed for general health, such as the incidence of upper respiratory infection, and work output, if feasible. A second physical examination would take place at the end of six months.

It may be that Sinclair had discussed this matter in person with Lord Nuffield sometime earlier; but it was not until a few days after his letter to

Morris Motors that Sinclair wrote to the Rt Hon. Viscount Nuffield, outlining the study and stressing the importance of good nutrition to the maintenance of industrial output during the war:

> Dr. Magee — who deals with nutritional problems in the Ministry of Health — and I called on Dr B at Morris Motors Ltd a few days ago, and he kindly offered to give us his help. We shall be assisted in the medical work by Dr V. P. Sydenstricker who is one of the foremost nutritional experts from the USA and who has come to this country on a brief visit to give us the benefit of his experience.

On 5 October Hugh Sinclair wrote to Hugh Smith, exaggerating the facts of the industrial survey:

> … Tremendous progress is being made with the factory experiment. I had an interview with Lord Nuffield last week, and he seemed very keen on it and offered us all facilities — in fact he seems pretty vitamin-conscious! Lord Nuffield has just sanctioned a considerable expenditure on statistical machines to be set up at Morris Motors for keeping careful records of all relevant things [including work output]. The new factory Medical Officer is a curious hypomanic individual who was thrown out of the Army because he wanted to introduce statistical methods. He is very keen indeed on helping us and will look after the question of getting volunteers and helping with the statistical side.

Unfortunately, members of the Morris Motors Works Welfare Committee eventually objected to the study because of its use of efficiency measures: many felt it was a means for managers to spy on employees. When mentioning this work in his next report to the Survey Committee, Sinclair stated that the Morris Motors factories were found to be unsuitable, but 150 factory workers at another engineering works were being studied by the Survey instead.

During the second half of 1942, the Rockefeller Foundation and Sir Jack Drummond, Special Scientific Advisory to the British Ministry of Food, began laying specific plans for the study of the nutritional status of civilians in liberated areas of the Continent. It was agreed to undertake a number of rapid surveys, beginning as quickly as circumstances would allow following liberation. The primary purpose of the surveys was to provide Allied forces entering an area with information about the food supply required by the local population. It was believed that as people regained their strength and health through better nutrition, they would be able to grow their own food and begin rebuilding their lives. In a letter to Dr W. A. Sawyer, his superior in New York, Hugh Smith wrote:

> As I reported to you before, the group working on nutrition in Oxford is keenly interested in the development of plans for Europe, and I have suggested to Drs. Sinclair and Meiklejohn that they put their ideas as to possible technical programs into writing. It will be useful to know their views as to what may be required in the way of equipment, laboratory apparatus, and personnel to form a well-rounded unit for nutrition studies in the field. Sinclair expects to be installed in the new headquarters in Oxford about the end of November. After that he will have ample facilities for the training of additional laboratory and field workers in small numbers there.[1]

Hugh Smith saw the rapid surveys on the Continent as a means to advance the work the Rockefeller Foundation had begun in Europe. Reporting back to Dr Sawyer, in New York, he wrote:[2]

> 8 October 1942:
> I gather that, since the British Isles are, and are likely to continue to be the principal center of research and teaching for the whole Commonwealth of British Nations, our best opportunity to inaugurate programs of far-reaching influence will occur here … there has been a tremendous impetus in Britain during the war toward the development of social medicine. The new Institute soon to be opened in Oxford has been warmly welcomed and aroused great interest. It will, if the right man can be found to direct it, have a great influence in British medicine.

Smith continued, discussing other contributions the Foundation was making towards improving the health status of Britain: up-grading housing conditions, helping the re-location of industries, and increasing the use of immunization. As for tuberculosis and venereal disease, he wrote:

> We can certainly teach the British health officers something about the control of the latter. Quite likely some action will be forced on them by the presence of large numbers of American and Canadian troops.

It must be mentioned that other investigators were discussed as well in regards to the postwar work, including Drs R. A. McCance and Elsie Widdowson, from the Dunn Institute, at the University of Cambridge. McCance and Widdowson were well respected for their work on the absorption and metabolism of nutrients, and were well known for their extensive research through self-experimentation. One of their significant contributions to nutrition was the publication, in 1940, of tables showing the nutrient composition of a series of foods. This work has been updated regularly, and remains an important tool in nutritional practice and research. In the light of things to come, it is worth noting that it was Professor J. A. Ryle, Regius Professor of Physic at Cambridge, who first invited Dr McCance to Cambridge as a Reader in Medicine. McCance took Elsie Widdowson and members of his staff with him. Ryle's respect for McCance and Widdowson would soon have an influence on Sinclair.

Meanwhile, Hugh Sinclair faced the problem of finding the equipment his staff needed to analyse the masses of data accumulating through the various activities of the Survey in Great Britain. As a high priority, he wanted a Monroe Adding Calculator, which at the time was not available through British sources. He asked Hugh Smith to obtain one through the American system, but U-boats continued to dog besieged merchant ships, and cargo space was increasingly preserved for the absolute essentials, such as oil and food. For one of the few times during the war, Smith gave Sinclair's work low priority.

Better luck came with Sinclair's request for a warming device for biological samples, known as a Barcroft Bath. Rudolph Peters loaned his department's bath to Sinclair on the condition that it would be returned in operating order on one month's notice. At great expense, the bath was modified to meet Sinclair's requirements. When the Survey finally moved to 10 Parks Road, items of equipment were taken from the department without Peters' permission: most irritating to Peters, an 'accumulator' [a crude computational device], and a microscope. Peters and Sinclair struck an agreement that these and other items would be purchased from the department by the Survey, but the damage to trust had been done. Bench Fees, which are the overhead costs of using laboratory space and equipment, remained a point of dissension, however. Until his death, Dr Walker had dealt with this matter; now Sinclair decided the previously agreed fee of £50 per person per year was too high, and refused to settle the Survey's account with the Department of Biochemistry. Eventually, at Hugh Smith's insistence, Sinclair paid the Bench Fees.

With Christmas approaching, Lady Place was readied for guests. Ration books were posted ahead to Rosalie, and Hugh was able to obtain venison to roast (probably from the Magdalen Grove) and excellent

South African sherry. (It is worth noting that it was not possible to slaughter an animal during World War II unless one had a licence to do so. For this reason, it is said, the Magdalen deer were registered as vegetables, and so needed no licence to be culled.) Rosalie also wanted a turkey or goose for Christmas dinner. There would be eight guests in the Dining Room, and five seated in the Staff Room. Hugh Smith and Dr Sydenstricker, among others, enjoyed a traditional British Christmas Eve and Day at Lady Place, where tables groaned with food, and a Scottish piper provided the evening's entertainment.

The Regius Professor of Medicine in Oxford, Sir Farquhar Buzzard, retired during the New Year, 1943, and was replaced by Sir Arthur William Mickle Ellis, previously Director of the Medical Unit at the London Hospital. Ellis lacked Buzzard's dynamism, and commanded less authority among the various factions within Oxford, who were warring over the structure and purpose of the new postgraduate medical school. Buzzard had seen a future in which the Institute of Social Medicine would influence the direction of both the Oxford medical school and plans for British social reform; Ellis was a less eager advocate of its merits.

Before Sir Farquhar retired, he wanted to see strong leadership at the helm of the Institute of Social Medicine: with support from men as influential as Sir Wilson Jameson, his choice was John A. Ryle. When it was learned that someone of Ryle's prominence was willing to leave a powerful position at the University of Cambridge to guide this new venture, the Nuffield Provincial Hospitals Trust underwrote the Institute. Ryle took up his post in April, 1943. He would be a Fellow of Magdalen College, and his rooms there were only a few doors away from those occupied by Sinclair.

Between the memorable Christmas celebration at Lady Place and Ryle's appointment, Sinclair began to feel under attack again. A letter from a faculty member of Vanderbilt University, in the United States, severely questioned the results of a report of Peter Meiklejohn's mobile survey in Merthyr Tydfil. It was suggested that samples used in the study were not 'random' as the report suggested, although that did not completely devalue the findings: it was a matter of understanding the source of the samples. 'It is possible that under the circumstances intelligent selective sampling may be preferable,' a reviewer wrote. And then there was the matter of the size of blood samples taken from children: was a drop enough? Questions were raised about the B_1 work in the survey: what was the 'correlation between your B_1 (Phycomyces method) and your other tests of B_1 nutrition including electrocardiograph?' it was asked. The Oxford report suggested that lowered levels of B_1 in the blood could be determined before clinical signs appeared. However, in a study at Vanderbilt, levels seemed not to drop until after clinical signs appeared. Finally, the Survey's conclusions were brought into question on what had been seen as clinical evidence of vitamin A and vitamin C deficiencies.

To doubt the results of the nutrition survey was to doubt Sinclair himself. He not only worked on the details of housing his research, but he also did much of the technical work. In the book, *Sinclair*, Brian Lloyd, a staff member at the time, remembers:

> … all the estimations of blood thiamine were done by him in person: the picture of a wryly-smiling but tired Hugh standing in a tattered lab coat for hour after small hour at the laboratory sink washing out horrid little tuffettes of Phycomyces Blakesleeanus under a dribbling tap is unforgettable. The curl on that massive forehead had long departed!

Sinclair did not limit work at 10 Parks Road to human subjects. Animal experimentation involving capillary fragility and vitamin C, and studies of the effect of ultraviolet light on rat cornea was also under way. Using the basement as an animal room, Hugh had borrowed or purchased cages from any source at hand. In the Department of Biochemistry, Sinclair worked under Peters's licence for animal experimentation, but when the Survey moved, Sinclair obtained his own. His skill with small animals was appreciated by those working with him, and some years later, in his contribution to the memoir, *Sinclair*, the notable Indian nutritionist, Dr V. Ramalingaswami, wrote:

> Hugh is also a superb experimentalist. I learned from him how to handle animals gently. It was a pleasure to watch him carry out slit-lamp examinations of rats for corneal vascularisation. The rats seemed to behave while he handled them.

Ramalingaswami did not join Hugh's laboratory until 1949, but this insight is important here because it testifies both to Hugh's use of animal experimentation, and the consideration with which he conducted this part of his work.

Hugh was deeply unhappy when John Ryle was appointed Director of the new Institute of Social Medicine. Had the Survey been more productive, Sinclair might have had a chance, but Sir Farquhar Buzzard and other university leaders knew Ryle commanded the academic prestige the new venture required. Hoping to smooth the way towards a good working relationship, Sinclair wrote to him with charm and welcome. His letter also showed his unwillingness to accept that the Survey would be an integral part of the new Institute. They could cooperate, but would be separate:

Dear Professor Ryle,
I should like to add my congratulations to the many that must be reaching you upon your appointment to the new Chair in the University. The importance of the Chair is reflected in our ability to attract from our rival University its Regius Professor and from clinical medicine one of its most distinguished leaders.

Those of us who are interested in Social Medicine have awaited the founding of such a Chair, and this is particularly true of some of the colleagues of my own age here who, in the last few years, have been ably directed towards these views by Sir Farquhar Buzzard. As you probably heard from Sir Farquhar, the experimental work that was being done here before the war on human and animal nutrition has been broadened into this Nutrition Survey as a war-time extension of the work. It is of course work that we hope will continue after the war and which, we hope, will be closely associated with the work of your Institute. Therefore it is particularly fortunate for us that we are housed in the same building, and will be able to collaborate with you and have the advantage of your advice.

Sir Farquhar has told me that he has asked you to attend the meeting of the Finance Committee of the Survey on Saturday morning, and that he hopes you may dine in Magdalen [our mutual College] on Saturday night …

John Ryle had known Sinclair since the younger man's days as a student at University College, where he was a member of the pro-peace student group, and wrote the contentious and anti-Semitic editorial published in the College magazine. It is difficult to believe that, during this time of war, Ryle and others did not think back on these events when they worked with Sinclair. After his arrival at 10 Parks Road, John Ryle wrote to Sinclair:

I have a number of things to thank you for — for showing me over the Institute; for supervising so much of the preparatory structural [work] and thereby saving me work and thought in anticipation; for the documents relating to your programmes of work undertaken by the Nutrition Survey; for your kind luncheon hospitality at Magdalen. I look forward very much to our association in the future and hope that I may be able to co-operate with some of the clinical work of the survey. I have had in mind a full Family Health Survey of samples from the main economic perspective with continuous follow-up over a period of years … For such a survey your work will clearly have prepared the way …

Several weeks later, Ryle again wrote to Sinclair, suggesting they should consider, as soon as possible, means of avoiding unnecessary duplication of personnel, equipment, books, and scientific journals. For example, an incoming statistician foreseen by the Nuffield Trust to take over the work of the Bureau of Morbidity Statistics might well direct the statistical work of both the Nutrition Survey and the Institute. They could build a common library, and clinical instruments and computing machines could be used by both. In addition, Ryle would like changes in the allocation of space: would it be possible to have his rooms, including at least one of the rooms on the second floor, cleared and 'spring-cleaned'? And, as the room set aside for his secretary was not adequate, could Sinclair's secretary work elsewhere? Ryle had not yet arrived on campus, and he was already trying to manage Sinclair's work.

Sinclair was in no rush to communicate, and responded to Ryle's requests two weeks later. Miss Hodgson, his secretary, would join him in his office, he said. Statistical resources could be shared, although he had already designed the type of Hollerith data punch cards the Survey would use. Hugh did not mention that he had secured money from Lord Nuffield for computation equipment, and had managed to purchase a calculator: he would resist seeing these items placed into a communal pot.

Like C. S. Lewis, the new Head of the Institute for Social Medicine was against all experimentation on humans — and particularly disliked the use of animals in research. His approach to discovery was through observation. Here then, was a growing seed of contention between Ryle and Sinclair, who believed that medical science could no longer be advanced by observation alone. Men like Buzzard and Jameson, who knew both men, might have anticipated that Ryle and Sinclair would not be able to combine their research efforts. Disagreements over the animal room became unpleasant and eventually Ryle insisted that it be moved, complaining of its disagreeable smell, and the likelihood it would attract vermin. In the end, Mrs Fisher was left to deal with the matter, and terminated the animal research. The loss of the animal room meant that Sinclair could not pursue some of the basic, or fundamental, aspects of nutrition that were necessary to justify him within the University's community of academic scholars. In months to come, this would become a significant problem.

While the University waited for Ryle's arrival, Sinclair decided to maintain a foothold in the Department of Biochemistry. He again told Rudolph Peters that he was anxious 'in view of the impending election to the Chair of Social Medicine' to continue fundamental research work in the Biochemistry laboratory. He also indicated that he would like to return to Peters's department after the war. Peters was surprised by this, and wrote to Sinclair:

> ... it is obviously sensible that there should be good reciprocal relations as a permanency between the human nutrition activities of the Survey (or its future developments) and the animal work in this department; hence, I feel that as soon as the war is over, this whole problem should be thoroughly explored ...

And so, it would be possible for Sinclair to return to the department after the war, if he wished.

Despite signs of increasing irritation over his lack of attention to the Survey work, and clear evidence that technical problems involving data were mounting, Sinclair decided to attend the annual meeting of the American Public Health Association. In large part, the trip was important because it offered opportunities to discuss post-war surveys on the Continent with Americans planning to participate in the work. Hugh Smith was in favour of Sinclair's trip, but wanted to make certain the Survey would be properly managed in Sinclair's absence. Smith asked Peter Meiklejohn to step in. Meiklejohn refused, saying he could better spend his time writing up results from the mobile surveys. As a second choice, Mrs Fisher took over the day-to-day direction of the work.

Some months earlier, Meiklejohn and Sinclair had had a major falling out, ostensibly over the method used by the mobile surveys to measure blood protein levels. Sinclair was incensed that the mobile work had received criticism in the United States, and disliked the fact that Meiklejohn had turned for advice to Magnus Pyke, in the Ministry of Food, rather than following Sinclair's method of choice. Meiklejohn accused Sinclair of not completing the work expected by Sir Wilson and the Ministry of Health, while taking on other responsibilities that would further his own career. Arguments between the men were bitter and long: Sinclair's notes record one in his rooms at Magdalen College continuing from 8.30 in the evening until midnight. Meiklejohn claimed the Survey now deviated from its original purpose, and he no longer felt sympathy for it. Data produced thus far were useless, he claimed. Instead of Sinclair going to the United States, Meiklejohn felt he should stay in England and complete the work Jameson expected.

Sinclair went to the United States at the invitation of Dr Wilbur O. Atwater, one of America's most prestigious nutritionists, with the express aim of his attending American Public Health Association meetings. As part of the arrangement, Sinclair was asked to participate in the scientific programme of the Section on a subject of his choice. He chose, of course, to speak from his experience of the Oxford Nutrition Survey.

It was agreed that in Hugh's absence the staff would write up each of the projects the Survey had undertaken; on his return, these documents, combined with a summary, would constitute a report on the first two years of the Survey's activity. Expectations were high. It was not enough that a general report of findings was produced, but data cards were to be prepared so that dietary intake information and clinical symptoms could be correlated: vitamin A and carotenoids with night-blindness, for example.

Sinclair would be away from Oxford for three months, from 1 August until 31 October. Hugh Smith had arranged a travel grant for the period, and Sir Wilson arranged Sinclair's travel. During his time in America, Sinclair was accompanied by Hugh Smith, who wrote frequently to Mrs Fisher, keeping her informed as to their whereabouts and matters that might be relevant to the Survey. His notes were informal, but many carried the same central message: the Survey was overextending itself.

During Hugh's absence, his secretary, Angela Hodgson, sent his mother reports of his whereabouts as information became available. To show her gratitude, Rosalie invited both Angela and Mrs Fisher to tea, saying that, as soap and hot water were scarce, they could have a bath beforehand, if they liked. Such were the realities of war time conditions in England.

CHAPTER 9

Eskimos and nutrition at Oxford

The United Nations Conference on Food and Agriculture was called to order on 18 May 1943, in Hot Springs, Virginia, USA. From then until 3 June, an international group of experts debated the importance of nutrition in the prevention and control of disease. Reading the conclusions of that meeting almost sixty years later, one cannot help but regret our failure to take up their challenge. The final Report states:

> Malnutrition is responsible for widespread impairment of human efficiency and for an enormous amount of ill health and disease, reduces the resistance of the body to tuberculosis, and enhances the general incidence and severity of familiar diseases; mortality rates in infants, children and mothers are higher in ill-fed than in well-fed populations; food consumption at a level merely sufficient to prevent malnutrition is not enough to promote health and well-being.
>
> A sound food and nutrition policy must be adopted by each Government if national diets are to be progressively improved, specific deficiency diseases eliminated and good health advanced.
>
> Given the will, we have the power to build in every nation a people more fit, more vigorous, more competent; a people with longer, more productive lives, and with more physical and mental stamina than the world has ever known. Such prospects, remote though they may be, should serve as stimulus in undertaking immediate tasks and overcoming immediate obstacles.

The United Nations Conference recommended that governments take steps to determine among their populations the prevalence of nutritional deficiency diseases, correct these deficiencies through dietary and therapeutic means, and provide means to prevent their recurrence. Further, it strongly recommended that governments should undertake the establishment of national nutrition organizations that would conduct research necessary to ensure a national food policy that would benefit all sections of society.

Such an atmosphere further encouraged advancing the study of human nutrition at the University of Oxford. A dynamic and well-funded Department of Nutrition had been established at the Harvard University School of Public Health, in Boston: why not achieve the same at Oxford? The form such an undertaking should take was another matter: considering the climate of change at Oxford, the debate was clear. Should human nutrition be undertaken as an extension of the war-time Oxford Nutrition Survey, which worked closely with governmental ministries? Or, should nutrition develop within a traditional academic department? The Department of Biochemistry, under Professor Rudolph Peters, had provided the foundation for the Survey. Just as biochemistry had been split off from the Department of Physiology to become a separate area of academic study, was it not possible to continue the study of Human Nutrition in the Department of Biochemistry, and separate it from that department when the time was right?

By late May 1943, the New York office of the Rockefeller Foundation was planning Hugh Smith's permanent return to the United States. He resisted, saying he wanted to remain in London at least until the end of the summer, when the nutrition work would be better established, and both Ryle and Sinclair would be established in offices at 10 Parks Road. He also wanted to ensure linkages between the work at Oxford and that developing elsewhere in Great Britain. The Medical Research Council was considering plans for a new department devoted to research in clinical and laboratory investigations of human nutrition, and the Ministry of Health was preparing for its post-war food studies.

About this time the Nutrition Society stated its wish to coordinate all nutrition surveys conducted in Great Britain. To discuss such a possibility, a meeting was held, on 25 May 1943, between the Director of the Society and the Oxford Nutrition Survey staff. Afterwards, Sinclair's group agreed the proposed arrangements for cooperation were chaotic and time-wasting. The next day Sinclair wrote to Peters, soliciting his help in stopping the Society, and suggesting that the Director was ill-informed and uninterested in basic survey methods:

> I have been told that Sir John Orr was opposed to the Nutrition Society
> undertaking the co-ordination of nutrition surveys ... Would it be possible

for you and Sir Edward [Mellanby] to take the matter up with Sir Wilson to see what arrangements could be made so that whatever co-ordination is desirable could be done without the chaos and waste of time that seem to be created by the present arrangement?

A few days later, Peters wrote to Sinclair, saying that he would give some thought to the matter, but that situations had reached a very complicated stage, and it was not obvious how to proceed.

About this time, Hugh Sinclair received a most welcome letter from his old friend, Major-General Sir Robert McCarrison, Emergency Medical Service, Oxford and District Group of Hospitals, who wrote:

> I send you herewith 104 Charts, together with 210 lantern slides, illustrating many of the figures included in the Charts. Most of the Charts, and lantern slides, are representative of certain aspects of my own work during the years 1913 to 1935. But I have included amongst them a number of others, prepared from papers published by other workers, which I found useful for teaching purposes. I ask your acceptance of these Charts and lantern slides in the hope that the Oxford Nutrition Survey may find them of use. I shall be happy to think that these records of an earlier day will have found a useful place in the School of Nutrition in Oxford.
>
> Yours sincerely, R. McCarrison

Sinclair responded with thanks:

> The Oxford Nutrition Survey has received very generous support from the Rockefeller Foundation, the Ministry of Health, and other sources, but the gift you made to it yesterday is perhaps the most encouraging that we have received. When I used to visit the Wellcome Museum, I often thought that the system of charts with photographs and scripts and microphotographs of pathological sections is the best possible way of teaching … I have discussed with Dr Smith and Sir Wilson Jameson the desirability of forming such a museum and in a small way we have started by collecting a few pathological specimens, a few slides and some photographs.
>
> Your magnificent gift of charts and lantern slides is a real foundation for such a collection and is by itself sufficient to give the Survey a permanent position in the University as a teaching centre. As you know, Sir Wilson is anxious that the Survey should continue in Oxford after the war as a centre for scientific research upon human nutrition, for teaching medical officers of health and others the latest knowledge about human nutrition and for

applying around the country and in close co-operation with the Ministry of
Health, the knowledge so gained. Such a centre would be impossible
without a museum; and it is particularly fitting that this museum should be
started with this collection illustrating your work … This collection will
always be associated with your name and will remind us not only of your
important contributions to this subject, but also of the support that the
Survey has received and continues to receive from you.

Yours sincerely HMS.

Sinclair's intentions were clear. Robert McCarrison's contributions to
the Survey would become the historical focus for the research and
educational centre he was creating. The basic understanding of human
nutrition, as set out in McCarrison's Cantor Lectures, would be the
foundation on which Sinclair would build his work. (It is a sad fact that
after Hugh Sinclair's death, in 1990, his scientific collection and papers
were divided, and McCarrison's contributions were distributed among
the Wellcome Trust, his family, and McCarrison's associates.)

In late summer of 1943, Sinclair left England to attend the annual
meeting of the American Public Health Association, as arranged by Sir
Wilson and Hugh Smith. The first leg of the journey was by Sunderland
flying boat to Ireland, from where he took a Pan American Airway
Clipper to New York. Once settled in his hotel, his first stop was the
offices of the Rockfeller Foundation. Bad news awaited him. He was told
that it was unlikely the Foundation would maintain its financial support
of the Survey after the war. It was also evident that the New York office of
the Foundation saw nutrition work at Oxford as part of the developing
Institute of Social Medicine, and not an independent department. With
this in mind, he set out with Hugh Smith on another marathon journey
around the North American scientific community. Just as he had done
during his earlier journey as a Radcliffe Travelling Fellow, Sinclair built
up a mosaic picture of nutrition research under way in the United States,
making precise notes concerning everyone he met, and what they were
doing.

First, he stopped in Princeton, New Jersey, to visit his old friend John
Fulton. He then travelled on to Washington DC, Pennsylvania,
Maryland, Alabama, Tennessee, Ohio, Wisconsin, North Carolina, Illinois, and Michigan. He also visited Minnesota, where he met Ancel Keys,
a man who would eventually come to Oxford and work with Sinclair.

Ancel Keys was then working on carbohydrates, vitamin B_1, B complex, vitamin C, and acute starvation; he would later become interested in fat and heart disease. Sinclair was impressed with Keys' staff of 15, his work in high temperature rooms, and the fact that he conducted both personality testing and biochemical investigations. On the day he met Keys, Sinclair encountered A. Hanson, who was working on fat requirements in dogs and children, and increasingly interested in essential fatty acids.

In Boston, Sinclair met Dr Fredrick Stare, the founding Chairman of the Department of Nutrition at the Harvard University School of Public Health. Hugh was impressed with Stare's research facilities, and was quite overwhelmed by the funding behind the operation. One project alone had a total budget of $250 000; another, an annual budget of $100 000. In 1943, these were enormous sums.

As Sinclair travelled with Hugh Smith, they talked of survey work to be done after the war. On 12 August, Dr W. A. Sawyer, Director of the International Health Division of the Foundation, met with Sinclair and made some notes of their meeting:[1]

> … HMS has definite ideas as to the war nutrition program in Europe. The group in England is considering first a rapid survey to obtain information of public health importance. The reports now available are obviously exaggerated and not based on good evidence. If a team could be sent to work first with and under the military authorities, it could quickly find out what the principal deficiencies are and hand over the information to the authorities. Second, the group could take the opportunity to study the significance of various lesions under conditions which might not recur for many years and could obtain information of great scientific value. A team going to Greece, for example, should have a small field laboratory, while a team in Holland or Belgium or Norway could send specimens to Oxford by air. The essentials of a mobile laboratory are a centrifuge and a refrigerator. The rest of the material is easily obtained and consists of chemicals and colorimeter. Surprisingly little can be done with clinical examinations alone. Errors in the dietists' examinations alone are enormous.

Before returning to Europe Sinclair travelled to Canada; first, to Montreal, where he made numerous notes on work for the Army at the Royal Victoria Hospital, and then on to Toronto. He viewed some of the work being undertaken by the Royal Canadian Air Force, and was most impressed by their research on the effect of riboflavin deficiency on the eye. While visiting with Group Captain Tisdall, in Ottawa, he saw a new photographic method for recording the degree of damage to the back of the eye caused by deficiency. Repeated photographs could provide an objective record of the effects of therapy. Hugh immediately wanted such a system for his laboratory in Oxford; along with standard split-lamp investigations this could prove a highly valuable tool in data collection. Hugh estimated the cost to be about $700.

A noisy bomber provided Sinclair's transport home. His discomfort was lightened by knowing he returned with a notebook filled with names and details that he could use to begin the serious work of preparing for post-war research on the Continent. He also knew he might have an opportunity to work further with Group Captain Tisdall.

31 October was Hugh's first day back in the Survey office. He exchanged greetings with Rudolph Peters, spoke with Peter Meiklejohn, who reminded Hugh he had known Stare while he was a Harvard Fellow, and settled down to a long conversation with Mrs Fisher. Nothing much had changed since he left: John Ryle had been difficult, and referred to the Survey staff as if they were his own; he had asked Mrs Fisher to become the administrative head of all in-house matters within the Institute, while remaining a staff member of the Survey. As Hugh had expected, Mrs Fisher had received unceasing demands for reports of Survey results. There were also the anticipated criticisms of some of the survey methods and reports, and one or two territorial disputes with other survey projects seemed possible. To someone returning from the stimulation of America, it all seemed tarnished. Few to whom Hugh spoke at Oxford seemed interested in post-war surveys, although there was interest in any information about how long the Americans would continue supplying food to Britain; without that help, rationing at home would certainly continue, and probably get worse.

As for the future of the Survey, soon after arriving in England Sinclair was relieved to hear that the Rockfeller Foundation would continue its financial support for the Survey for another year. The re-naming of the Survey had been discussed, and it was generally agreed that it should be known as the Division of Human Nutrition, Institute of Social Medicine. Hugh was in adamant disagreement on this last point: he was now firmly committed to establishing an institute of his own, and he boldly defined such an institute in a report to the Survey Advisory Committee:

It is impossible for an Institute of Nutrition to study the scientific problems without any interest in or contact with their application to public health; it is impossible for a ministry to conduct a public health nutrition programme without recourse to scientific research. Both fail. Food without drink descends slowly; and drink without food goes to the head. Various Universities already have connections with animal nutrition. Cambridge has the Dunn Nutritional Laboratory and the Animal Nutrition Institute; London has the Division of Nutrition in the Lister Institute of Preventive Medicine; Reading has the National Institute for Research in Dairying …; Aberdeen has the Imperial Bureau of Animal Nutrition. It is odd that so much emphasis has been placed on animal nutrition and none on human nutrition … There are strong reasons why the establishment of such an Institute of Human Nutrition in Oxford should not be delayed … There is a mass of information on nutrition, and more is rapidly accumulating particularly as a result of the war; a central place is needed to collect and sift this information. The occasion is instant because the nutritional problems are at hand and challenge solution … After the storm of war, a new era dawns. Blood, toil, sweat and tears are exchanged for food, health, work and homes. Physical science has provided the comforts that man needs, but medical — and above all nutritional — science provides the right food, which is the basis of health … We need a more intelligent and effectual provision, use and management of food. The new science of nutrition will shed much-needed light on one of the most fundamental, widespread, urgent and inadequately managed problems in the whole domain of human welfare.

A few months after returning from North America, Sinclair received a letter from Canada that greatly excited him. Owing to his contacts with Tisdall, the Medical Division of the Indian Affairs Branch of the Department of Mines and Resources was asking him to participate in a study concerning the nutritional status of the Canadian Indians. The goal of the study was to collect information of value both to the Indian population and the outside world. In conjunction with the Millbank Memorial Fund, the Hudson's Bay Company and the Royal Canadian Air Force were now planning another in a series of scientific expeditions into northern Canada. As Hugh Sinclair had expressed interest in this work (particularly that related to dietary therapy using riboflavin, and vitamins A and C) the Canadian Government wanted to know if he would like to participate in the next expedition. Hugh would need to make his own way to Winnipeg, Canada, after which the Department of Mines and Resources would look after him. The group would leave Winnipeg between 10 and 15 March, he was told, and would return eight to ten days later. Data collection would be limited to a few days.

Hugh began making plans for the journey at once, and wrote to Sir Wilson for help with travel funds to Canada. His excuse for going pertained to a scientific dispute he was having with Group-Captain Tisdall:

> There is one particular reason why I wish to join the expedition: Group-Captain Tisdall, Dr Kruse and I have rather different views about some of the signs of nutritional deficiency, and I think the best way to settle our differences is to see together a large number of cases of nutrition deficiency, and the effects of specific therapy upon them …

On his birthday, 4 February 1944, Sinclair received word from the High Commission for Canada saying he had been given priority passage to Canada. All he need do was find the cost of ocean passage, about £40, to be paid in cash. He would also need funds to cover transport from London to Liverpool, and to cover living expenses in Liverpool prior to embarkation.

Sinclair brought the Canadian invitation to Rudolph Peters' attention. Peters immediately questioned the value of the trip, in light of Hugh's other obligations. Hugh knew that the growing mound of unanalysed data caused Peters and others to question the value of his work, but he was unrepentant. The men argued, and Sinclair went on the offensive, changing the subject from Canada to John Ryle. Hugh said he would not work under John Ryle's direction, and that if he were forced to do so, he would leave. Over the next weeks Sinclair repeated this warning; Sir Arthur Ellis, who was soon to replace Sir Farquhar Buzzard in the powerful position of Regius Professor, was among those who heard such threats.

It can be documented that by this time Sinclair had developed an interest in dietary fats, and on a schedule of work for the Survey jotted down: 'Essential fatty acids deficiency in man. Estimation of unsaturated fat in blood in Germany.' What inspired this interest is unknown, but visiting laboratories in the United States must have played a role. This was a new, but still secondary, research interest as he prepared to leave for Canada.

With Mrs Fisher, Sinclair detailed the work to be done in his absence; if anything went wrong while he was away, she was to contact Hugh Smith.

[Meicklejohn planned to finish his data collection and then take a few weeks off.] On 21 February 1944, Sinclair left Oxford for London, from where he travelled on to Liverpool. For security reasons, relatives and friends were not allowed to travel with passengers to the port, and were forbidden on the station platform. Authorities warned:

> All passages are subject to cancellation without notice at the port of embarkation in the event of an emergency … All persons travelling overseas do so at their own risk … Due to war emergencies, applicants are warned that ocean travel accommodation is very inferior to that available in peace time. Discomfort and hardship must be expected …

Hugh's ship, HMS *Andes*, left Liverpool at 7 a.m. on 27 February and arrived in Halifax six days later. After arriving in North America, he went first to Ottawa, where he stayed briefly and met the other participants in the expedition, and then travelled south to the Rockefeller Foundation New York and to Washington DC for meetings concerning the impending post-war relief work in Europe. At last, satisfied that his position with regard to the post-war surveys was secured, Sinclair left for Winnipeg. He had with him a slit-lamp and 850 pounds of scientific gear and luggage. After arriving in Manitoba, the members of the expedition travelled by dog-sled to a base camp Rossville, where details of the research work were agreed. For the better part of eight days, teams of investigators examined and surveyed the Eskimos they had come to see. In his diary, Hugh jotted down details of a camera he so very much admired, and planned to use in his Oxford laboratory. He also made notes on various analytical methods used during the scientific investigations. Unfortunately, Sinclair failed to include in his diary details of any clinical or experimental observations, and this author could find none among the papers available to her. However, in a paper published much later in his life, Sinclair wrote:[2]

> In 1944 the Royal Canadian Air Force asked me to join its nutritional Consultant (Dr Tisdall) and Dr Harry Kruse of New York in an investigation of riboflavin deficiency as a cause of snow-blindness. Tisdall and Kruse maintain that Canadian Indians had severe riboflavin deficiency, but I disputed their criterion of superficial corneal vascularization. I agreed to take part provided I could also examine Eskimos, since they had the highest intake of fat in the world (but very rich in EFA], and allegedly no heart disease or other non-infective diseases of civilisation. In fact, the

'vascularization' in the Indians was the result of lime burns on the cornea
some years previously.

I went on to examine Eskimos, using a slit-lamp microscope with cross
Polaroids to detect very early arcus senilis as an indicator of atherosclerosis.
I found none, even in old Eskimos, although I could detect it in ten per cent
of my Oxford students.

During Sinclair's absence from Oxford, Mrs Fisher and the Survey staff
coded and analysed as much data as possible, until they were suddenly
confronted with an almost insurmountable problem: new food composition tables were being prepared by the Ministry of Food, and would not
be generally available for some weeks. Without accurate food composition tables the Survey's work was impossible. The procedure of the
Survey was to obtain information about food consumption from its subjects by interview. Then, using standardized tables listing the composition of individual foods, staff members would laboriously calculate the
total intake of nutrients consumed, based on the types and weight of
foods reported as having been eaten. All British nutritional scientists
used the same tables, thus establishing comparability in their work. Most
unfortunately, changes in the food composition tables meant that all calculations based on previous tables required correction. The additional
staff effort and time required to make these changes would be enormous.

After hearing of the changes, Mrs Fisher contacted Dr Bransby, in the
Ministry of Food, and urgently requested copies of the new tables.
Unfortunately, she was told, the new tables would not be circulated until
all alterations were completed. Using inaccurate draft tables would only
compound the problem of correcting calculations. Work on the Survey
data ground to a halt.

When Sinclair returned to Oxford, in mid-April, and heard the state of
affairs, he immediately arranged to go to London and meet with Sir Jack
Drummond and Magnus Pyke at the Ministry of Food to discuss plans
for the proposed post-war surveys on the Continent. While there, he had
an opportunity to stop by Bransby's office, which was in the same building. Finding that Bransby was not in, Sinclair convinced his secretary to
give him two copies of the draft tables. Bransby was furious when he discovered the copies missing, and wrote to Sinclair at once. He wanted the
draft tables back. Hugh held on to the papers, and told Mrs Fisher to use
them judiciously.

The food composition tables proved to be the least of Sinclair's problems. About a week after his return, he learned of staff changes within the Rockefeller Foundation. Dr Sawyer had retired from his key role as Head of the International Health Division, and been replaced by Dr George K. Strode. Hugh Smith was to be the new Assistant Director. Sinclair contacted Smith, asking about the future of the Survey. In his notes of the discussion, Sinclair states that Smith said that he did not want the Survey to die, but saw it supported with British money. The Foundation would continue financing it for a time, but Hugh should find other sources of funds as quickly as possible.

With the exception of Smith's promotion, none of this information was new, but with Smith in his new role, it put pressure on Sinclair to begin active fund-raising. Equally important, with John Ryle now an antagonist within the University, getting agreement on securing the Survey's future was becoming more difficult; without Hugh Smith's support and timely interventions, agreement might be impossible.

A meeting of the Survey Advisory Committee had been held during Sinclair's absence. Soon after returning to London, he met with John Ryle, Chairman of the Survey Advisory Committee, to discuss its outcome. Ryle reassured Sinclair that the University saw within it a place for the study of Human Nutrition. But, whether that should be as a Chair, a Readership, or as part of the Institute of Social Medicine was unclear. Although some members of the Committee saw continuation of the Survey in a limited form, perhaps continuing to work with the Ministries of Food and Health, there was a growing sentiment that Sinclair should write up his data, and then focus full-time on fundamental research. As for funds, perhaps the Foundation could support him for more than the time agreed, and perhaps other sources of finance could be found.

Sinclair prepared the proposals necessary for fund-raising. He gave the expanded Survey work a new name: what was once the Oxford Nutrition Survey was now the Institute for Human Nutrition. When members of the Survey Advisory Committee learned he had changed the name without their discussion and agreement, they were furious. After considerable debate, it was eventually decided that Hugh's nutrition work should be known as The Laboratory for Human Nutrition. Despite this, Sinclair did not drop the word 'Institute' from his fund-raising documents. [Fortunately, Hugh Smith agreed with Sinclair's decision, and used the term in his communications from the Foundation, thus strengthening Sinclair's hand.]

It was agreed that Sinclair would seek funds from the Nuffield Trust; however, members of the University community decided the scope of Sinclair's research should be carefully considered before such an approach was made. With Hugh Smith's help, Sinclair set about preparing a proposed outline of research and, for the first time in such a document, mentioned dietary fats. The scope of research described was more ambitious than anything previously prepared.

The proposal outlined an approach to integrated research and teaching which Sinclair pursued until his death, almost half a century later. It harked back to his earlier proposal to the Rockefeller Foundation, in which he wrote about the three-legged stool of nutrition research, and it formed the skeleton of a plan for a great institution that encompassed both the ideas of his most important mentor, Sir Robert McCarrison, and the early work of great agriculturists, including his ancestor, Sir John Sinclair. To fulfil Hugh Sinclair's plans now, the subject of nutrition would be given its own lexicon, its own utility within the community, and its unique and leading place in medical teaching. Now, he spoke of 'nutriture' when describing nutritional status. The complete programme of the new Institute of Human Nutrition would eventually include the study of the epidemiology of malnutrition, human dietary requirements, and patterns of food consumption, absorption, metabolism and excretion of nutrients, food analysis, induced deficiencies in man, deficiencies observed in certain clinical conditions — such as liver disorders and diseases of the nervous system, a study of the effect of improved diet upon physique, health, mortality, and morbidity, and the development of new scientific methodologies for the study of nutrition.

These research activities would be accompanied by programmes of teaching, and a museum, library, and information bureau that would support research and the dissemination of information concerning nutrition. The research programme would also have an international component. Sinclair's time at Harvard had impressed him and he was particularly interested in work the Department of Nutrition had undertaken in Peru. Oxford should work with Harvard, Sinclair thought, by establishing a similar programme in West Africa.

In addition to the proposal for his new institute, a second document now competed for Sinclair's time: it was the Oxford Nutrition Survey's

Army Nutritional Intelligence Units (Preliminary Suggestions).[3] Using information gathered both during his recent trip to the United States, and through his various contacts, Sinclair devised a plan to distribute food on the Continent under the combined authority of the Army, the national governments of the territories released from enemy occupation, and the United Nations Relief and Rehabilitation Administration [UNRRA]. Created in 1943, and discontinued in 1947, UNRRA was an extensive social-welfare programme designed to aid nations devastated by World War II. After the war, these agencies faced the problems of determining the areas where malnutrition existed, who in these areas was in need of food, and which specific foods were required. Once these questions were answered, priorities could be established for distributing supplies.

Sinclair suggested that the Army Nutritional Intelligence Unit should be involved in the work directly under Corps Headquarters, thus enabling it to move into occupied territories as quickly as possible. He wrote:

> An advancing army cannot afford to look over its shoulder at the civilian population it leaves behind, and even if malnutrition becomes temporarily worse in territories immediately after they are liberated from the enemy, the knowledge that food is being distributed in the best possible way should do much to make the population content.

Sinclair recommended that a typical post-war survey team should include: a medical officer; a junior medical officer [ideally one trained as a pathologist]; a biochemist; a nurse-dietist [preferably one from the country concerned]; two technicians; several driver-mechanics, who could double as assistant technicians, photographers, and wireless operators; and one *aide-de-camp*, who would be concerned with the organization of the work and would be responsible for liaison with the military authorities.

On 30 May 1944, at 11.30 a.m., the Advisory Committee met to discuss the future of the Survey. Professor John Ryle was in the Chair. Also present were Sir Farquhar Buzzard, Sir Wilson Jameson, Professor Liddell, Sir Robert McCarrison, Professor Peters, Drs Hugh Smith and Hugh Sinclair. The Registrar attended, but Sir Jack Drummond was absent. Mrs Fisher was asked to take the minutes; in the past, Sinclair had done this. After a lengthy discussion of funding and equipment requirements, the

Committee turned its attention to its principal responsibility. Chairman John Ryle said the Committee must consider this to be the 'winding-up' year for the Survey in its present form, and the staff who would be released from the Survey should be so informed as soon as possible, and warned that no large new projects should be launched. A heated debate ensued. Sir Wilson Jameson said he knew the form of the Survey would change, but he wanted to see it continue as the work at Oxford provided guidance in nutritional work that would otherwise be lacking. Ryle said that human nutrition at Oxford should be part of a permanent department within the University to investigate the scientific aspects of the subject. Jameson questioned whether or not this meant paying little or no attention to the social aspects of nutrition. In response, Ryle assured him a means could be found to link the work with that of the Institute of Social Medicine.

The University Registrar spoke up, stating that work on human nutrition was 'absolutely essential' for the work of the University. If the Survey were to become defunct, it would be necessary to create some other institutional body to carry on investigating nutrition. Professor Peters agreed. John Ryle said that no one thought that the study of human nutrition should be dropped from the University; the question was, how it could be established independently and maintain its connections with related work, such as physical anthropology, biochemistry, and social medicine.

Robert McCarrison wanted to move matters on, and asked whether on not they should put forward definite proposals as to how the programme should be run. Putting the notion of proposals for programme development to one side, John Ryle said he thought it unlikely that either the Nuffield Foundation or the University Grants Committee would support a nutrition programme as a separate entity.

Rudolph Peters spoke on the dual nature of nutritional research. First, he pointed out, there was the practical application of present knowledge; for example the administration of iodine to populations where goitre is prevalent. Second, was the advancement of knowledge by the discovery of new information through fundamental research. It was his view that these two aspects of the subject required separate discussion. Sir Wilson responded that both were being done, and could continue to be done, well at Oxford. Peters agreed, but argued that the discovery of fact was separate from the application of fact. The discovery of facts, he emphasized, should be in the hands of 'a man with the status of a Reader'.

The argument continued. At last, Sir Robert McCarrison spoke: he understood, he said, that Professor Peters wanted to separate the science and the art of nutrition. Considering the current situation, he suggested that an Institute of Human Nutrition should be established in which fundamental scientific work could be conducted: the results of this work could then find application in the Institute of Social Medicine. Every department in the University should have an interest in nutrition, and these interests could be integrated through such an Institute.

Members of the Committee scoffed at McCarrison's vision, saying it was too grand, and that while Institutes with similar goals had proven successful in the United States they would not take hold in England. Rudolph Peters said such an Institute should evolve very gradually, in any case. Jameson countered, saying that nutrition would be an important subject in the future, and a centre was needed for the correlation of research and information. As the discussion continued, John Ryle picked up on Sir Robert's comments, and suggested that all the science departments in the University have a connection with such an Institute of Human Nutrition, and through it, anyone interested in a subject related to human nutrition could consult the appropriate person. Sir Wilson liked the idea. The mood of the discussion was beginning to change in Sinclair's favour.

Rudolph Peters intervened again: things should begin on a limited basis. The important matter for the University was pure research, and he suggested that a storehouse of scientific knowledge concerning Human Nutrition was needed. This should include information about, and access to, the special apparatus necessary for nutritional research. He suggested a place where, for example, biochemists could apply for knowledge, or advice, or equipment. He also said that his department worked on nutrition because of the light it threw on biochemistry. If, at some time in the future, he found that this no longer was the case, he would lose interest in nutrition.

In his comments to the Advisory Committee, Peters stressed that real research workers were artists, 'men of artistic temperaments, and that such men were not easy to work with and cooperation between them was difficult.'[4] His experience with Sinclair most certainly had influenced this comment.

Hugh Smith, representing the primary source of funding for the Nutrition Survey, brought the conversation around to the practicalities of the issue. Should the present organization, and its equipment, remain as applied medicine, and become part of the Institute of Social Medicine; or, should it remain in the university as part of its basic research effort? [This question illuminates the difference between a 'Department' and an 'Institute'. The former is a part of the body of the university, whereas the latter serves as a link between the university, governmental bodies, and the community as a whole.] Hugh Smith said he felt strongly, particularly after his recent trip with Sinclair to visit laboratories across the United States, that fundamental research in human nutrition was what was most needed.

Sir Wilson Jameson did not object to this idea, but wanted selected, practical surveys to continue as well. These, he thought, could be conducted by the Institute of Social Medicine, assisted by government funds, since this type of work was not appropriate for a university department. In addition, the clinical aspect of nutrition must be pursued. 'Those concerned with this must have an active knowledge of the subject, and must be in touch with its development all over the world,' Jameson said. Surely he must have had Sinclair in mind as the man to lead such a wide-ranging responsibility.

Peters argued that one person should look after the surveys, which should be done under the auspices of the Institute of Social Medicine. The fundamental research should be put in the hands of a Reader, attached to one of the departments for purely administrative purposes. He thought it possible, but not desirable, that the same person could hold both positions, devoting one-third of his time to social application, and the remainder to fundamental research.

It was almost two o'clock, and time to break for lunch in Magdalen Senior Common room. The Registrar suggested the discussion be brought to a close. Sir Farquhar put the question: Are we generally agreed that there should be an Institute of Human Nutrition as a University department in Oxford? Peters replied that it depended on what else existed: McCance had a unit in Cambridge; the Medical Research Council was establishing a nutrition unit in London. In the end, he agreed there was room for a further Institute of Human Nutrition in Oxford. His views were underlined by Hugh Smith, who stressed that fundamental research in human nutrition was absolutely essential.

It was done. The University Committee most intimately involved in Hugh Sinclair's work had agreed an Institute of Human Nutrition should be established at the University of Oxford.

After lunch, Sinclair reviewed his existing and planned research activities, distributing copies of the Memorandum on Army Nutritional Intelligence Units. Peters found the university's position regarding the work 'difficult'; Sir Wilson, on the other hand, supported the work, saying the most important thing now was to provide good nutritional advice. Sinclair suggested that, in light of the funding difficulties surrounding the Survey, if any outside agency such as UNRRA, was interested in its help, and would supply funds, he felt inclined to accept the responsibility.

The Committee warned him not to go ahead with the proposed post-war surveys. Sir Farquhar was sympathetic, but suggested that engaging in large-scale work out of the country could jeopardise the entire Survey. John Ryle advised Sinclair to keep in close touch with the planning process just placed in motion by the Committee: Sinclair should not become 'too readily detached' from the primary work in Britain, except for a short time.

During that fateful meeting, Hugh Smith knew his return to the United States was imminent. Armed conflict would continue for some time in Europe, but for Smith the war was over. On a dark night of 2 June 1944 Smith listened to the terrifying sounds of anti-aircraft fire over London: he did not learn until later that the first German V1 flying bombs had been launched against Britain. In the morning, he took a bus to the airport, boarded the plane on which he was ticketed, and flew back to the United States. Like other Americans who had come to help save Europe, his going home left a terrible void that would not easily be filled.

CHAPTER 10

An end to the Survey

An Institute of Human Nutrition would be concerned with the advancement of nutrition science, and with the collection and integration of existing knowledge. Although it might be concerned with teaching within the University, it would not 'educate professional and lay thought', or 'direct legislation'. However, since man cannot be investigated in cages like laboratory animals, the Institute might study the nutrition of man in three ways: by the study of and by experiments upon normal healthy volunteers; by the examination of patients; and by field work.

Hugh Macdonald Sinclair, 26 June 1944

In the autumn of 1944, Sinclair applied to an American philanthropic organization, other than the Rockefeller Foundation, for funds. He wanted to conduct a one-year depletion study on a group of conscientious objectors: his subject was vitamin C. This potential donor contacted Hugh Smith in New York, and inquired as to Sinclair's clinical ability to direct such a study. Smith assured them that Sinclair had an excellent clinical background. A second letter arrived, asking Smith's opinion on whether or not Sinclair's proposed study was a peace-time or war-time investigation. Special consideration had been given to research work in Europe as part of the war effort, they said, but peace-time studies outside the United States and Canada were beyond their mandate. Smith responded, saying that any study which Sinclair might carry out in the future would come too late to affect the troops in the present war, but that did not mean the study was not of war-time significance. Sinclair's proposed study was valuable, and Smith thought the Oxford group was better equipped to conduct the work than any other group in existence. His only concern was whether the study outlined would interfere unduly with the programme of work already under way.

Information about Sinclair and his work was sought from other sources; in the end, the project was turned down on the basis that it fell outside peace-time policy. However, this case strongly suggests concern regarding Sinclair's ability to complete projects in a timely fashion was being voiced, even by his robust supporters.

The failure to secure the funds for the vitamin C project was a blow. It would have paid staff salaries, and given Sinclair work similar to that undertaken by Hans Krebs at the University of Sheffield, thus providing a possible link between their laboratories. Of all the groups investigating nutrition outside Oxford, Krebs impressed Sinclair the most. Krebs's studies on the citric acid cycle would result in his receiving a Nobel Prize in 1953, which he shared with Fritz Lipmann.

On 9 November 1944, the new Regius Professor of Medicine, Sir Arthur W. Ellis, asked Sinclair to prepare new documentation justifying the continuation of the Survey. He wanted summaries of projects undertaken since the inception of the Survey; a summary of conclusions arrived at as a result of the work of the Survey; copies of reports and published work of the Survey; and a programme of projected work and expenditure. Sinclair and his team had prepared several such documents in the past, but these were brief and provided little analytical content. Ellis wanted details.

Two days after Ellis's letter was posted, Sinclair sent two letters to Hugh Smith. The first, a long and melancholy message, was marked <u>Personal</u>. In it, Sinclair recognized his debt to Smith, offered explanations for the Survey's failure to fulfil its potential, attempted to vindicate his part in the matter. In the end, however, the message is one of loneliness.

> … A strong … realization of my inability to express adequately my thanks to you has overwhelmed me with inhibitions and prevented me from writing a personal note earlier. I hope you have not interpreted my silence as ingratitude.
>
> For I am very fully aware of the support you gave us both on the stage and behind the scenes, and how you acted as our ambassador when others were critical …
>
> The Survey passed through many storms in the three years. I know that I was responsible for them all, mainly for two reasons: if there is a principle involved I will fight for it to the last ditch, and if one feels that in wartime one must work to the limit of one's ability there is little time to sit back and meditate about the most tactful way of doing things. For instance over the dreary business of bench-fees … Then there was the most unhappy quarrel with Peter [Meiklejohn]; the desire of Platt and Bransby to sit in armchairs and

tell me what to do. And almost worst of all, I felt there was a period — now happily far distant — when you and Sir Wilson were very dissatisfied with our work.

I was, as it happens, aware of several of the faults of the Survey. In one sense we took on too much — for two reasons. First, I felt that the main difficulty of the work was the organization: getting people (who were very busy) to come voluntarily at an awkward time to an inconvenient place (for our dark-room for the dark-adaptation test has varied from a lavatory to the altar of a prison chapel!) … I felt that if all this organization went all right, then one should make the maximum possible use of the person and her blood particularly since I had throughout the work been skeptical of most of the allegedly standard methods for surveys. Secondly I felt — as I know you did — that those of us who were not in uniform and risking our lives should work to the utmost limit. It has made me very weary and care-less, for I have not had a holiday since the summer of '38 — except for two grand weekends on the last two Christmases and the visit with you to the USA. I now long — though infinitely less than my many friends in prison camps — to get back to my books and test tubes, and away from a desk.

But the black patches were overwhelmed by the generosity and support we received. There was the sound advice that one could rely on from Walker, and the friendly support from Veale and from Buzzard. Peters was surprisingly good-tempered and generous … Above all, I felt that if you and Sir Wilson were behind us nothing could go really wrong, and I felt profoundly grateful to both of you.

Now I miss your advice and friendship, and my colleagues in Magdalen miss you too … To some slight extent James Mackintosh is taking your place, for he is becoming a frequent and very welcome visitor for weekends. But — and this I have told to no one so <u>please do not repeat it even to him</u> — he has an ulterior motive in that he wants me to take a chair of Human Nutrition in the School of Hygiene … But I feel I must get back to some work and stop signing cheques and sacking charwomen!

… I also look forward to another visit from you as soon as the war is over …

Yours ever …

Formally written on Oxford Nutrition Survey letterhead, the second letter had a different tone. Sinclair spent no time apologizing for the past, nor the present state of the Survey data; instead, he promised books, data, and future reports, which by now Hugh Smith must have realized were not to be forthcoming in the near future, and probably not at all.

> … In your letter of 22 June you asked me to send copies of all our forms. I purposely delayed doing so because we are anxious to circulate to all those who may be interested not only the forms (which without explanation convey little) but also a sample set of forms filled in and a written description of their use. This is now nearly complete, and I hope to send you several in the very near future. I have also designed a small form for the quick and simple recording of clinical findings in rapid nutritional surveys, and we are discussing the possible international use of this in London next week with representatives of various nationalities. I shall also send you a copy of this after that meeting …
>
> You will recall the rather chaotic preliminaries about co-ordination of nutrition surveys … Chaos continues. A conference … was called … in the School of Hygiene at which everyone talked about nothing in particular … From the point of view of the Survey the only satisfactory thing is that any question that arises gets referred to us for our opinion as few of the other people present seem to have any ideas about the problems involved …
>
> The Confusion at these meetings underlines the need for some statement about the methods that are available for use in nutritional surveys of all types. When I was in the Foundation Offices last March Dr Strode asked me if I would put together all the methods in the form of a book; I have started to do this and plan to enlist Dr Sydenstricker's help in the clinical section and Lloyd is taking part in the biochemical section. I hope that all aspects will be covered, including the computation of data, and although no doubt there will be nothing original in the book it may be useful to have a summary of the available methods, their detailed application, and personal views about their limitations.
>
> … We are hoping to obtain some of the standard nutritional textbooks from the USA … We feel that this is particularly important since Sir Wilson has repeatedly stressed to us recently that he wishes the Survey to become a centre in this country for the correlation of all the literature regarding nutrition …
>
> P. S. All our results are now nearly in order; there has been great and very successful activity in the way of computation.

Professor Rudolph Peters had clearly stated his support for the Institute of Human Nutrition only as long as it conducted research that enlightened the study of human biochemistry. Sinclair knew that Peters was on the Committee overseeing the increasingly impressive work on vitamin C being directed by Hans Krebs. Failing to obtain funding for a vitamin C study of his own, Sinclair now tried to find other ways to cooperate with Krebs's work, possibly in an attempt to please Peters. Over the past several years, Brian Lloyd and others in Sinclair's group had worked on methods to measure vitamin C in various blood fractions, and had accumulated quantities of data on the subject. Hugh decided to present himself as a potential participant in Krebs' research.

> 13 November — HMS to Peters:
> I understand that the vitamin C experiment at Sheffield is about to start, and wonder if we could be of any help to you and your Committee in it. There seems to me to be two ways in which we might offer help.
>
> First, the clinical examination is of course in the hands of highly competent clinicians with wide experience of scurvy. But certain techniques, of unproved reliability, have been introduced for the diagnosis of early deficiency … Also we have apparatus here for testing the ability of the gum to resist pressure without pain or bleeding, and for determining the capillary resistance of the gum … I first examined the gums of several of the subjects in July 1942, and have followed under magnification the appearance of the skin since that date …
>
> Secondly, as you know, we have spent a great deal of time upon the estimation of vitamin C for clinical purposes and are, I believe, the only laboratory with considerable experience of the estimation in the leukocyte-platelet layer. Lloyd has been doing this since June 1941, and has done about 2500 estimations. He and Webster have done about 12 500 estimations of vitamin C in different parts of blood. We would of course, be prepared to cover our expenses in examining the subjects and doing occasional estimations of vitamin C in the different parts of blood and in urine; and to place our results at the disposal of your Committee …

Peters took time to reply, and when he did, the response was negative. He had placed Sinclair's offer before the Committee on vitamin C, but several members of the group were against it. The legacy of Sinclair's failure to complete work and cooperate with other investigators was growing. Peters wrote to Sinclair:

> … the idea that you should help has not been viewed favourably so far … I did not feel prepared to urge it, because I had regretted asking you to

participate in the Heythrop trial, and your participation in the A deficiency trial does not seem to have been appreciated.

In regard to the specific points which you raise. Firstly, J. D. King is looking at the gums ... this was arranged some time ago. Have you reported any results of the pressure experiments? The second point is being met by analyses in two laboratories, and — so far — no discrepancies of significance between the analyses have turned up. I know that Lloyd has been at work for some time and presumably is efficient at the vitamin C estimations now; but so far as I know he has no chemistry degree and there has been no report on the series of estimations from which one can judge their value. Mere numbers alone are not particularly impressive ...

This was lukewarm support, to say the least. Sinclair was now deeply worried about the future plans for an Institute. There were rumours that the November meeting of the powerful Committee of the Faculty of Medicine, where the decisive decision was to be made about the future of the Survey, had been postponed. Under the pressure of these circumstances, Hugh wrote a rambling letter to the Regius:

29 November   PRIVATE

Dear Regius,

I hesitate to trouble you when I know you are so busy, but I have heard a rumour that your SubCommittee of the Board may not report this afternoon on the Survey Committee's recommendations because it may wish to await a report of a SubCommittee of the Committee of the Institute of Social Medicine. I feel surprised to hear that rumour, since both the other two members of the Institute of Social Medicine Committee (Sir Farquhar and Professor Ryle) were present at the meeting of the Survey Committee and supported the recommendations to the Board ...

I should like to underline what I said at the last meeting of the Board about the need for some decision at this afternoon's meeting. Three of our staff have just left, and two more are, with my support, contemplating leaving very soon. Sir Wilson Jameson would not permit us to take part in work in Europe on a large scale because, he said, his Ministry needed our work in this country. To continue that work until 31 July next. I should have to replace those who leave but could not honestly do so while the future remains completely undecided. ... Therefore I should have to tell Sir Wilson that I could not carry on the work of the Survey.

Replacements, temporarily, would be very difficult to make in any case, because UNRRA are offering jobs and because it is widely known that the Senate of London University have been considering for some time creating a degree in Nutrition; the syllabus was discussed in London last Saturday at a meeting attended by four of our staff. [I have myself been offered by Professor James Mackintosh a Chair of Human Nutrition in the School of Hygiene — and also a similar Chair in the USA — and shall have to make a decision very soon, but this is for your ears only as the former is very tentative and in any case I wish to keep my own personal position as far as possible divorced from the principle (obviously accepted by London) that Human Nutrition is a University scientific subject and should be studied as such.] …

May I suggest that the Board considers this this afternoon, and either passes on to the Council (or the General Board) or rejects the broad recommendations of the Committee of the Survey. Surely the details can be filled in … if they are rejected by the Board, then the Survey will close at once so far as the University is concerned and we shall be free to work for the Ministry of Health [or] elsewhere if we wish …

Sinclair soon learned that the meeting of the Committee of the Faculty of Medicine had been delayed at Ryle's request, as he was travelling on the day. Hugh wrote to Ryle at once about the budget and staff leaving, and then called upon the name of Sir Wilson Jameson. Hugh wrote that 'He [Sir Wilson] found it extremely difficult to discuss the "winding up" of the projects [which you had envisaged] and dismissal of staff …'

On 12 December, Hugh Smith answered Hugh Sinclair's personal letter. As before, he was supportive and willing to share Sinclair's vision. Perhaps he should have been less kind:

… Anyone … who was in Britain during the dark periods of 1941/1942 will realize the doubts and uncertainties that beset those responsible for the public welfare. It was clear to me shortly after I arrived that the easiest course for you and many others would have been entering the armed services where the responsibilities could be pushed off on those with higher rank.

… I believe when the war is over and one can look back over the course of events with a detached point of view that the role of the Oxford Nutrition Survey will appear as a very bright one in comparison with the confused utterings and publications emanating from other sources … No doubt this period of trial and error, difficult as it has been, will prove most valuable in the evolution of proper facilities for the study of human nutrition. I sincerely hope that you will have a future role in Oxford commensurate with your great abilities.

I shall be thinking of you especially at Christmas and wishing that I could join in again the happy celebrations at Oxford …[2]

Such kindness only fuelled Sinclair's growing sense of isolation within Oxford. He had previously confessed to feelings of exhaustion; now, his tiredness and frustration were clearly evident. He was convinced Rudolph Peters wanted to discontinue work on human nutrition after the war, despite Peters' protestation to the contrary. All this made it impossible for Sinclair to accept as genuine the facts and criticism in Peters' recent letter concerning the vitamin C investigations (see page 136–7), and on 13 December, the anger, the hurt, and all the old resentments boiled over. Sinclair wrote to Peters.

> My dear Professor:
> I have delayed answering your letter of 28 November because I was to see Sir Edward Mellanby yesterday and thought it best to wait until I knew his attitude towards my letter to you in view of the information you gave me that he was against our helping in the vitamin C experiment at Sheffield. I think it might be best if I answer your points at some length putting my thoughts straight onto the typewriter.

Sir Edward, Sinclair claimed, was in favour of the Survey participating in the vitamin C experiment, and would speak to Peters about it. Also, Sir Edward said that he had heard that the Survey had failed to deliver some vitamin A results to Sheffield on time: an allegation that Sinclair denied. The letter to Peters continued:

> I am glad to learn from your second paragraph that my second point — namely, following vitamin C in whole blood, red cells, white cells and plasma in depleted subjects — is being met. Lloyd showed O'Brien the method for white cells, but I did not know that either O'Brien or Krebs have been estimating vitamin C in whole blood up to now. It is technically true that neither Lloyd nor Webster has a BA; but each has passed all the necessary examinations and could get the actual degree by signing a cheque — a formality that I had regarded as being unimportant! …
> I wonder, Professor, if I may take this opportunity of making some general comments in seeking your advice about the future. I realize, of course, that you think very little of my ability as a biochemist … I also think very little of it,

and realize that I suffer too from the disadvantage of having no ambitions: I do research because I am intensely interested in the problems and anxious to use my hands and brain in solving them, but I am not interested in broadcasting any results, particularly before I am quite certain of them. I know that this policy has disadvantages — that, for instance, you did not believe the very consistent catatorulin effect that I got with pyruvate whenever I tried in 1932–33 and therefore in my thesis I adopted a very lame explanation of the discrepancy between my results and those obtained in the Department during the previous year. More recently, I spent August 1940 searching the chemical literature for substances that formed firm combinations with arsenic and found that the most likely ones were reduced hydron blue, chlorodimercaptomethylene violet, and dithiol. Philpot said that he had the latter but could not find it. Walker rightly would not order it because he knew that there was some about in the lab. When, two months later, Thompson came to the conclusion independently that dithiol should work, he obtained some from Philpot and at once found that it was very promising. I was absolutely content to leave the field because to remain in the work might have led to competition, and I immediately got to work on the nutritional survey. Of course I wanted, and deserve, no credit for first having suggested dithiol — in fact, I would normally have forgotten about it; perhaps if the Philpot filing system had worked the Nutrition Survey would not have started; but perhaps you will therefore regret the failure of his filing system!

I hope very much indeed that you will advise me about the future. It is quite simple for the Survey to disband completely if you think that is the right course — indeed it may have to; and it happens to be easy for me to leave Oxford and work elsewhere if you would advise me to do so. I do not want to leave, and have no ambitions about holding Chairs (which are extremely cheap now because I have been offered two in Human Nutrition) but merely want (in peacetime) to do research and teaching, and above all to rely, as before, upon your friendship and advice. To take a very small example, I have been asked by Churchills [publishers] to write a volume on 'Recent Advances in Nutrition', and could not do so without the benefit of your guidance on the most significant recent trends. I know that, partly I hope through not having had a holiday since August 1938, I have allowed friction to arise all too often and on occasion to become bitter, and for that I am very sorry indeed. I have written this letter spontaneously without much preliminary thought and it will not I hope kindle spontaneous combustion! I thought it might be useful to set things down honestly and frankly, for the heaviest shower often heralds the end of the storm. I remain, as ever, deeply grateful and indebted to you for all the help and encouragement you have given me in the last twelve years, and I am most anxious to continue to look to you for advice in the future. So I hope that you will forgive both the tedious length of this present letter, and my past shortcomings.

I remain yours very sincerely, HS

Only a man, like Peters, who believed scientists share with artists a limited ability to work with one another, could have received such a letter with grace. Peters took time before responding:

18 December

My dear Sinclair,

Thank you for your long letter of the 13th December. Just now, I cannot spare time to deal with all the points raised, though many have a simple answer. In the meanwhile I am glad that you have written all this and for the kind sentiments expressed. In regard to your last paragraph, I hope that I shall always be willing and able to give an old pupil like yourself advice when asked for it, and will certainly give this careful thought — it is most important. Somehow I do not really feel that the friction has been on my side; at the same time I think also that some of the difficulties have been accentuated by interference from outside.

With kind regards, yours very sincerely, R. A. Peters

Some weeks later, in the New Year, Rudolph Peters wrote two brief letters to Sinclair, both marked 'Personal'. In the first, he discussed Hugh's comments concerning the dithiols, and suggested that Hugh had gone about matters in the wrong way: he should have asked Peters for the chemicals needed.

… I was the director of the team and responsible for keeping them all at work, and actually had a hard fight to prevent the closing down of the work. If you wanted this compound, why did you not come to me? …

In his second letter, Peters dealt with Hugh's future:

In reference to the last paragraph of your letter, I cannot say what decision the SubCommittee are likely to make about the future of the Survey. So far as _your_ future is concerned, I have thought a good deal about this, and how your special capabilities are likely to be best and fully employed. It seems to me that though you are much interested in biochemistry, it does not hold all your attention in the way that your nutritional work does. This leads me to the view that the career which would suit you best would lead though nutrition to social medicine in some form. I have felt for long that you wanted work which had some definite contact with humans. Of course it is so difficult to know what structure things scientific and medical will take after this war is over, when much is sure to go into the melting pot.

Yours very sincerely, R. A. Peters.

Professor and pupil, father and son in science, had made an uneasy peace.

23 December 1945 — Illustrated London News:
OXFORD DON SURVEYS WARTIME DIETS
Even in wartime Dr Hugh Sinclair's work has taken him to out-of-the-way places. Last year the Canadian Government asked him to report on malnutrition among the Red Indians in the North. With two other experts be worked at 60 degrees below zero, but was warmed by the friendliness of the Crees, who called a 'band' meeting in their honour and presented him with a soapstone pipe. The pipe, unfortunately, leaks. But his report should be of great practical value.

The Indians, apparently, were better nourished before civilized foods became available. The wild beasts they trapped and ate, entrails and all, supported life more adequately than the white flour, white sugar and pure lard now supplied them by missionaries and traders.

Young Dr Sinclair, DM, MA, BSc, Fellow and Tutor of Magdalen College, Oxford, University Demonstrator and Lecturer in Biochemistry, is already an international authority on the relation of food to health. Half-expecting a bewildering academic discourse on the incidence, cause, severity and distribution of malnutrition, I visited him to learn why he started the Oxford Nutrition Survey which has been called 'revolutionary'.

First impressions were not reassuring. Sinclair, a lofty, bespectacled, ramshackle Scot with a high, bald forehead, looked every square inch a scientist. He eyed me as he might a vitamin under a microscope.

'I understand you are only 34,' I ventured nervously, 'but you look at least ten years older.'

Sinclair smiled wanly, 'It's a great advantage to look older …'

The article went on to highlight Sinclair's ancestry — Sir John Jackson, Sir Archibald Sinclair — Air Minister at the time, and his great grandfather, Sir John Sinclair, founder of the Board of Agriculture, who wrote over 600 books and pamphlets on agriculture and health. It then stated:

'Agriculture has always been considered before nutrition in Britain,' said Sinclair, in a quiet, precise voice … 'That's putting the cart before the horse. Agricultural policy should be linked more closely with the <u>nutritional</u> needs of the people.'

For example, he wants a post-war increase in production of milk, vegetables and eggs. He thinks, too, that the scientific aspect of nutrition has been neglected …

'We've been more concerned with the scientific feeding of animals than of men. There are two institutes of animal nutrition in England, but none of human nutrition. …'

On the walls of Sinclair's office I noticed brightly coloured photographs revealing in unpleasant detail some of the more advanced results of vitamin deficiencies. Two pictures of protruding tongues, one scarlet and the other magenta, vividly illustrated the lack of two different vitamins present in both bread and beer …

I had been saving up a very important question to ask him. 'Are turnips good for me? I don't like them.'

'No. You can't digest food you dislike. The palatableness of a meal is important.'

That answer alone made my trip to Oxford worthwhile.

CHAPTER 11

Brigadier Sinclair on the Continent

My darling Hugh — It will be nice to see you again — Come as soon as you can before lunch Saturday … John is full Colonel in Vienna. I wonder if you got any of my letters … Mother

As the war ground to a halt during 1945, much of the Continent lay in ruins. The people of Great Britain suffered terrible hardship and loss during the war, but did not starve. In fact, under Lease-Lend, American food aid produced a healthier population than had been seen for years. Now, the people of these islands would be called upon to help feed the starving on the Continent, and this would eventually include the Germans. Hard times and hard feelings were certain to follow.

During his most recent trip to the United States, Sinclair positioned himself to play an important role in what lay ahead. Highest priority was given to aid for people in liberated areas, including those in German prison camps. Once word arrived that Sinclair should immediately join the medical team chosen for this task, he turned his attention almost fully to that cause, leaving the matter of his Institute to linger in the hands of others.

While working on the Continent, Sir Jack Drummond, Special Advisor to the Ministry of Food, and a friend of Hugh's from his days as a medical student at University College Hospital, was Sinclair's closest ally. Drummond was one of the first medical experts to assess conditions after Allied Forces entered and secured what had been enemy territory. The situation he found in the concentration camps was dire. In April 1945, Allied Forces liberated Belsen. Unlike Auschwitz, this was not an extermination camp. Here people did not die from gas, but from neglect, disease, filth, and starvation. When Drummond first saw this horror, Sinclair was probably with him.

To ease the introduction and coordinating of civilian scientific experts with the activities and personnel of the Allied military community, military ranks were assigned. Hugh was made Hon. Nutritional Consultant (Brig.), CCG (BE). Secretly, Hugh must have been well pleased to outrank his brother.

Brigadier Sinclair's first assignment was to help examine starving children brought to Britain from Holland. He believed he had been given responsibility for organizing this effort. However, in early January, Hugh received a telephone message informing him that Rudolph Peters had been asked to bring together all the people who would be interested in helping examine and care for Dutch children being evacuated to England. As this would have diminished Sinclair's authority, he did not wait for Peters to act. By himself, he at once met with the Committee of the Medical Research Council, representatives of the Dutch Government, and staff members from the British Ministry of Health. On 7 February Sinclair wrote Peters a letter stating that he and a member of the Royal Canadian Air Force had been asked by the Committee to make a rapid clinical examination for nutritional deficiency of a large number of Dutch children soon to arrive in England. Peters appears to have accepted the slight.

Soon Sinclair was assigned work on to the Continent. Civilians in Holland were in grim condition, followed by those in Belgium, and then in France. The general picture was the same everywhere: except for potatoes, there had been no vegetables in the markets for many months. Some unfortified margarine was available, but the total fat ration was low, and there was no butter or milk. Dietary protein levels were low; average total caloric intake was low, and vitamin A deficiency was a major concern. However, depending on the length of time of occupation, and pre-existing conditions, starvation levels differed from place to place. Detailed information about nutritional deficiency was required before food distribution could begin in any effective manner. To provide intelligence about the nature of starvation in specific areas, Sinclair's proposal for mobile nutrition surveys was now activated.

The mobile teams worked under the auspices of Supreme Headquarters, Allied Expeditionary Forces of World War II [SHAEF] and UNRRA. Sinclair informed both Peters and Ryle he was about to leave for the Continent, and that during his absence Mrs Fisher would oversee the Survey in Oxford. Both agreed, although they again cautioned Sinclair that this work should not detract from his commitment to complete the analysis of existing Survey data.

It was decided that the Survey would need a mobile laboratory, equipped along the lines Sinclair had proposed some months before (see page 127) Two large transport vehicles were required to form the physical base of the laboratory. But these were expensive, and when questioned about the eventual use of the laboratory Sinclair said that, after the post-war work was completed, the mobile unit would revert to University of Oxford, probably for use by John Ryle's Institute of Social Medicine. The Rockefeller Foundation agreed to cover the necessary expenditures.

Once Sinclair knew money was available for the mobile unit, he purchased a five-ton X-ray lorry (without the X-ray equipment) from The Ministry of Health, and a National Fire Service Hose van from the Ministry of Supply. He then personally supervised the refit of the vehicles. The entire team worked to secure the supplies and equipment needed on the Continent: loyalty was tested to the full. Staff stayed far into the night (once all night) but usually left Hugh still at his desk when they decided it was time to sleep before the next day's travail.

On 18 April 1945, Hugh Sinclair, Jack Drummond, and others travelled to Eindhoven, Holland, where they stayed at the Hôtel du Commerce while they and people from other Allied countries planned a special feeding programme sufficient to treat 70 000 cases of severe starvation. It was during that time the medical team first visited Belsen.

They went to the concentration camp at the request of Brigadier Kennedy, 23rd Army Group, who had overall responsibility for nutrition surveys. There, the men met with Colonel Johnson and other officers to discuss diet and supply questions affecting liberated prisoners. The 21st Army was rushing in supplies of hydrolysed protein with glucose and vitamin supplements for a liquid feeding programme, and about 100 London medical students were scheduled to arrive within days to help administer this treatment. The liquid feed was tried on the worse cases, but seemed to have few advantages over the simpler treatment of a milk diet. Most of the prisoners were semi-starved and very weak; for these, a special light diet was devised, which was as good as could be provided with the supplies available. A few individuals had not suffered severe malnutrition; some were recent detainees in the camp, and others had been obtaining extra food in return for working for German camp personnel, or serving as prostitutes.

Just prior to Drummond's arrival at Belsen, unpleasant scenes developed when the Russians, who were in a dangerous frame of mind, demanded, with threats, that the British provide full British scale army field rations to a large group of liberated Russian prisoners. Eventually, cooler heads took control and the matter was resolved through negotiation.

The SHAEF group returned to Brussels on 1 May, and proceeded to Tilburg on receipt of news that an agreement had been reached between the 1st Canadian Corps and the Reichs Kommisar for the Occupied Netherlands to permit four members of the Advisory Committee to pass through the German lines for discussions with the Netherlands Public Health Authorities [NPHA]. The next day, the group's controlled passage through German lines was blocked. It proved to be impossible to get the members of the Advisory Committee and the vehicles containing the necessary stores and equipment through: a postponement of 24 hours was granted. That evening news arrived of the liberation of Holland. On 5 May, under a flag of truce, the convoy passed through German lines and moved on to Biltoven. On the following morning, the Advisory Committee travelled to Hilversum, where the Senior Germany Army Medical Officer of the occupied area authorized them to travel anywhere in the German-occupied area other than special military zones. At the Committee's request, they were also detailed a Begleits-Offizier to accompany them on all their visits. Later the same day, 5 May, they moved on to Amersterdam, and reviewed the feeding problems there: it was agreed it was time to call forward four special feeding teams. Monday, 7 May, the Committee moved on to the Hague and Rotterdam, where they made contact with the appropriate Public Health authorities. It was agreed to call together six feeding teams and establish at the Hague the Test Ward and Laboratory.

In general, the adults looked well, and not overly thin; the children, however, appeared undernourished and pale. For more than four months, intelligence from that region reported people in the large towns living on 400–800 calories daily. Drummond wrote:

> The very first sight of the cheering crowds in the streets of the liberated towns was sufficient to dispel that impression. … Comparison with the general impression formed on seeing the populations of Brussels and Eindhoven soon after their liberation suggested that a large proportion of the Dutch cities had by hook or by crook been able during the past few months to get food representing about 1500 calories daily.[1]

But impressions from street scenes were misleading; the old and infirm suffered the war in a very different way from those able to deal and find means to shop on the black market. Cases of starvation were not uncommon. Hunger oedema was prevalent in many towns, and as many as 700 cases were reported in one day in Rotterdam. For those forced to live on the official ration, of 500 calories a day, the story was disastrous. In a well-equipped and modern asylum near Amersfoort which housed 1600 inmates, Drummond said most were reduced to skin and bones and were dying from starvation because no food other than the official ration was available.

Certain inconsistencies were identified between the intelligence reports and what was found on the ground: before entering Holland, the Committee was led to expect grave conditions among infants under 1 year of age: reports had said that few born during the previous year had survived. In reality, the situation was not this dire, although the infant death rate had increased three- to four-fold in many towns. This increased death rate was not, however, entirely due to poor nutrition. For months, it had been almost impossible to pasteurize milk and baby food, and there had been an acute lack of hot water and soap. Diarrhoea killed infants who might otherwise have survived on the nourishment available to them.

The collapse of German authority made it possible to expedite planning for the feeding programme, and between 7 and 8 May, feeding teams entered Rotterdam, four went into Amsterdam, and two set up in the Hague; the Test Ward and Laboratory teams went with this last group. Over the next four days, Survey teams made 'snap' appraisals of the nutritional situation in numerous towns and cities. On 8 May, the war in Europe officially ended, but the nutrition work had just begun. There was no way Hugh Sinclair and others could turn their backs on reality and return home.

By 14 May, Drummond was able to provide the appropriate authorities with a broad picture of the food situation and the operations of the special teams. As the operation was functioning well, Drummond left for London on 15 May, arriving there the next day. Sinclair stayed on to work with others on detailed plans for assistance and evaluation of the starving population. He would remain until the end of June.

He was to deal with two immediate problems: first, to identify cases of true starvation and get them into hospitals or treatment centres; and second, to improve the general food supply at once. The latter was a problem of logistics — supply and transport. Subjects in Sinclair's diary from this time move between notes on methodology, information about local contacts (a particular professor at Leiden University was interested in nutrition, for example), lists of supplies to be shipped from England with the vans, and jotted memories of social events: Tea at 8 p.m., Liebfraumilch (some very pleasant) 1937. Hugh's diary notes on diagnosis mingled with notes on the wine: 'Cataract due to scattering of light, hence it does not occur with macular degeneration. Lipoid dystrophies (cholesterol). Yellow colour of macula seen direct.'

Back in England, Brian Lloyd, Mrs Fisher, Angela Hodgson, and the other members of Sinclair's little team struggled to meet his requests for information and supplies. Other work fell to the side, although Lloyd continued working on reports for publication: by this time, Brian Lloyd was the closest scientific colleague Sinclair had on his team.

In Sinclair's absence, Professor John Ryle held a meeting of the Survey Advisory Committee. As its Chairman, he reported to the Registrar saying that the Survey had received £3500 to purchase and equip a mobile laboratory for the Continent. Further, with the support of Sir Wilson Jameson and Sir Jack Drummond, Sinclair had agreed that expenditures would not exceed this sum, and Mrs Fisher was collecting bills and receipts to confirm this. The Red Cross covered the expense of moving personnel and equipment, and the expenses of the units while in Holland were the responsibility of SHAEF (Netherlands) and the Government of the Netherlands.

That same day, Ryle also wrote to Sinclair:

28 May
We had some informal discussion about the rather difficult prospect for the Survey from now on. The major contributions come to an end this year. The Institute decided not to continue their allocation beyond the 31st December. There seems no likelihood of an early decision about the proposed Readership and special Laboratory of Human Nutrition. This means that we should be considering seriously what to say to the members of the Survey staff. You will hardly be able to keep on more than a small nucleus for continuing current work and working up the material of the past three years. It seems to me that we owe to those members of the staff whom you are not likely to retain to notify them as soon as possible …

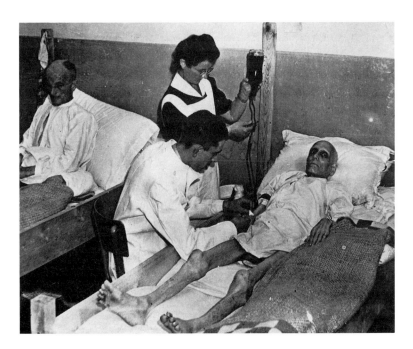

Allies Aid Starving Holland. Two victims of the Nazi-created famine in Holland receive treatment in a Rotterdam hospital. The emaciated man at right is receiving intravenous feeding. *May/June 1945.* [This caption, typed by Sinclair, appears on the reverse of this photograph kindly supplied by Professor Christine Williams, University of Reading.]

While Sinclair was on the Continent he had no means, nor apparently much inclination, to alter the course of events within Oxford that affected the Survey. As indicated in Ryle's letter to him, various groups within the University had diluted Sinclair's concept for an Institute of Human Nutrition to a special laboratory of human nutrition. Several weeks after Ryle's letter, Sinclair received a report regarding a meeting of the Board of the Faculty of Biological Sciences: Committee on Laboratory of Human Nutrition. It stated that the Board was:

> … in favour of the establishment of a laboratory of human nutrition in Oxford. It holds the view strongly that the laboratory when established should be concerned with the fundamental problems in human nutrition which have their basis in physiology and biochemistry. It is also of the opinion that the laboratory should be under the control of the Board of the Faculty of Biological Sciences, which should be advised by a standing committee appointed by the Board. The appointment of the scientific staff of the laboratory should be made by the Board on the recommendation of

Baby Clinic in Amsterdam, Holland. One of the nurses holding a young patient. *May/June 1945.* [This caption, typed by Sinclair, appears on the reverse of this photograph kindly supplied by Professor Christine Williams, University of Reading.]

that committee … In order that the Reader may be able to keep in close personal touch with the research carried out, it is desirable that the laboratory should not be too large. In the first instance the income of the laboratory should not exceed £5000 per annum, exclusive of the salary of the Reader …

Dr Harriette Chick, now of the Ministry of Food, was particularly interested in how Sinclair's work progressed on the Continent, and asked Mrs Fisher to keep her informed. On 1 June, Dr Chick received a letter from Mrs Fisher:

We have received more news from Leiden during this week … The vans arrived at Calais on 22 May, and Keith Taylor and a Dutch driver met them there and drove them straight up. The Nutrition Unit, which consists of thee teams, the Canadian under Dr McCreary, the American under Dr Fred Stare and our own, are based at Leiden. They are working in six towns … and each team covers two towns … With the Dutch drivers, typists and dietitians the unit contains some 50 persons: as this was too many for the original SHAEF HQ our team has moved to a nearby gynaecological clinic, requisitioned from a Quisling. This provides living accommodation for 20, and ample office space for typists and computers. All-day clinics, of 150 to 200 people, are being held, and the results worked up immediately. Everyone is most anxious to help, and the difficulty is to keep the numbers down. They say that conditions are very bad indeed: that whether there are numerous cases of starvation and of famine oedema, but whether there are specific signs of lack of any particular constituent I do not yet know.

Considerable effort was needed to get the mobile laboratory to Holland. Sinclair had gone on ahead, with Brian Lloyd and the others following on a priority flight. Once there, they first had to locate Sinclair, who was moving about with Drummond. Then, they had to locate the lorry and van that made up the mobile laboratories: this presented serious problems. The ship transporting the vans did not have the right type of cranes to off-load the heavy vehicles. Worse still, Dutch ports had received heavy bomb damage, and there was no place able to take the ship or manage the removal of the van and lorry. Lloyd spent much of his time on the telephone, trying to trace the equipment. They were finally off-loaded in Calais and driven north to join Sinclair, first to Eindhoven, and finally to Leiden, where Sinclair had set up headquarters. A house was commandeered, and the process of weighing people, collecting blood samples, and investigating dietary intake began. The work went on for two months, until the worst of the famine had passed. Data in hand, the Survey staff flew back to England, but the vans remained on the Continent; they would soon be needed in Germany.

True to form, before leaving Leiden, Hugh Sinclair wrote letters of thanks to all those who had helped him and his team while in Holland. His gestures were generous. One of the reasons Sinclair enjoyed his time in Leiden was its medical history; especially its connection with the great eighteenth-century physician, Hermann Boerhaave, the inspiration for medical education not only in Holland, but in Edinburgh, Scotland, and other parts of Europe as well. When Sinclair found the university owned no original copy of Boerhaave's work, he arranged to provide them with one from his growing private collection of rare books. And, from his personal supply of alcohol, probably purchased from the American PX, or military store, he laid on a party well lubricated with gin, whisky, brandy, and liqueurs. Forty-five people were invited. Seven bottles of whisky and three of gin were consumed.

Results of the Survey's investigations were provided by verbal and brief written reports from Sinclair to Jack Drummond; no major published account of the work in Holland would be available until 1948, when the Ministry of Social Affairs, of the Netherlands, published a two-part document, *Malnutrition and starvation in the Western Netherlands*. The Editorial Committee was composed of Dr G. C. E. Burger (Chairman) of the Netherlands, Lt Col. H. R. Sandstead, of the United States, and Sir Jack Drummond, of Great Britain. The story of that report appears later. However, it should be noted that Sinclair was criticized for not publishing data from this survey soon after arriving back in England, although this would have been against military orders.

On 2 June 1945, an entry in Sinclair's diary records a meeting in which it was agreed that there would be no private publication of work before the publication of an official report, or before 1 June 1946. All other reports would be confidential, and a Central Publications Committee would give approval for the use of figures. Even if Sinclair had data ready for publication, the opportunity was denied him.

While Hugh was on the Continent, in mid-June 1945, he received a letter from his mother: John was home working in the garden and Catherine was also home for a time; they had enjoyed the theatre together in London. She also mentioned that Angela Hodgson and her mother came to tea on the previous Monday. From another source, he also learned that Peter Meiklejohn had been seen in the Gezira Club, in Cairo, in January 1944, and that had gone on to provide medical care in the Belsen prison camp. Meiklejohn had played no role in the survey work with Sinclair.

CHAPTER 12

Germany

During the summer of 1945, the University of Oxford was settling the financial affairs of the Oxford Nutrition Survey in anticipation of its being replaced by a peace-time programme. It was found that Hugh had charged some personal expenses against the Survey, and he was asked to settle the debt. Hard pressed for cash, he turned to Rosalie; specifically, he asked his mother to sell a large diamond necklace. She considered the idea, but, under Catherine's influence, decided against it. Her lack of support infuriated Hugh and his resentment simmered for months. For a time, his debt would not be paid.

After returning from the nutrition surveys in the Netherlands, Hugh remained in Oxford less than a month before travelling back to the Continent with Jack Drummond. On 26 July, they boarded a Dakota, in Northolt, England, and flew through a thunderstorm into Frankfurt, Germany. Meetings began that night and continued throughout the next few days. The Army wanted a nutritional assessment done in the Ruhr Valley; devastated by heavy bombing, little of the infrastructure was left to support any forthcoming relief effort. Terms of reference and definitions were agreed for a study of the nutritional state of the people, covering the British, French, and American zones. The Russians were reluctant to participate in the work, saying they could supply food to their own sector.

On 30 July, the group began what Sinclair referred to in his diary as the Grand Tour, starting in Frankfurt. Sinclair's diary from this time is rich with notes concerning the opinions of the members of the international group assembled, the problems they expect to find, and the logistics of getting food from one Allied zone to another. To complicate their work, the black market was already well established, and 63 vehicles loaded with illegal food had been stopped on the previous day alone.

By the time the group moved on to Stuttgart, Sinclair had formed the opinion that information recorded about the severity of the starvation seen was inaccurate. Problems were being missed, and the general nutritional status of the people overestimated. As an example, he wrote in his diary about a child who was pale, had cheilosis [an eruption of the skin at the sides of the mouth, often associated with riboflavin deficiency], and was obviously undernourished: there was nothing written beside her name in the file other than 'poor appetite'. Procedures were sloppy; men were seen being weighed carrying their shoes. Physical examinations did not include clinical indicators Sinclair believed were important. A number of physical parameters and diagnostic signs had been agreed by the Allied group of medical experts before arriving in Stuttgart; once there, Sinclair found these lists had been altered. According to people already working in the city, everyone was well fed, with the possible exception of the elderly. However, Sinclair was told that the medical teams there were not visiting institutions, where he knew the real effects of a restricted food supply would be seen.

Sinclair noted that after the meetings in Stuttgart the group enjoyed a 'magnificent dinner', no doubt washed down with fine wines, and plentiful amounts of beer for those who preferred it, in what had formerly been a German General's home. From Stuttgart, the group moved on by plane to Augsburg, where before beginning work they were fed a 'large lunch of chicken and ice cream'. The men who came to study starvation ate well.

In Augsburg, Sinclair again worried about the accuracy of the medical data recorded. He felt the laboratory work was being done badly; in his estimation, not enough blood was taken for serum, for example. Men and women were weighed and measured in the same room, and so the women remained in their dresses. Goitre was 'extremely common', according to Sinclair, but not much was made of it. That night, after a day's work with the malnourished, there was another dinner party, with good food and excellent wines, at which the Allied teams were entertained by a popular musical trio. The contrasts between Victor and Vanquished were unceasing.

The next morning, Sinclair and the others flew to Bamberg, over the devastated city of Nuremburg. Their first stop was a clinic being held in a factory. Here, they were told, the food situation was the best in Bavaria, but condensed milk was needed for the children. About a third of the population had access to a garden for fresh vegetables, and the fat and protein supply seemed adequate. However, there was concern about the coming winter. Unless there was a radical change in supplies, there would not be enough food.

The next afternoon, in yet another location, there was a nutrition conference, at which Sinclair again met Brigadier T. F. Kennedy, OBE. The group compared notes on the availability of food and types of nutritional deficiency observed. Everyone was concerned about the general condition of the people in the Ruhr. Pellagra had been seen in Düsseldorf, and there were higher rates of oedema in men than was expected. Mining had almost stopped, and there were no stores, so the threat of winter became even worse. There was no hope of any food other than flour and possibly some fish from Sweden and Norway. Some potatoes could be made available from the British Zone. As for transport of food, the whole of the canal system was out of action in Germany, and the railways were operating on a very limited scale. That night, there was a conference with Kennedy in his rooms: Drummond, Magnus Pyke, and Sinclair attended. Sinclair recorded: 'Kennedy wants the [nutrition] teams. The General wants to know the position in the coal-mining district. We shall be going to Essen. He wants factual information about how they are getting along.'

Plön, Düsseldorf, Essen, and Saarbrücken: — everywhere the story was the same. Medical services were mixed and limited. General upheaval caused by the destruction of property, lice, the run-down of food and fuel stocks, badly disrupted transportation, and a population largely depleted of nutrients, were ingredients for a difficult winter ahead. A plan was agreed by the Allied nutrition experts, outlining the contents of the report to be written after a detailed investigation of the food problems in the region. Only a short time was available for the work. Sinclair detailed the discussion in his diary, noting that: 'The following report is objective and we, as scientists, are not concerned with policy.'

In early August, Sinclair returned to England. He was badly shaken by the intense suffering he had seen. The situation in Holland, seen in his previous assignment on the Continent, was terrible; but what he had just witnessed in the Ruhr Valley was worse. For Sinclair, the destruction and waste of the war was not lightened by the sweet knowledge of victory. He had seen first-hand the bitter herbs of the Vanquished: hatred and mistrust. He talked to friends and colleagues about suffering in Germany, but few were receptive to his views.

16 August from T. F. Kennedy, CA/Mil Gov HQ 21 Army Group BLA
Dear Sinclair
As a result of a meeting yesterday with Major-General G. W. R. Templar,
Director of Military Government 21 Army Group, and Sir Jack
Drummond, I am writing to ask if you and your two Nutritional Survey

teams could be made available for work on nutritional surveys in the British Zone in Germany, probably Ruhr and Berlin areas.

It is not known for how long the teams will be required, and probably you yourself will be in a better position to judge this, once you are over here I imagine, though, that it would be for not less that 6 months and probably a year … We have approached the British Red Cross here, and they see no objection at all to your coming over under their aegis …

On 16 August, Sinclair wrote a long letter to John Ryle; there is no record of a similar letter to Rudolph Peters.

… I have just taken part in an extensive tour by air of the three zones of Germany along with Sir Jack Drummond [and others] … I have been asked by the British Control Commission and by 21 Army Group to make a survey in the Ruhr and possibly afterwards in Berlin; an American team has reported that in Berlin the nutritional situation is … deteriorating rapidly. I understand that an official request will be coming from Brigadier Kennedy … in a day or two; and Sir Jack Drummond … tells me that it is suggested that outstanding questions such as finance will be settled as soon as we get out to Germany.

Preparations for his return to Germany dominated Hugh's time, and he was rarely seen far from his office in the Institute of Social Medicine, or from his rooms at Magdalen College, where he now lived most of the time. The sheltered life at Lady Place was too emotionally distant from what he had seen in the Low Countries and Germany. He found the usual academic business of the university wearing; the debates and struggles seemed trivial beside the problems he knew faced millions of people during the coming winter. His reserves of energy and humour had worn thin. A few days after receiving word he was to return to Germany, Hugh wrote to his mother:

My darling Mother,

Now that the war is at last over, I wonder if I may make one or two suggestions even though I have made them before. As you know, I try hard during weekends to say as little as possible … I should very much have liked to have told you — as I have my friends — about Germany … But somehow it always seems impracticable. One feels that whatever one says will be met with a lot of complaints.

I know and greatly appreciate how hard you work for us, and your difficulties. But I wonder if you are right to kill any suggestion we have to make. I came back for the weekend hoping that we might take our tea onto the Downs but refrained from suggesting it because I did not want to be told once again that you haven't a cook (which is very irksome particularly as I have been in Holland and Germany where there are no servants, no fuel and no soap — but in Holland I heard no word of complaint anywhere) … I am sorry to have to write all this, Mother, particularly as it is not new. But I may not be back next weekend and do not know how long I shall be in Germany. I always come back tired out at weekends and just want a rest, quiet time and to feel at home — to feel I can help myself at meals instead of having to wait to be asked. Perhaps, now that this war is over … we can rebuild things. For there is nothing I want more than to look forward to my weekends and enlarge them so that I can spend more time at home and see more of you. The more we can understand one another the happier we'll be. And compared with so many people, we've had the greatest luck in this war. Best love from Hughie.

Rosalie responded at once:

I began to read your letter before breakfast and when I saw the bit about taking our tea on the Downs I just broke down entirely — it would have been just heavenly to get away for a few hours — to peace and quiet and a good blow and I feel weepy whenever I think of it. You aren't <u>quite</u> fair! … I tried to get you to tell me about Germany at 2 or 3 meals … Try to say as much as possible at the weekends and don't shut me out of your life. I wanted to join you in the garden after supper but thought you wanted to be alone as you shut the door and did not suggest my going out … I don't want to kill any suggestion you make (I did leave Kent and come here for you) but I have to think of 4 of you … I long for the time when you can be at home more with me now even if just for coffee and a smoke after dinner, I only seem to see you at meals and then I try to talk — you forget too I have got <u>very</u> deaf … and find it hard to hear what you say — By the way your pyjama jacket is unmendable, can you get some new ones. … Come home as much as you can and I want to hear about Germany — I have not time to read this so tear it up. I keep thinking of how lovely it would have been on the Downs.

Lots of love and kisses Your loving Mother.

Just before Sinclair was scheduled to leave for Germany he received a letter from John Ryle, who remarked that Hugh seemed to be 'spreading more widely'. In response, Hugh asked for a meeting with the Survey

Committee before his departure. Ryle said he thought it would be difficult to arrange such a meeting in the near future, and he doubted if it would be likely to alter the programme into which Hugh had launched himself and members of the Survey staff. Ryle wrote to Hugh:

> … I think, however, that you should summon a meeting definitely for the last week in September… or in the first week in October and be present yourself (a) to report on the situation and to hear the views of the Committee … (b) to discuss seriously what is to become of the Survey. You might ask Miss Shaw if there has been any reply to my letter to the Registrar about the description of the cars [the vehicles used in the mobile survey unit] and their ownership … Failing a decision you had better state in your report that the negotiations are incomplete in this respect …

In response, Sinclair wrote still another report to the Committee detailing again his plans for the study of human nutrition, requesting more space for his work, and asking that decisions on the matter of the future of the unit be delayed until after his return from Germany.

For days, Hugh and his staff worked almost without sleep, writing reports for the Survey Committee and preparing the various items needed for the work in Germany. On the night of 31 August, he was to have gone to London with his mother to see a show, but instead, at midnight, he found himself still at his desk. He wrote a note to Rosalie, saying he was leaving the next morning by car, and then flying to Germany via Holland. He expected to return in early October, although he had been told to expect to stay longer.

By 3 September, he and thirteen others, including Brian Lloyd and several other members of Sinclair's Oxford Nutrition Survey staff, were temporarily attached to a unit in Düsseldorf.

Sinclair's four teams began work weighing members of the public and conducting limited physical examinations. He travelled to a number of locations, noting in his diary that major laboratories and medical facilities were badly damaged. Visits to clinics where data on the general health of the population were collected underlined the need for the imposition of stricter procedures: people were examined for skin lesions in poor artificial light; all left their shoes and socks on; women were examined fully clothed; gum probes were not used; only right legs, below the knee, were examined; and blood pressure equipment was incorrectly read.

On 19 September, Brigadier Sinclair met with Brigadier Kennedy, who was about to go on leave for almost two weeks. Kennedy was in charge of the work Hugh had undertaken, and most decisions now went through him. They discussed the best use of the Survey teams, and Sinclair gave Kennedy a memorandum suggesting that five Army doctors were required. Kennedy then told him that Peter Meiklejohn had been cabled for, and should be in Germany soon to start survey work. Another physician had been asked to stay on, so there would be adequate cover when Sinclair returned to Oxford in early October.

While in Germany Sinclair worked countless hours, making sure all his team were properly fed and billeted; German personnel assigned to him often received rough treatment, a situation which he tried to amend. On one occasion, Sinclair made certain his staff ate well, while he went without. Matters concerning the treatment of the Germans were eventually resolved, but in the meantime Sinclair saw first-hand the rejection by the Allied forces of educated and talented men whom he had come to like and respect.

A Hollerith machine was obtained in Germany for the tabulation and collation of Survey data. Brian Lloyd, who was specifically responsible for the biochemical work, was assigned the task of managing both the installation and use of the machine. Lloyd managed to find a place to set up the machine in Düsseldorf, The Finanze Präsidium, which had been used for Customs' investigations. Among the previous tenant's exhibits of roulette wheels and diagrams of people smuggling things on the railways, Lloyd and a team of five German women set up a data processing laboratory. When operating, the machine vibrated so violently Lloyd feared for the structure of the building. All was well, however, when the German Hollorith team cut a hole in the cement floor of the basement and placed the machine on the earth beneath it. Lloyd was impressed; in England such a feat would have taken months; here, the task was done in days.

Problems multiplied. Survey team members were slower than expected in their work and general activities. Equipment problems plagued Sinclair: at one point while travelling, all the capillary tubes needed for blood work were left behind at a previous stop. On the morning of 24 September, Sinclair left his staff to unload one of his trucks. It was accidentally driven into a river, smashing its radiator. Still partially loaded with necessary equipment, it had to be towed to a garage the next morning for repairs. While other members of the Survey team took time from work to enjoy what was left of entertainment in the area, Sinclair stayed behind to work, making notes of the nutritional conditions he witnessed.

On 7 October, Sinclair motored to Bünde and lunched with Brigadier
Kennedy. Peter Meiklejohn had sent Kennedy a report suggesting a
different staffing pattern for the survey teams from the one Sinclair
wanted. Sinclair and Kennedy discussed the matter, and in the end
reached a compromise. Sinclair was unhappy with the outcome, and in
his notes of the meeting wrote:

> Show him copy of the memorandum I gave him previously (which he has
> forgotten) … Give him an account of the surveys. He wants to know how
> long they can continue on present rations without disaster — I guess Berlin
> [one month], Düsseldorf [two months]. He is very interested in the vitamin
> A situation, particularly as regards its (alleged) anti-infection properties.
> He considers giving A to all Ruhr and Berlin 6 000 000 persons! <u>Body
> weights</u>. Give him memorandum. He is against the second part — B —
> because he doesn't trust the Germans. The arrangement is that the
> Provinces will send me the results and I'll get them punched, sending the
> results … back to him through the provinces.

Hugh and several of his team returned to England on 12 October with
samples, data, and more plans for work in Germany. He left behind strict
instructions as to what information was to be obtained from subjects, the
methodologies to be used during the collection process, and how the
resulting data were to be coded and recorded on Hollerith cards.

Once back in Oxford, Hugh's schedule was hectic: in addition to checking
on the progress of the surveys in Britain, there was demonstrating for Peters
in the Department of Biochemistry, and students to tutor at Magdalen
College. John Ryle pressed Hugh's team into increasingly less space within
the Institute. He irked Sinclair even more by claiming items belonging to his
Institute were finding their way among possessions of the Survey.

In the evening, after a day of fighting both staff incompetence and
arcane bureaucracy, Sinclair would turn to writing reports about work
completed and work he believed needed to be done. Finally, there were
plans to make before he could return to Germany. Acerbic arrogance
helped him survive. A profile of Sinclair, published in the *Oxford Medical
School Gazette*, Vol. X, No. 2, Trinity 1958, gives clues about his
demeanour during the months following the war:

> … He became a controversial figure in the field of nutritional and metabolic
> problems. It was quite acceptable that he should suddenly reappear, from

Germany it might be, as a Brigadier, requiring to be saluted by humbler brethren who had plodded through the field.

'What are you doing in London?' I once asked him, not looking at all like a Brigadier, and added, 'and in those clothes.' 'Oh,' he replied, 'this is my clinical suit, and I've been being admitted a Member of the College of Physicians.' 'By the back door?' 'I believe some ignorami refer to admission by published work as such, but it's by the side door.' In a few moments he reappeared, ducal, with vast courtesy, in evening clothes … M. P.

Hugh missed a meeting of the Institute Committee, on 20 October. Four days later, John Ryle wrote to Sinclair:

> PERSONAL
> … At the meeting of the Institute Committee held on Saturday, October 20th, the future arrangements for the administration of the Institute building came up for discussion. You will recall that at a meeting of the Committee held on January 13th, 1944, it was agreed that Mrs Fisher be appointed as administrator, and that her salary should be shared between the [Institute of Social medicine and the Oxford Nutrition Survey], and that the appointment should be reviewed at the end of one year.

In view of the fact that the Survey would cease to operate after 31 March 1946, according to Ryle, the Committee thought it was time to make other arrangements for the administration:

> They are anxious that you should express to Mrs Fisher their gratitude for the assistance she has given and also the hope that she will be able to continue the work until 31 March, 1946, but feel it proper to give you due notice that the Institute's portion of Mrs Fisher's salary will not be available thereafter.

At the same meeting, the Committee decided to refer the matter of space allocation to the Hebdomadal Council of the University. Hugh and his staff were on their way out of 10 Parks Road.

On 1 November, Hugh responded with a long report to the Joint Scientific Advisory Committee of the University. Laboratory space was equipped primarily with funds from the Survey, he pointed out, and while bench space had been allocated to members of the Institute of Social Medicine, the Institute had no laboratory workers.

The Control Commission contacted Hugh again on 2 November, and told him to return to Germany on 8 November for an urgent meeting; it was expected that he could return to England on 12 November. People 'at the highest' levels were anxious to receive accurate information about the food situation in Germany, and, based on information in Sinclair's diary, Sir Wilson Jameson wanted the Germans 'fed better'. The work Sinclair now felt was truly important was in Germany, not England, and he was anxious to return.

Earlier, in mid-October, Sinclair and Peter Meiklejohn had met in Oxford, and over dinner at Magdalen College discussed Kennedy's plans. Fred Stare, from the United States, had been asked to arrange the next tour of Germany cities, which Kennedy asked Meiklejohn to join. Kennedy had also arranged for an UNRRA nutrition team to conduct surveys in Hamburg, Münster, and possibly Hanover. According to Sinclair's diary, over dinner Meiklejohn said that he had suggested to Kennedy that Sinclair should meet the tour in Düsseldorf, but Kennedy refused. Sinclair tried to confirm that he was still supposed to form four teams to conduct nutritional work as he had planned. Meiklejohn's response was vague: perhaps it was three teams; or, perhaps, two. Meiklejohn said he was to conduct the work in the cities now covered by UNRRA. In preparation, he had found a laboratory in Hamburg, and would soon have a Red Cross van to use as a mobile laboratory.

Sure that he would find conflict on his arrival, Sinclair returned to Germany on 9 November. He may have escaped the tumult of Oxford, but he had not altogether escaped conflict. His diary from that time indicates trouble was brewing between himself and Brigadier Kennedy, who had military responsibility for the administration of the nutritional surveys. On reaching Germany, Sinclair met with key military personnel, and was assigned to their mess. One contact warned him that Kennedy was 'hurt' because Sinclair had said Meiklejohn had displaced him in the work. Kennedy, Sinclair was told, had thought Hugh was not available. Two days later, Sinclair met with Kennedy in the Senior Officers' mess, where talk over tea extended well into the dinner hour. In the presence of another Brigadier, Kennedy gave Sinclair orders as to what he was to do, and what reports and budgets he was to prepare. Sinclair accepted Kennedy's orders, and within ten days returned to England, planning to recruit four teams for the nutrition surveys.

Back in England, Sinclair learned there were numerous errors in the data from a 3-day study of Dutch diets. These data had not been included in a report already prepared for SHAEF, and Sinclair decided to leave them until later. It also appeared that calculations had been made without checking published Medical Research Council standard values. Since these tables were the standard by which all dietary studies in the UK were conducted, this gave rise to a major error in methodology. Mrs Fisher wanted to go through the data cards and ink in values. Sinclair looked at a sample card and found it incomplete — yet Mrs Fisher claimed she was finished with it. Sinclair delayed the inevitable longer, and decided to leave the cards until after the next set of MRC food values had been added. Then, he said, Mrs Fisher and one other person would review them — separately.

> 20 November   Douglas Veale [Registrar of the University] to Hugh
> Sinclair
> Dear Sinclair,
> You are, I think, aware of the recommendations that were made by the Board
> of the Faculty of Biological Sciences in favour of the establishment of a
> laboratory of human nutrition in Oxford. This question has now been
> further considered by the Joint Advisory Science Committee. On the
> recommendation of these two bodies, Council has decided that a Readership
> and laboratory of human nutrition be established on the following conditions:
>   (1) it shall be concerned with the fundamental problems in human
> nutrition which have their basis in physiology and biochemistry;
>   (2) its finances shall be wholly met by a special permanent endowment in
> such a way that no burden can fall upon the University either immediately
> or in the future;
>   (3) the laboratory shall be under the control of the Board of the Faculty of
> Biological Sciences which shall be advised by a Standing Committee
> appointed by the Board.
>   An overriding condition shall be that adequate accommodation for the
> laboratory shall be provided.
>   Council further considers that the expenditures should not exceed £5000
> a year, though discretion should be given to the General Board to allow this
> sum to be exceeded if money is forthcoming. Should the required
> endowment be forthcoming the necessary legislation will be promoted.
> Yours sincerely ...

Was this victory? Or, a poisoned chalice? The wording of the University's decision left it to Sinclair to obtain outside funds for his nutrition work; and, he was initially restricted both in the type of research he could undertake and in the size of the laboratory's future budget. As usual, Sinclair took the circumstances as presented and described them to his best advantage. He wrote at once to the Rockefeller Foundation, saying how Oxford University had agreed to:

> … create a Readership and Institute of Human Nutrition provided that a permanent endowment of £5000 per annum and a capital sum to provide adequate accommodation can be secured for this purpose. … It is therefore probable that, if funds are forthcoming, an Institute of Human Nutrition, the first in this country, will be started in the University and the work on nutrition done by the Ministry of Health will be intimately linked with it. I believe it is possible that there may be a suitable site available for this joint enterprise in the very near future, well situated between the University and the Churchill Hospital. … I write … in case you think that The Rockefeller Foundation, which has already financed so generously the scientific work in general of this University and the work of the Oxford Nutrition Survey in particular, might be interested in this natural development. I had a talk with the Vice-Chancellor recently, and I am sure that if you are interested and wish it he would write to you a more formal letter …

Hugh's letter to the Rockefeller Foundation was passed from the Division of International Health to the Division of Medical Sciences: with the end of the war in Europe, the priority for funding projects had changed.

Sinclair also sought support from Sir Wilson Jameson, and on 11 December wrote:

> I have been informed by the University that it wishes to create an Institute of Human Nutrition provided £5000 a year can be secured and a suitable sum for accommodation. I have therefore taken immediate steps to see if this sum might be forthcoming, and Sir Henry Tizard suggested to me that the Prudential Assurance Company might be interested. Some time ago he spoke to the Chairman of this company regarding the possibility of a sum being guaranteed by it for, say, ten years, and received a favourable reply, and he is anxious to pursue the question again now. … Sir Henry Tizard thought it was possible that the Chairman of the Prudential Assurance Company might wish to consult you, and I was therefore anxious to keep you informed about the present developments.

This letter was a serious blunder. Time would reveal that Sinclair misstated the case regarding the Prudential, and this led to embarrassment for Jameson.

22 Dec

My darling Mother — This is to wish you a very happy Christmas, and all health and happiness in 1946 — the first full year of peace. I am very sorry that you will be alone at home and wish I could be with you … I was asked if I would form two more Nutrition Survey teams — making 4 in all — to cover the whole British Zone, and Germany including Berlin. Of course I brought out two teams in September but had to take back some of the people for the work in England. I got most of the people together for the four teams, but the interminable delays and inefficiencies of the Army and Control Commission caused things to move very slowly, and so I said I would not go back to Germany but was going to Austria. The immediate result was a telegram asking me if I would return to Germany at once to help them, and there is so much to be done here that I decided to do so. I travelled by train as that was the quickest way — fog had held up air transport for 2 days — and went to Bünde, which is in the middle of the B … near Minden. This rather pleasant town is Headquarters of a lot of the Control Commission and I have conferences there with many people about nutrition things. Then I came down to join the remnants of my two teams here in Düsseldorf. Tomorrow morning most of the rest of the teams are arriving.

We have very pleasant quarters here. As soon as I came out in September I requisitioned 4 houses — modern blocks of flats — on the outskirts of Düsseldorf near the hospital. It was not very pleasant turning even Germans out of their houses at 24 hours notice, particularly when almost all the big town of Düsseldorf is a heap of rubble and people live crowded together in cellars. But it is the only and correct thing to do, and the German official assured me he would find accommodation for them elsewhere. Since then the Army have given us what extra furniture we need. Airmen, servants and chef, coal and all our food etc., repaired the houses and laid on military and civilian telephones. I have quite a fleet of cars: our two big mobile laboratories which we brought from Holland, 3 big Army staff cars, an ambulance and a truck. The Army are getting me a further German car for my own use which I do not really want as I like the staff cars.

… The Germans make a lot of Christmas — more than we do — but it will be incredibly miserable for them this year. My main German interpreter was a University student when war broke out and had to join the Wehrmacht; he comes from a good family, speaks English excellently and is a most loyal and efficient person. He has just got married and I gave him 80 cigarettes and some matches for Christmas. He was so pleased he did not know what to do; the Germans get about 1 cigarette a day if they are lucky and matches are practically unobtainable for them. I shook hands with all the German staff and wished them a happy Christmas as they went off this evening — many people will not shake hands with Germans, but all my staff, like all Germans employed by Military Government, have been very carefully examined to make sure they had no connections with the Nazis and it seems stupid and unnecessary to ostracise them.

One cannot describe the havoc and misery in Germany — brought by the Germans upon themselves. Every large town, especially in the Ruhr and Berlin, is completely devastated — a few walls here and there, the rest rubble. Wherever there is a tottering wall, a cellar or a piece of a roof, people live there. There is practically no fuel, and almost nothing in which they can burn what wood they collect from the debris. Their clothing is in general very poor — although there are quite a lot of silk stockings stolen from France; shoes in particular are dreadfully poor. But worst of all is the food situation. The diet is now almost entirely bread and potatoes — with very little bread and almost no potatoes. They get about 12 ozs bread a day per adult, $^3/_4$ ozs meat (not always available), a little fat ($^1/_2$ oz) and 9 ozs of potatoes (on the ration, becoming unobtainable now). Add to that the difficulties of cooking (no fuel, no utensils) and you can see that life is pretty miserable. I have recommended that the German police should be given extra rations — they are unarmed, work long hours and are out in the cold most of the day. One cannot but feel sorry for the children: they did nothing to bring this devastation upon themselves — I suppose it is just bad luck to have been born a German in the last few years. Yet one has to be very stern about it. I found when we were first out here that half-starved children would come around and help the sergeants to clean the cars in the hope of getting a bit of bread; but not only is it slightly illegal to give Army rations to Germans, but also one would have dozens of children around if one once gave a bit of bread to one child.

I have a large organization for following the weights of the Germans throughout the Zone. One way we do it is to have weighing teams (Army Sergeants) at busy places in towns weighing all the people who come past. We have collected and analyzed thousands of weights — we have installed at our HQ here vast mechanical machines worth thousands of pound for analyzing all the results.

The Director of the Economic Division for Austria — a very important person — has invited me to visit Vienna to advise on the situation there. Sometime later I hope to go to make a proper survey there, but now there is not time as I want to get back to England about 12 January. But I have made a tentative arrangement to fly there on 30 December. I hope therefore that John and I shall be able to drink your health together on New Year's Eve! I am looking forward to seeing him. I expect to be about a week in Vienna, then another week here getting the new teams properly started, and then go home. If the teams run properly I should not have to come out so often in future, particularly if, as I hope but do not expect, the food situation improves. I am afraid that thousands of Germans will die of starvation this winter, as well as of cold.

What do you want for Christmas? There is absolutely nothing to buy out here … would you like to buy yourself something? … P.S. Would you like to take some of my wine from the cellar for Christmas. I think it is all labelled.

(Such an offer was generous, as Hugh always made certain his wine racks were full of the best that was available, and by this time had bought several cellars as an investment.)

29 December

My darling Hughie — I have heard nothing of you since you left and hope you have got my letters … Xmas Eve a telegram came saying John has got the D.S.O. — which he fully deserves and X-mas Day he rang up from Vienna about 9.30 p.m. I could not hear well but think he will come on leave next month … I had a letter from Catherine [too] X-mas Day, and Salie's presents — Catherine is probably in Borneo now … Theobald went away Mon–Wed. I hope to get the geese moved soon, then the ducks and then the hens. I expect you heard Sir Farquhar Buzzard died — write some time if you can even if it is just a short letter …

Now, another of Sinclair's most devoted champions, from his father's generation, was gone.

30 December

Dear Mrs Sinclair,

I don't think Dr Sinclair will have a minute to write to you just now, so I thought I would send you a line to say he was catching a plane for Vienna at 7.00 am this morning. … I think he expects to be away a week or 10 days.

I arrived a week ago after a rather hectic journey by land, sea and air! Since then we have had a very frivolous Christmas, including amongst other things, a party for all the small children in our two streets — about 20 or more of them. It was very good fun and well worth doing.

Very best wishes to you, yours sincerely Angela Hodgson.

On the morning of Saturday 29 December 1945 Hugh Sinclair wrote a note to a staff member who had just resigned, placed it on her desk, and went back to bed. He recorded a temperature of 102 °F, and a pulse rate of 82 in his diary. Late that afternoon he received word that a plane was available to take him to Vienna, where he had hoped to see his brother, John. At 7 p.m. Hugh made his way to Bünde, where he spent the night. In the morning he motored to Bückeburg, boarded a Dakota and tried settling his long feverish frame into the discomfort of its military simplicity. Three-quarters of an hour into the flight the plane's left engine failed and it turned back. After a night in a Transit Mess, Sinclair again boarded the plane and it headed for Austria. Once there, he waited two and a half hours before transport was available to take him to his hotel. Before he could unpack, Magnus Pyke called, suggesting they do some clinic inspections while Sinclair was in Austria. Despite having a streaming cold, Hugh agreed. Minutes later his brother called. John was appalled by Hugh's obvious illness and moved him to a nearby villa. That night the Sinclair brothers dined and drank well, celebrating the first New Year's Eve of peace in Europe.

After food and rest, Hugh's flu seemed better; and so on 2 January he and Magnus Pyke visited some local clinics. In his diary, Sinclair noted down dietary information, clinical observations, and comments on what he thought were the poor methods of clinical examination used by local medical teams. More clinic visits consumed most of the next day, and that evening Hugh and John celebrated their reunion in Vienna by dining at Kinsky Palace, and attended the opera. But Hugh was still unwell: by the evening of 4 January his temperature had reached 107° F, and the next morning he was taken to hospital. On the ninth day after admission, the specialists diagnosed pleurisy, and he was moved to a private ward and X-rayed. The middle lobe of his right lung had collapsed.

When strength allowed, Hugh wrote first to Peters, then to Brigadier Kennedy, and finally to Mrs Fisher, explaining his delay in returning to England. On 19 January he received a warm letter from Ryle, suggesting that he should not rush back, as he would do his best to mitigate the upheavals pending in Oxford regarding the Survey. Peters wrote several days later, saying that he was sorry to hear about the flu, and curious that the immunity Hugh had enjoyed for so long had suddenly broken down. They had planned the term without him, so he need not hurry to return to England.

Perhaps stimulated by these suggestions that he remain abroad, Hugh ignored doctors' advice that he stay in Austria to recuperate and returned to England, directly from Austria, as quickly as possible. This left the German Survey work in chaos. His teams collected as much data as possible, abandoned the two vehicles they had used as the mobile laboratory, and followed Sinclair back to England. Unfortunately, it later appeared that they left a considerable amount of data behind as well.

In his description of his life during this time, Sinclair's comments about his period in Germany are minimal, although he sarcastically hinted at deep problems between himself and another Brigadier almost twice his age: this must have been Kennedy. An unpublished document, which Sinclair called *Curriculum Vitae*, describes surveys in the Netherlands, then states the following:

> … Then the laboratories went on to the British zone of Germany where Hugh, with rank of Brigadier, formed five teams to make nutritional surveys for the Control Commission. The head of public health was an aged regular colonel, promoted to Brigadier, who thought Hugh should take his orders. As Hugh regarded him as incompetent, the Minister in charge of Germany (Lord Pakenham) put Hugh (aged 35) in Medical Officers Mess A in Berlin with the Major-Generals, the Brigadier (nearly twice his age) being in Medical Officers Mess B. This did not produce harmony. But the US Government awarded Hugh its Medal of Freedom with Silver Palm (the Highest civilian award) for his achievement in bringing about the cooperation of the UK, US, USSR and French officials.

Having access to some of the facts about Sinclair's activities in Germany, it is possible to see why a career soldier like the Brigadier would have expected to be in charge. Sinclair was constantly in transit between Oxford and the Continent. He was more willing to tell people how work should be done than to explain what he was doing. Sinclair was always ready to boast of his contacts, both in Britain and America, and he lived well while in uniform; what Angela Hodgson described as '… a very frivolous Christmas' may have been resented by members of the regular army.

The United States of America did indeed award Hugh Sinclair the Medal of Freedom with Silver Palm. This medal was also given to Sir Jack Drummond and the other senior members of the international team overseeing the study of post-war nutrition. Hugh's personal charm, and his impressive comprehension of the history and scientific basis of the study of nutrition, had won him admirers among the group of international experts brought together to study the nutrition of starvation. The Oxford University Nutrition Survey had gained an international reputation, and in 1944 was mentioned by the Truman Committee in a document supporting the idea of an Institute for Human Nutrition in the United States, in which it was said:

> The need in the United States today is for a nutritional testing group such as the Oxford Nutrition Survey in Great Britain. It has been stated that there exists no method by which the nutritional well being of the public can be periodically tested, although this is known by nutritional authorities to be greatly needed.

At this time, Sinclair was vulnerable on counts other than his minimal scientific output from the Survey. He returned from Germany convinced that errors in the recorded weight and physical condition of the civilian population led to an underestimation of their degree of starvation, and that this flawed reporting was intentional. But these observations and conclusions were not well received by many at home. Food shortages were now a part of British life, and bread rationing, unknown during even the darkest days of the war, deeply embittered many. The British people were now being forced to sacrifice even more than during the war: and for what? To help save the hated Germans. Both Jack Drummond and Magnus Pyke reported starvation among the Germans, and along with many Allied officials worried about the consequences of failing to correct the situation. But in most parts of the country, anyone suggesting that the Germans should have more food at Britain's expense was despised. [Brian Lloyd recalls Hugh telling him that the nutritional surveys in Germany were more for political than practical reasons.]

Was it possible that Hugh Sinclair sympathized too openly with the German civilian population? Years before, as recounted earlier, while a student at University College Hospital, Hugh Sinclair had inflamed the Jewish community in England by publishing in the College newsletter an anti-Semitic article, which he had thought clever. Old memories of such an act do not die quickly. There is no document to establish the effect Hugh's early anti-Semitic outburst had on his career following the war, but it must have been a fact of his life remembered in some circles.

CHAPTER 13

Finance and accommodation

In late February 1946 the Trustees of the Wellcome Foundation gave provisional approval for a grant of £25 000 towards establishing a Laboratory of Human Nutrition at the University of Oxford. Hugh Sinclair had been the moving force in obtaining the money, but there were conditions that he may not have appreciated. The grant stipulated that the scheme of the Laboratory must fit into the Government's plans regarding nutrition, and be in accord with the policy of the Medical Research Council. The University Grants Committee must make a recurrent grant to the University of Oxford to support the work of the Laboratory. There was one final consideration: space must be secured for the Laboratory.

Sinclair was obliged to vacate the premises of the Institute of Social Medicine, at 10 Parks Road, by the end of March, and at once began a campaign to raise the sum of £10 000 to purchase a house, known as The Shrubbery, near the Department of Zoology. Unfortunately for Sinclair, new academic programmes were thriving in the post-war atmosphere of enthusiasm, and competition for premises was keen. Senior members of the University supporting Sinclair's nutrition work did what they could to help. After a meeting on 27 March 1946, the Committee of the Oxford Nutrition Survey met to vote its support of the new laboratory. The Chairman, John Ryle, wrote to the Vice Chancellor:

> At the final meeting of the Oxford Nutrition Survey Committee yesterday, it was made apparent that unless the suggested accommodation at The Shrubbery, Woodstock Road — which would be particularly suitable — can be allocated to the Laboratory of Human Nutrition, the whole work of the organisation must come to an abrupt end. At the best, it would become suspended for an indefinite period. This would not only mean loss of employment, by disbanding, of most of Dr Sinclair's staff, but also a cessation or serious interruption of research activities and of the analysis of accumulated results, which may prove to have a large human as well as scientific value.

I am not, of course, in a position to know anything of the rival claims to this building of other departments in the University, but as Chairman of the Oxford Nutrition Survey, I would like to submit this special plea on behalf of the Laboratory of Human Nutrition …

J. A. Ryle

Despite the pleas from such a prestigious group, the desired property was allocated to the Institute of Statistics before the end of the month. Ryle immediately informed Sinclair, saying he realized the difficulty this presented, and offering to continue housing the Survey at the Institute for Social Medicine for another three months, until the end of June. Hugh, grateful for this temporary arrangement, continued the search for permanent premises.

On 9 April Mrs Fisher received a letter informing her that the House Committee of the Radcliffe Infirmary had agreed that a part of Nissen Hut 37, outside the Churchill Hospital, could be loaned to the Survey for a period of six months. Meantine, Hugh also heard that another property, Summertown House, might be available for his work, and began lobbying the proper authorities at once. He wrote to Hugh Smith, painting a somewhat glossy picture of the situation, and defining his view of how his developing design for an Institute for Human Nutrition would be organized:

You will recall that towards the end of 1944 the Committee of the Survey recommended to the University that an Institute of Human Nutrition should be established. The University is always rightly hesitant over establishing new institutes, and therefore the matter was fully discussed by all the relevant Boards … at the end of November 1945 the Hebdomadal Council decided to create a readership and Institute of Human Nutrition provided that a permanent endowment of £5000 per annum and a capital sum to provide adequate accommodation could be secured for the purpose. Planning the future was also not made very easy by my frequent absences first to Holland and then in Germany … However, I was fortunate in finding that the Wellcome Foundation was very interested in our work and, although such work did not fall within the plans that its scientific trustees had laid down, they were keen to support us. The Trustees fortunately made full enquiries from all the relevant persons, including Sir Wilson Jameson and Sir Edward Mellanby, to see if our proposed work fitted in with the national plan, and then offered the University at the end of last month £25 000 for purchasing and equipping premises for an Institute of Human Nutrition. I believe that there is a reversionary interest on the estate so the Trustees could not consider giving a recurrent sum but I think they were assured that this was likely to come from the University Grants Committee; they therefore stipulated that the University should provide £5000 a year for five years.

Yesterday Council had its first meeting of the term and I understand that it decided to accept the offer with gratitude and to purchase a house on which I secured an option. It happens to be extremely difficult to secure suitable accommodation in Oxford at the present time, and building a new Institute would be out of the question for at least five years. However, there is a large house near here, with $2^1/_2$ acres of land, which has been requisitioned by the Army and recently came up for sale. It is extremely suitable for the purpose, particularly in view of the land which will permit the Ministry of Health to establish its public health nutritional work in close association with the University's experimental work. I am of course taking up this question with Sir Wilson at once. The plan is to have a threefold organisation: experimental work on human nutrition done by the University, its application to public health being done in surveys conducted from Oxford by the Ministry of Health, and a Bureau of Human Nutrition which will I hope be common to both organisations, and be a place from which education and teaching can be conducted and to which Medical Officers of Health can come for advice. We hope to take over the requisition of the house from the Army as soon as the Army can find alternative accommodation, which should be towards the end of June. We shall take with us the benches and much of the equipment which we have here and which was of course purchased with the generous capital grant that your Foundation gave us.

A few weeks later Hugh heard from John Ryle that the Secretary of the University Chest had written to the Rockefeller Foundation, acknowledging that a Laboratory of Human Nutrition was to be established (if funds were available) and that Dr Hugh Sinclair would direct this. He further said that, from 1 August 1945 to 31 March 1946, the Survey accumulated a deficit of £2528.17.0; he was writing to the Foundation to ask if they 'should see their way to allotting any grant for the period mentioned above'.

More than a month passed before a response arrived: the answer was negative. The Foundation praised the Survey and Sinclair, and expressed pleasure in the fact that Oxford was converting the Survey into a Laboratory of Human Nutrition, and were sorry to hear that a deficit had occurred subsequent to the termination of their support. However, they could see no way they could cover the outstanding sum.

> 14 May 1946 — letter to Sinclair from the Secretary of Faculties, University Registry
> Dear Sinclair,
> (1) Will you hold yourself in readiness to attend a conference at the Ministry of Health on the subject of Human Nutrition which is to take place after the return of Sir Wilson Jameson at the end of the month.
> (2) A Committee of the general Board has instructed me to ask you to state in writing what staff and what accommodation you will need for the proposed laboratory of Human Nutrition as distinct from the Survey. It occurs to the Committee that the present staff consists mainly of survivors from the Survey rather than of persons suitable for the laboratory.

During all this negotiation in Oxford, Hugh returned to Germany several times, and was there when Mrs Fisher relinquished her duties with the Survey. Angela Hodgson, now Hugh's main administrative support, contacted him for advice on how to respond to the letter. Hugh failed to reply until after he returned to England.

> 27 May
> Your letter of 14 May has just reached me upon my return from Germany where I have been representing the Health Branch of the Control Commission on the Tripartite Nutrition Committee. Since the Committee has been working all day and travelling by night for the past fortnight, and since I have to submit a report tomorrow to the Controller-General, you will perhaps forgive me for not answering in detail: and circumstances which I shall mention below would in any case have made it difficult for me to do so.

He said he would hold himself in readiness for the meeting.

On the afternoon of 13 June a meeting was held with Sir Wilson Jameson in the Ministry of Health concerning the Survey. The Registrar, Douglas Veale, Howard Florey, and Hugh Sinclair were present to support the University's position. It was agreed that the University no longer wanted to conduct surveys but, instead, wished to establish a Laboratory of Human Nutrition; however, the accumulated data from the Survey were still useful, and steps should be taken to see that they were properly reported. Sir Wilson Jameson argued that the necessary analytical work should be undertaken in association with a University department in which research on human nutrition was being done. Several staff members were required for this work, and Douglas Veale and Sir Howard Florey agreed that the University would pay the necessary salaries, which the Ministry could then reimburse. The Ministry would also apply some of its own staff and resources to data analysis.

Accommodation for the work again appeared as a stumbling block. Summertown House was now empty and for sale. It was generally understood that the Local Authority would have no objection if the property was purchased by the University and used for nutritional research. Sir Wilson was asked if the Ministry would approach the Local Authority on the matter. In the course of the discussion, the Ministry agreed to pay the salaries of data analysis staff for five months: from 31 July to 31 December. They would also pay for accommodation, if necessary. Two Ministry employees would take part in the work. If called upon, the Ministry would also help in the matter of securing permanent premises for the new Laboratory of Human Nutrition at Oxford. A Note of the meeting reads, in part:

> The Ministry attaches value to surveys, and hopes to continue them with its own staff and in its own premises. Proximity to a University will be valuable but not essential. It hopes that some University, preferably Oxford, will establish a department for fundamental work on human nutrition. The purpose of such a department would be solely the advancement, and not the application, of knowledge; but the Ministry would hope by the methods usual in such cases to obtain its help from time to time on special problems, and would, of course, make use of its published results.
>
> As to the immediate problem, the Ministry agrees that much material, which may be of scientific value, has been collected in the course of the surveys already made in this country and on the continent. This material ought to be worked over and the results published in scientific journals. Although a good deal of the survey work was done in Germany and was not the responsibility of the Ministry of Health, that Ministry is prepared to take a broad view in order that the results may be published without loss of time …
>
> As regards Summertown House, if the University wished to buy it and use it for nutrition work, the Ministry would, if asked by the City Council for advice, urge that permission should be granted, but would not intervene otherwise. It could not keep the house under requisition and allow work on the survey to be continued there.

On 17 June the Registrar wrote to the University Secretary of Facilities, informing them of the meeting with Sir Wilson. He asked that the General Board advise him by 24 June as to whether or not they accepted the agreement. Six days remained after that date before Hugh Sinclair would be forced to vacate John Ryle's Institute.

Gaining support from the Ministry of Health had drawbacks. Dr Bransby, for example, now felt he had much greater say over how the data would be handled, as his employees were involved in the matter. Sinclair saw Bransby as an intruder, and often felt his requests and actions wholly outside the agreement between the University and the Ministry: one case of disagreement over the matter stands out. Soon after the 13 June meeting, Bransby informed Sinclair that he was working on a scientific paper, and asked to quote some of the Survey data in it. Specifically, he wanted to combine the results of the three war-time surveys conducted on lactating women. Hugh responded that the staff was re-calculating all their dietary results, and would prefer to send the data after corrected calculations were available.

In October 1946, the gift of £25 000 from the Wellcome Foundation was accepted by Oxford University. It was agreed that if the Institute for Human Nutrition should cease to exist, the house and equipment purchased with these funds would be applied to research in those sciences which contribute to the well-being of man and animals.

A Readership in Human Nutrition was established, and the occupant of that post was to be elected by a Board consisting of the Vice Chancellor, the Regius Professor of Medicine, a person appointed by the General Board of the Hebdomadal Council, an appointee of the Board of the Faculty of Medicine, and three persons appointed by the Board of the Faculty of Biological Sciences. It was clearly understood that the Reader would reside at least six months during the academic year within the University. He was to lecture and give instruction in Human Nutrition, and conduct research within the Laboratory of Human Nutrition. Hugh Sinclair was eventually appointed to the position.

After leaving Ryle's Institute, in late June, Sinclair and his staff moved into Summertown House and made what progress they could on the data analysis. In late November, the Oxford City Council decided against Summertown House being used as a laboratory, and Hugh's staff was to vacate the building before the end of January 1947: a building recently used as a hostel for labourers was offered as an alternative. Funds from the Ministry of Health for Survey staff and accommodation ran out on the last day of December 1946. Unfortunately, arrangements had not been made to charge the University for the extra month's rent. Feeling certain that the Ministry of Health would cover the cost, Hugh personally guaranteed meeting the expense. In the end, he accepted the offer of the Nissan hut near Churchill Hospital, and on 28 December, he and his staff moved in.

Some decades later, when describing this period in his life, the story differed considerably from the facts presented here. In *Sinclair*, the memoir published by the McCarrison Society, Hugh stated:

> The war ended, the University professors all fought for money from the University Grants Committee and the promise of a department of Human Nutrition, formally made during the war by the Regius Professor of Medicine, Sir Farquhar Buzzard, was renounced, Buzzard having retired from the Regius Chair in 1943. However, the UGC sated it would give the recurrent sum needed, and in March 1946 the University gave me ten days to raise £25 000 for capital expenditures. I did it with one day to spare: Sir Henry Dale, as chairman of the recently created Wellcome Trust, offered the University a Wellcome Institute of Human Nutrition. The Vice-Chancellor summoned me on 23 April and told me the University Council had been advised by its Joint Scientific Advisory Committee (Peters, Florey and Professor J. H. Burn) that the offer should be refused because in ten years' time there would be no human nutritional problems to study and 'the Wellcome Institute would be a white elephant'.
>
> I was then, I believe, the only person in the world interested in coronary heart disease as a problem of EFA, on which I had done some work during the war. The Vice-Chancellor himself regretted this decision, and I told him I would form my own Institute, work for ten years to try to show there were problems of EFA and disease of civilisation, and then review the situation ...
>
> Exactly ten years, almost to the day, after the Vice-Chancellor had told me that in ten years' time there would be no human nutritional problems to study, I published in the Lancet ... its longest and rudest letter suggesting that a relative deficiency of EFA is the main cause of various diseases of civilisation ...[1]

Diaries and correspondence available to this biographer provide no record of this exchange between the Vice-Chancellor and Sinclair, and there is very limited evidence that, at that time, he would have chosen essential fatty acids as the nutrient that he would champion.

Hugh believed Oxford University had let him down, having failed to accept his comprehensive approach to nutrition research, and limiting the scope and scale of his endeavour. There is some truth in this. However, one must acknowledge that Farquhar Buzzard, Rudolph Peters, and others tried to support and to protect Sinclair, even during those most difficult times when they were protecting him from himself.

Nissen huts and honours received

It was 1947: Princess Elizabeth Alexandra Mary, the elder daughter of King George VI, married her third cousin, Philip, the Duke of Edinburgh; Captain Chuck Yeager flew the Bell X-1 rocket plane at supersonic speed; and India gained her independence from British rule. Perhaps telling more of the time, the fashion house of Dior enjoyed instant success when it introduced the long and swirling skirts of the New Look, flaunting the extravagant use of fabric in the face of lingering rationing.

Soon after the war, old family resentments re-established themselves at Lady Place, and the Sinclairs resumed their peculiar way of living separately together. Catherine and Rosalie had the house and the servants to themselves most of the time, but at the weekend, Hugh tried to get home for at least a day, and John, who now worked in the War Office in London, also came home as frequently as possible. Notes were still pushed under doors, and each of the protagonists maintained a personal sitting room for privacy.

Hugh's personal financial problems increased. His travel and life-style during the war had drained his private capital, and much of the money he earned was now quickly spent entertaining influential friends who might help finance his Institute of Human Nutrition. Other funds were spent expanding his collection of rare books and maintaining Lady Place and its ground. But with financial help from his mother, Hugh filled the rooms and gardens of Lady Place with items purchased at auction and house sales. These stood proudly beside fine, inherited pieces, and Sinclair did nothing to dissuade people from assuming that all were part of a wealthy past that deserved a fine future.

By the late 1940s, the plan Sinclair had developed almost a decade before for the study of B vitamins and diseases of the nervous system was expanded and re-written as a proposal for an Institute of Human Nutrition. In this plan, nutritional surveys as previously conducted played an important role. Next, he needed a great library. Beginning with his own books and his growing collection of reprints, he would build a great library from which students, government officials, medical clinicians, scientific investigators, and members of the public could all obtain information about food and its relevance to health. This information would not be limited to the composition of food, but also to the economics of food, agriculture, processing, and social behaviour affecting food choices. Sir Robert McCarrison's gifts of books and charts and pictures would be the basis for a museum of exhibitions pertaining to nutritional disease.

With money from the Wellcome Foundation the University of Oxford established the Laboratory of Human Nutrition, with Hugh Sinclair as its Director, but it was decided that the apparatus and equipment which formerly belonged to the Oxford Nutrition Survey should be left behind in the Department of Biochemistry. Animosity between Sinclair and important members of the University community was palpable by this time, and his personal and scientific belongings were treated with little regard when they were removed from the Department of Biochemistry. According to Sinclair, on the day he and his staff moved from the Department boxes of specimen slides and files were haphazardly emptied and the contents scattered and lost. Data and samples were destroyed. Someone on his staff protected his microscope, but he claimed that much of the remaining equipment was apportioned among members of the department. What he could save was moved to his rooms at Magdalen College, or to his personal laboratory at Lady Place. The day of the move was marked by animosity and Sinclair's sense of loss.

Money was available to start again, but the task was Herculean. He was Director of a potentially powerful organization that could change the future of research into human nutrition, but in reality all he had was a dream and two Nissen huts.

Over the cold weeks of winter, Hugh and what remained of his staff moved into two Nissen huts in the grounds of the Churchill Hospital. Having been formerly used by the American Forces as hospital units, they were ill equipped for basic scientific research. (Somewhat later, a prisoner-of-war washhouse was converted into an animal house, but at the start all animal experimentation was done at Lady Place.) Sinclair and colleagues cleared the huts, broke up toilets and basins with sledgehammers, and installed basic laboratory benches, which he purchased from the University of Birmingham with his personal funds. Sinclair lamented that the huts were too small to house volunteers for nutrition experiments. The conditions were primitive, and embarrassed Sinclair when visiting scholars arrived. When Brian Lloyd later referred to the Laboratory, he described the Cinderella-like characteristics, typified by its mode of space-heating: 'Dickensian coke-burning cast-iron stoves with tin-plate chimneys wriggling their dusty way up through the roof.'

In fact, cast-iron stoves were the normal way to heat Nissen huts. During the early post-war years many scientists at Oxford and other British universities were forced to work in Nissen huts, and in even less attractive surroundings. But as Sinclair saw it, his new work environment was totally inappropriate for his important work on human nutrition. It was an insult to him, and to his subject.

Once work in the Laboratory settled down, and he had secured more staff, Sinclair returned to work on the role of trace metals in diseases. He was particularly interested in multiple sclerosis. This led to collaborative work with Dr Ritchie Russell of the Department of Neurology, and with R. B. Bourdillon, of the Medical Research Council. Sinclair soon developed a need for new and better equipment.

At about this time John Marshall D.Phil, a Magdalen College chemist, developed a means of conducting spectrographic analysis of trace elements in tissues, thus opening new avenues of investigation. This methodology excited Sinclair, and for a time it appeared that trace elements would become his central interest. Unfortunately, that interest lagged. Hugh Sinclair was never one to spend great amounts of time investigating a problem: he became bored as soon as he thought he saw its solution. Also, he had lost most of his most dependable staff members. Brian Lloyd had returned to the University after returning from the survey work in Germany, and planned to take his Physiology Finals during the coming year, 1948. Mrs Fisher had accompanied her husband to America, and Angela Hodgson had left to get married. Sinclair realized that, until he could find new money to expand his research and administrative staff, he should apply himself to building the Institute's library

Over the years Sinclair collected, bought, borrowed, and begged for scientific reprints on a wide range of subjects pertaining to human nutrition. He devised a remarkable system to cross-reference these. As reprints were obtained, Sinclair or a trusted member of his staff wrote out a description of each on a slim strip of index card, which was then assigned a code for identification and cross-referencing. Before the end of his life, files containing tens of thousands of these little cards were accumulated. Many of these were later lost, which was regrettable as they not only detailed the historical foundations of nutrition, but they reflected how Sinclair conceptualized the subject.

The nutritional surveys in Germany dragged on. In early April 1947 Hugh received a letter from Dr Harriette Chick (who two years later would become Dame Harriette Chick in recognition of her work on human nutrition). She would soon visit Germany, she said, and wanted to see some of the survey work underway; would Hugh give her permission to visit some of his units? He gave his permission quickly and graciously, and to facilitate matters, sent her a letter of introduction to the Administrative Officer of the Düsseldorf Survey team. In the cover note he wrote:

> Advice is unfortunately just a little difficult at this precise moment because of a rather delicate situation with the Director of Public Health Branch … He has been most anxious to show that there is almost no nutritional problem in the big towns of Germany, and during the last tour of the Combined Nutrition Committee Dr Youmans and I were in very sharp conflict with him over the incidence of nutritional oedema in Hamburg …

A few weeks later Hugh received a hand-written note from Dr Chick:

> When in Germany recently I visited the Survey Unit at Düsseldorf on two occasions. The first time things were very hectic and all members of the staff were very busy preparing for the visit of the Tripartite Commission which was imminent. On the second visit, we spent longer and saw quite a bit of the work of the Unit and were presented with copies of the Report they prepared for the Tri-Part-Com. We were much interested to see the results of the enquiry showing to what extent the official caloric and food rations were actually being honoured! It appeared that the ordinary adult ration was being provided only to the extent of about 1100 calories. Dr. W, who was showing us around demonstrated her own ration book, with the unused bread coupons, which could not be honoured for the existing month and the previous month.
>
> At the same time it was quite obvious that the people thronging the streets were not subsisting on 1100 calories daily and of course we heard and experienced a good deal about the ways and means that more original and more fortunate people use to get extra food. That is undoubtedly true and the regular barter of things and cigarettes between the people and

farmers is known to everyone. At the same time there are many classes who get no advantage that way — people who live alone and have full-time jobs and no one to forage for them, old and invalided people and the very poor. At [one clinic] … there seemed to be about 40 fresh cases of light hunger oedema a week and those must be an index of many more who did not get to hospital. I met several people among my own acquaintances, e.g. hard-working High School Teachers, women living alone, who suffer very badly from lack of proper food and often even more from lack of appropriate housing. For if they are fortunate enough to retain a small flat they usually have to take in many unsuitable 'lodgers'.

These are only the impressions of a very short period in Germany, but we tried to make the most of our opportunities. I am glad to be able to thank you once again for your letter of introduction and to tell you how very … hospitably we were received …

On 30 May, Sinclair responded to Dr Harriette Chick, agreeing with her observations and saying that he thought the nutrition survey in Germany would soon be reorganized. He returned to Germany once more, at the end of the year, but had little authority. In later years of his life he wrote and talked more of survey work in the Netherlands than of that in Germany, where he spent more time. A memorandum written by a staff member of the Rockefeller Foundation furthers our understanding of the situation:[1]

> … tea at Magdalen College, with Dr H. M. Sinclair: Then on to see his department of Human Nutrition (The Wellcome Foundation) which has recently been separated off from Ryle's Department of Social Medicine and is now housed in two Army huts at the Churchill Hospital … S. has a great collection of statistical material, some of which has recently been put in shape for publication; unfortunately, the completed manuscripts were lost last week when S. suffered the theft of his baggage in London!!
>
> The Mobile Laboratories used in Germany and Holland are still on the Continent, presumably in Düsseldorf; 4 nutrition teams are at work in the British zone of Germany, and some $8000 worth of laboratory equipment is in the various laboratories erected by S. during his nutrition surveys in Germany. S's textbook on Nutrition, which he expected to publish in 1946, has not yet been completed; however, he is beginning to get his statistics into shape, to determine the results of his surveys; these have been reduced on to a quarter of a million cards, to be analysed by his Hollerith machine.

For his work in Europe, Hugh Sinclair received both the Order of the House of Orange from the Government of the Netherlands, and, as already noted, the Medal of Freedom with Silver Palm from the United States of America. The citation that accompanied the Medal read:

Dr Hugh M. Sinclair, British Civilian, for exceptionally meritorious achievement which aided the United States in the prosecution of the war against the enemy in Continental Europe, while serving as Honorary Nutrition Consultant and Coordinator of Research in the British Zone, Control Commission (British element). His outstanding professional knowledge and ability, his excellent scientific training and his energetic and judicious activities as a representative of the United Kingdom has materially contributed to the military effort as a whole, both in the liberated countries and occupied Germany. He further contributed to the detailed plan of operations by stimulating a spirit of mutual scientific cooperation, interest and assistance between the British, French and American public health officials, which spirit has prevailed throughout the entire public health program to the present date, thereby meriting the praise and recognition of the United States. Official Headquarters European Command

CHAPTER 15

Reports on the Western Netherlands and dietary fat

In the post-war climate of political idealism, the British Government sought to heal the social woes of the country. The National Health Service, initiated in 1948, was the greatest legacy of this time, setting out to provide medical care to all, from the cradle to the grave, free at the point of delivery. To meet changing demands, medical education at Oxford accelerated its transformation, and Hugh Sinclair was asked to participate. In February 1949, Leslie John Witts, the Nuffield Professor of Clinical Medicine at the Radcliffe Infirmary, and a supporter of Sinclair's, had an idea for remodelling the medical lectures, devoting the first term to dietetic, metabolic, and renal disease. He wrote to Sinclair, asking him to comment on the plan.

It would have been politically desirable for Hugh to help define the new curriculum; instead, he failed to comment on Witts' ideas until April. He had needed time to entertain important visiting scientific experts, he explained, and other academic matters had required his attention. Sinclair assured Witts that he would be delighted to help in any way with the new curriculum, and that the Laboratory of Human Nutrition had a good selection of slides to support any lectures they undertook. However, when it came to the matter of content, Sinclair did not respond directly to Witts' proposal, but presented his own views on the reorganization of medical education.

As shown in his work a decade before, Sinclair believed the understanding of human nutrition was fundamental to the understanding of human illness; therefore, it was fundamental to medicine, and should be taught at all stages in the course of a student's education — not just one term.

For instance, [he wrote] he [the student] should learn quite a lot while he studies Botany. When he does Physiology Finals, the clinical aspects of certain nutritional deficiencies are quite properly briefly mentioned by way of illustration — but only briefly, of course. Therefore, when he studies Clinical Medicine, it seems to me reasonable that 5 or 6 lectures should be specifically devoted to nutrition … Clearly nutrition will come in over and over again during the lectures on metabolic disorders … Rickets and infantile scurvy presumably are covered in paediatrics; toxaemia of pregnancy in obstetrics; pre- and postnatal operative dietetics … in surgery.

Sinclair believed that the social aspects of nutrition (such as the effects of culture on dietary patterns, environmental influences, effect of religion, education, economic circumstances, and rationing) would best be dealt with in lectures on public health. To this end, he willingly participated in a course offered by John Ryle and his staff, entitled 'The Nutritional Requirements of a Working Population'.

At about this time, the Oxford bacteriologist Arthur Duncan Gardner became the Regius Professor of Medicine at Oxford, and therefore the leader of medical education within the university. His appointment did little to support Sinclair's views on human nutrition, as Gardner had scant interest in the subject, and probably saw it as a part of Rudolph Peters' domain. Unfortunately, tension between Sinclair and Peters remained high, and in February, Sinclair noted in his diary:

State of some confusion — papers flying around. Prof. said that I glowered at him whenever we met as if he had done me dirt; I said that I regretted we never seemed to meet! … Prof. then stressed the need for immediate and early publication of our results; we had had a long run and very little to show for it. He was most anxious we should get work done and published as soon as possible.

The University acquired property as it became available. One possible solution for housing the nutrition studies was Nuneham Park, a large property close to Oxford on the Henley road. It consisted of several grand buildings connected by passageways, and was big enough to house Hugh's information bureau, provide quarters for volunteers used in nutrition experiments, and accommodate a large animal research facility. However, the size of the property was its greatest drawback. University authorities wanted to give Hugh's Laboratory, or Institute, as he referred to it, a wing of the main building, housing other programmes in the remaining sections. Hugh's previous experience of sharing space with the Institute of Social Medicine made him cautious, and he thought it best if he could choose the other tenants for Nuneham Park. Unofficially, he wrote to the Ministry of Food, suggesting they might want to rent a portion of the building from the University. He received little encouragement.

In the Ministry's view, if Hugh planned to pursue 'fundamental studies in human nutrition', and that was the reason for the Ministry being invited to rent space at Nuneham Park, this fell within the responsibility of the Medical Research Council, not the Ministry of Food. If, on the other hand, Sinclair wanted to conduct surveys on food consumption, this fell under the authority of the Ministry of Food, and would be done under interdepartmental agreement with the Ministry of Health. If his intention was to determine the effect of nutrition on human health, he was told to contact both Mellanby and Jameson, and see what cooperation they were willing to provide.

While the search for accommodation continued, so did the struggle to cope with the masses of information collected by the Oxford Nutrition Survey. Thousands of data cards remained to be corrected, punched, sorted, and tabulated. The work would slowly grind on; eventually most would be completed at the Ministry of Food's facilities in Stanmore, Middlesex, where the staff took on the task with little enthusiasm. The volume of work was greater than originally estimated, and questions were soon raised as to who would cover the mounting costs: the matter went as far as the Treasury. While the debate over funds continued (between the Ministry of Food, the Ministry of Health, and Oxford University) data management at Stanmore fell victim to the demands of seasonal crop variations, changes in Ministry of Food personnel, and wrangling over budgets. At last, one kind woman, responsible for work on the Hollerith machines for the Ministry of Food, slipped Sinclair's work in around her other duties on a no-charge basis and, to the extent that it was possible, the work was concluded after transferring some of it to the Potato and Carrots Division of the Ministry. The few people left in Hugh's staff struggled to complete the analysis of the sorted and tabulated information; meanwhile, he found time to lecture on survey methods and instruct others in the use of the rapid survey data system his team had used in Düsseldorf and Germany.

By the end of the surveys, Sinclair accumulated above a quarter of a million punched Hollerith cards filled with nutritional data. Information from this source was in great demand, but no one Sinclair worked with was able to cope with the size and complexity of the analytical tasks involved. Dr Andrew Neil, of the Institute of Health Science, the University of Oxford, has pointed out a great irony in this situation. The Colossus computer that was built during the war at Bletchley Park, about twenty-five miles from Oxford, would have been capable of analysing the Survey data, but this vital tool was deliberately smashed after the war on orders of the War Office.[1]

The international team of nutrition experts who had worked in the Netherlands was anxious to complete its report, to be known as *Malnutrition and starvation in the Western Netherlands*. The report and its accompanying book of appendices were finally published in 1948 in The Hague by the General State Printing Office, and paid for by the Government of the Netherlands. A facsimile of one of Sinclair's Hollerith cards appears in the report's Appendices, and shows the variables collected and how they were positioned on the card.

Most of the Oxford work reported was produced by Brian Lloyd [wrongly listed in the final publication as Mrs B. B. Lloyd] and his three colleagues: Miss C. M. Wood, Mr E. V. Quinn, and Mr K. B. Taylor, a Magdalen medical student and pupil of Sinclair's. In addition, there is a statement acknowledging the Oxford Team's preparation of the mobile laboratory and their analysis of samples flown back to England. Typically, Sinclair's contribution to the publication fell behind schedule and was late.

The task of completing the report was complicated. Independent, national teams of investigators had used different methods to collect and analyse data, making comparative observations almost impossible. The Editorial Committee agreed it was better to include at least a part of each group's report, more or less as it was submitted. In the end, it was decided to prepare a strictly factual document with general conclusions, as opposed to the comprehensive analytical volume originally planned. In its final form, the report describes the seriousness of the famine in the Netherlands, but fails to serve as a scientific guide to others confronted with the starving victims of war. The Introduction to the report states:

Perhaps the most striking impression carried away by those who attended the meetings at Brussels and Eindhoven in February 1945, when the original plans were drawn up for dealing with the rapidly deteriorating situation in Western Netherlands, was the deplorable paucity of knowledge regarding the treatment of individuals suffering from grave starvation. In living memory millions had died of starvation in catastrophic famines affecting the Far East and the USSR, yet none of the experts who were consulted could say with assurance how such a famine should be met ...[2]

The gravity of the situation in the Netherlands is best conveyed in the Appendices, which include an open letter from a group of unnamed Dutch physicians to the Reichskommissar of the Occupied Area of Holland. It is a brave cry against the atrocity of starvation. In part, it states:

> Once again the Dutch physicians feel obliged to address you. You will realise that even without signatures this letter expresses the opinion of thousands of physicians, who in December 1941 made clear what they considered to be their professional duty, and who since then have, by their firm will, left no doubt as to their attitude ... Certain measures taken by the occupation authorities ... constitute so serious a threat to the Nation's health that they cannot remain silent.
>
> They hold your Administration responsible for the dire shortage of even the most necessary foodstuffs ... The Occupying Authorities are to blame for these conditions ... The occupiers attempt to blame the Dutch organisations, but this charge must be rejected, as every measure taken by these organisations to improve the food situation has been opposed ... All this makes it increasingly difficult for Dutch physicians to maintain the human attitude of their profession towards the Germans ... You must recognise that, if in future you wish to call upon the help of Dutch physicians, they will give their assistance solely because they respond to this sense of professional duty; they do not wish to degrade themselves to the same level as their opponents ...
>
> They protest against the abuse of power over the Dutch people by the Occupying Authorities. They consider the pronouncement of the German Führer that his people are the bearers of a higher culture to be brought in ridicule by events. Finally, they point out that by continuing to maltreat the Dutch people the German Occupation Authority lose every claim to the respect of humanity.[3]

The authors of the report on the Western Netherlands agreed general conclusions, and a statement of principle was included:

> The Editorial Committee wishes to record its considered opinion that there is no justification for the ruthless sacrifice of a civilian population in such circumstances and that the United Nations should devise a Convention of international scope of which the object would be to protect civilians

subjected to an occupying power from the suffering and grave injury to health as a result of inadequate nourishment.

No similar report was generated from the survey work in Germany, although it held more promise because of the size of the effort and the geographic mix of locations studied.

During the surveys in Germany, as Hugh and his team moved from place to place their data and equipment were also relocated. Inadvertently, documents were scattered between Düsseldorf, Hannover, and Hamburg. Sinclair knew that unless his notes and original data could be retrieved, the British work would be useless. When the surveys were almost completed, a representative of the American Government wrote to Sinclair, saying that the United States Army was planning to send a mission to Germany to gather the data their survey teams left behind. With Hugh's encouragement, Bransby, from the Ministry of Health, joined the American effort, and travelled to Germany intending to retrieve a large number of Sinclair's Hollerith cards, files, charts, graphs, and a series of notebooks containing details of estimations completed. Before Bransby left England he received a note from Sinclair:

> … I very much hope that you will be able to obtain <u>all</u> the records and data since we have information that a number of the clinical signs were wrongly recorded and that a number of the biochemical estimations were erroneous; and there are errors in the body-weight survey data … it is absolutely essential to go back to the original data …

Bransby returned to England with as much information as possible, but found little stomach within his Ministry to pour more resources into the study of details from a past war. Large quantities of data were packed into cases and put into storage as money for analysis ran out. The collaborative report of work done in the Western Netherlands had been criticized because the research was poorly coordinated and few conclusions could be drawn; but, when compared with the chaotic situation surrounding the Control Commission nutrition surveys in Germany, the earlier work appears highly successful.

Nonetheless, Sinclair began working with Bransby on reports and papers based on the available survey data, including one containing a section on anthropometric measurement. For this, Sinclair designed a set of callipers useful in the rapid appraisal of the body from anatomical landmarks. Physical form and its changes during the life process and under the influence of nutritional change had interested Sinclair since boyhood. (Several series of photographs of himself, taken nude at various ages, along with detailed measurements for comparison were found in his archive.) Perhaps influenced by this work, or perhaps as a coincidence, about this time Hugh Sinclair joined Alastair Hamish T. Robb-Smith, the Nuffield Reader in Pathology, and a friend of Hugh's, in writing *A history of the teaching of anatomy in Oxford*, published by the Oxford University Press in 1950.

Despite his prodigious workload, Hugh followed his father's advice: he kept a certain balance in his life, and often found time to enjoy himself. The Athenaeum and the Worshipful Society of the Apothecaries were his London bases for entertaining, which he did frequently and with great pleasure. In Oxford, his well-furnished rooms in Magdalen College often rang with the laughter of friends, and Hugh continued to enjoy entertaining at home, organizing impromptu afternoons of cricket, croquet, and the occasional tennis match. Periodically, he took advantage of his lifetime membership of the Marylebone Cricket Club, and also managed to arrange cricket matches for various groups within Oxford. After Hugh's death, one of his students recalled, with gratitude, that Sinclair inspired him to follow a simple rule: work hard, but find time for physical sport and friends.

Sinclair certainly had no fear of hard work, but in his professional career, his efforts were damaged by a lack of continuity. Sinclair would work on a project until he thought he saw its answer, and then became bored with the effort. Now, as he approached the age of forty, he urgently needed to find a focus for his work. All the university committees reviewing his work stressed the need to concentrate on the fundamental biochemical problems of diet and disease. But which problem should he choose? Which issue was complex enough to sustain his attention for an extended period of time?

World War II marked the time when the greatest advances were made in curing infectious diseases. Now, scientists turned their attention to the prevention of disease caused by social patterns (including diet) and the environment. The importance of dietary fat was one issue under debate. Is dietary fat a cause of serious illness, and should it therefore be avoided? Or, is it a substance we must consume to maintain normal growth and health? Its general importance was already known: fat makes food pleasant in the mouth, it is the most concentrated source of energy in food, and it is necessary for the absorption of vitamins A, E, and D during the process of digestion. But what about the need for fat itself? Could it be that certain forms of fat have biological significance other than as highly efficient means of storing energy?

About this time, Magnus Pyke, now a Scientific Advisor to the Ministry of Food, reported that the amount of fat in the natural diets of different cultures varied widely. Pyke questioned whether, on the basis of epidemiological evidence, dietary fats influence longevity. In a review of the literature,[4] Pyke noted that the consumption of a high-fat diet was found among many populations without evidence of impaired health. The Maasai obtained 46 per cent of their calories from fat; Finnish students (43 per cent), Dutch high-school teachers (40 per cent); and middle-class Scots (40 per cent): even higher intake levels (51 per cent) had been recorded among Eskimos.

It had also been observed in Westernized cultures that the percentage of calories consumed as fat increases with rising income: in 1936, Boyd-Orr found that the poorest groups studied in England consumed 28 per cent of their calories as fat, while the comparable figure for wealthy groups was 38 per cent, for example.

Paradoxically, Pyke pointed out, many populations live healthy lives on low levels of fat intake; many Oriental diets were examples. In 1905, Oshima concluded that only 7 per cent of the total calories were from fat in the diets of professional and business Japanese, and heavy labourers consuming more than 4000 calories per day received only 5 per cent from fat. In China, figures ranged from 8 per cent among the poor and up to 19 per cent for middle class families. Pyke wrote:

> While the general well-being of, say Italians, as indicated by figures for death rates, shows that a low proportion of fat is compatible with a reasonable state of health, there is no direct evidence of experiments on human subjects to fix the magnitude of the optimum intake.
>
> It appears, therefore, that although the Western European likes to derive 30% or more of his calories from fat and, if permitted by economic circumstances to do so, will increase his fat intake still higher, there is no scientific evidence to show that the proportion of fat in the diet is within very wide limits of physiological significance as a nutritional factor in adult human diet.

Unlike proteins and carbohydrates, the two other classifications of major nutrients, the role of dietary fat in human development and health remained unclear during the 1940s. In part, this difference was due to the primitive technology then available to study fats and lipids. Animal fats and plant oils were recognized as major components of food from the time of the Egyptians. However, they were impossible to examine scientifically until 1815, when the French chemist, Michel Eugène Chevreul, defined their molecular structure.[5] Eight years later, Chevreul published a classical examination of fats, describing the most common form as a combination of glycerol and fatty acids.[6] It would take almost a century and a half for knowledge to advance much beyond this point.

The basic structure of the most common form of fat is a 'triglyceride', and consists of three fatty acid molecules attached to one glycerol molecule. Glycerol is a stable substance with little biological activity, but fatty acids can have significant powers. Fatty acids are made up of a chain of carbon molecules studded with hydrogen atoms, and vary from one another in two main ways: the number of carbon atoms in the chain, and the extent to which hydrogen atoms occupy all the locations on the chain available to them. Each carbon atom can hold, or attract, two hydrogen molecules. When the hydrogen complement is complete, the fatty acid is referred to as being 'saturated'; when hydrogen atoms are missing, the fatty acid is referred to as 'unsaturated'. Saturated fatty acids have little or no biological activity: unsaturated fatty acids can be highly active, depending on their length and the number of hydrogen atoms missing. On a carbon chain, when hydrogen atoms are missing, adjacent carbon atoms form a double bond. The placement and the number of double bonds within each chain determine the exact nature of its biological activity. Saturated fats are rigid, and are found in very high concentrations in butter, and tallow. In contrast, because double bonds are flexible, unsaturated fatty acids form oils. Saturated fats dominate in animal tissues; unsaturated fats are primarily found in plant products, such as seeds and nuts. With time, unsaturated fatty acids would become very important to Hugh Sinclair.

Early chemists interested in fatty acids found it very difficult to separate and quantify different species of molecules. Methods used were based on the fact that, when prepared in solution form, molecules of fatty acids with differing length and saturation travel at differing rates over prepared media — such as a treated glass plate, or chemically packed column. These methods are slow, and separation is poor, making research difficult. The answer came with the development of gas–liquid chromatograph. This equipment uses the same principle, but heat and gas separate the fatty acids with great rapidity along a core of prepared medium. As the molecules separate and pass an end point, a mechanical tracing is made that allows identification of specific molecules, and, by calculation, an estimation of their relative amounts. There is one other characteristic we should remember about fatty acids: saturated varieties are stable, whereas unsaturated molecules are highly active and easily enter into chemical reactions. Combining with oxygen, or oxidation, is one of the most common reactions, and this is why cooking oil exposed to the air becomes rancid. Unlike vitamins, which are biological catalysts and remain physically stable during biochemical processes, fatty acids change form at speed. Therefore, a series of changes, or a 'chain' of metabolic events, can happen before accurate measurements of these reactions can be made.

By the late 1940s a stream of visiting scholars began pouring through the doors of the Nissen huts occupied by the Wellcome Laboratory of Human Nutrition. Professor V. Ramalingaswami, from India, a close friend of Robert McCarrison's, was one such individual. Sir Robert agreed with Sinclair's comprehensive approach to the study of nutrition, and arranged for the young Indian doctor to join the growing international team of biochemists and clinicians working in the huts outside the Churchill Hospital. Ramalingaswami was already an accomplished scientist. He had worked in the Nutrition Research Laboratories, Coonoor, South India, and before coming to England had described the symptoms of the protein-deficiency illness, kwashiorkor, in several Indian journals. He had also described the symptom of nutritional diarrhoea due to vitamin A deficiency. He was training in pathology in Poona, at the Armed Forces Laboratory of Pathology, when he received a telegraphic message saying that he had received a place in Sinclair's laboratory.

While at Oxford, Dr Ramalingaswami completed his doctoral research degree: his subject was the effects of deficiencies in vitamin A, essential fatty acids, and pyridoxine (vitamin B_6) on the production of skin lesions. This work was interesting for two reasons: first, it was the earliest published work on essential fatty acids from Hugh's laboratory; and second, the role of fatty acids was compared with that of, and in combination with, other nutrients.

Hugh Sinclair is said to have been reserved towards the Indian scholar at first, but soon invited him to both Magdalen and Lady Place, took him on clinical rounds, and motored him to meetings of the Nutrition Society. By Ramalingaswami's account of that time, we know that Sinclair's extensive collection of scientific reprints both facilitated his work and broadened his scientific understanding. In his contribution to the memoir, *Sinclair*, he wrote that Hugh was 'ahead of his time ... marshalled epidemiological evidence from data on dietary intake of fats to establish a relationship between the two', and was '... a physician scientist of the highest distinction'. 'Courage and candour characterized his nature ...' He praised Sinclair, saying that he had: 'nurtured my growth and development. The sharpness of his intellect, his analytical skills, the depth of his understanding of human biology and his mastery of literature and history, make him outstanding as a rare and exceptionally gifted scientist. Biochemistry was his forté but he [was] also a fine physician.'[7]

While working in his Nissen huts, Hugh was expanding his circle of influential friends and wanted to make certain Hans Krebs was among them. It had not escaped Sinclair's notice that the Sheffield scientist was a strong candidate to replace Rudolph Peters as Professor and Head of the Department of Biochemistry when he retired. Sinclair made certain they exchanged greetings at scientific meetings, and that Krebs was entertained and flattered whenever possible. '... It has been my pleasure lately to sit rather often on Committees with you,' Hugh once wrote to Krebs, 'and to experience — as others have done over long periods — the wise advice you give backed by scientific knowledge ... I often wonder how you have found time for so much careful scientific work ...'

During the late months of 1949, another important named appeared in Sinclair's files: Dr Lawrence W. Kinsell, of the Division of Medicine, University of California Medical School, San Fransico, California. Sinclair wrote to Kinsell, requesting a reprint of a recent paper on the clinical aspects of essential fatty acids. This was the start of a friendship of like-minded people that would carry Hugh through difficult times ahead.

At home, Hugh's family maximized their income from Lady Place wherever possible, and increased the number of garden and flower shops they owned to sell their goods in and around Sutton Courtenay. John Sinclair still worked in London at the War Office, but had suffered his first heart attack, and was now unable to do much without help. Rosalie maintained the property as best she could, but Hugh had to do more than usual, and resented the fact that the business consumed much of the time he had previously enjoyed spending with friends in Oxford and London.

CHAPTER 16

Biochemistry, Hans Krebs, and essential fatty acids

Hugh Sinclair was appointed the first Reader in Human Nutrition at the University of Oxford in 1951; he could reapply for the post in 1958. Although tainted by the sour history of the Survey and burdened by limited funds, he began his new position determined to make nutrition an important and independent science within the Oxford community. To help achieve this, he secured staff for his Laboratory from every possible source, putting each person to work using the scientific skills he or she had already mastered. True to his concept of programme design, their topics were diverse, involving both clinical and experimental work.

Ideas were picked up and put down. Some would have led nowhere, while others, given a little time, may have had profound consequences. One of the more promising of the ideas that failed to get beyond the early stages of development was the *Atwater*, a nutritional measure named in honour of the American scientist Wilbur Olin Atwater. In the late 1890s, Atwater meticulously identified and described the energy (or caloric) content of protein, carbohydrate, and fat. Hugh Sinclair's *Atwater* was an attempt to find a universal system to measure nutritional value. It was described only once, in very brief paper published in 1949 in the *British Journal of Nutrition*, although it was mentioned in other publications. The idea was based on a concept, developed in the late 1920s, of a common system by which one could quantify all the nutrients in a food and relate them to some agreed standard requirement. The number of *Atwaters* in different food would vary. For example, if the total requirement for nutrient X was 12 *Atwaters*, an individual would need to eat sufficient portions of the right foods to supply 12 *Atwaters*. Hugh's idea was to substitute one standard measurement, and thereby eliminate the confusion of unrelated weights and measures that plague the subject of nutrition: milligrams, international units (i.u.), calories, and so forth. As a baseline, Sinclair used the nutritional requirements of a moderately active woman, as these, he said, were closer to the average requirements of the population as a whole. The total of all nutrients needed by such a person was calculated for seven days, not a single day, to even out fluctuations in both intake and metabolic demand. The sum total for each nutrient, whether measured in milligrams, grams, or i.u., was determined and this sum was then divided into 1000 equal parts. With this tool, individual foods could then be described in terms of how many *Atwaters* they contributed to the total nutrient requirement of an individual. A food containing little protein, such as an apple, would have a very low *Atwater* number for protein, but a slice of ham would be assigned a much higher value. Conversely, the apple would have a higher *Atwater* for carbohydrate than the ham, and so on. By looking at a label on a tin or bottle, a shopper could see at a glance the extent to which a specific food would fill their nutritional requirements. As example, according to Sinclair, 100 grams of 'summer milk' contained 4 *Atwaters* of calories, 6 *Atwaters* of fat, 23 of calcium, 14 of phosphorous, and 17 of vitamin A. (Respectively, the accepted measurements used to describe these same nutrients were 65 calories, 3.6 grams of fat, 120 milligrams of calcium, and 100 i.u. of vitamin A.) It is easy to see how the *Atwater* would be an easier means to judge the nutritional value of a food. Further, the *Atwater* contained no decimal points, making interpretation even easier.

Changing knowledge about nutritional requirements could be taken into account by altering the value of the *Atwater*, without confusing the consumer. In the same way, different *Atwaters* could be devised for populations with special needs: expectant mothers could be told they require more than 1000 *Atwaters* of folic acid during the first months of their pregnancy, for example.

The concept of the *Atwater* was never developed, but it is easy to see how it could be useful today, when consumer information about nutrition is becoming ever more complex, and ever more important in the fight against illness.

Hugh and the people working with him published on a range of subjects, and although he allowed each to develop their own ideas, he would inject a measure of his own wit, along with sound clinical advice, whenever possible. As a case in point, in a paper titled 'Oral manifestations of deficiency diseases', Sinclair wrote:[1]

> One of the greatest adornments of the medical profession, Frances Rabelais, said that the boy Gargantua, between the ages of three and five, always looked a

gift-horse in the mouth. A patient should be similarly treated. Careful in spection of the lips, tongue, buccal mucosa and gums may give early indications of nutritional deficiencies; at a glance the simplest and most highly differentiated tissues may be seen, and since these relatively unprotected tissues are constantly insulted by various forms of trauma, they tend to show structural evidence of deficiency before changes occur in other regions.

Dr Ancel Keys, an American scientist, joined Sinclair's Laboratory in 1950, and worked there for two years. In the beginning, the two men were friends, travelling together to scientific meetings and sharing ideas; but rivalry grew with time. Sinclair wrote some years later that while Keys was at Oxford he thought all fats, including corn oil, raised plasma cholesterol levels. This was a matter of real irritation for Sinclair, because he and a member of his growing staff, Dr V. Basnayake, were showing that a relative deficiency of essential fatty acids caused deposits of cholesterol in avascular tissues such as the epidermis of the rat. Another matter that irritated Sinclair was the obvious esteem in which Rudolph Peters held Keys. Nonetheless, Sinclair and Keys produced interesting work together. 'Real nutritional deficiency', a paper authored jointly by Keys and Sinclair, and published in the *British Medical Bulletin*,[2] 1952, was one such piece of collaboration. It reflects contributions made to our understanding of nutrition during the period between the start of World War I and the end of the World War II. The source information is that obtained through the systematic study of the starving.

For the next few years Sinclair and his team continued to publish widely, many papers touching on his earlier work. For example, based on work done with V. Ramalingaswami, Joyce Buxton, and Agnes Mayer, a young German student, Hugh Sinclair published a paper on the nutritional aspects of pyridoxal as a coenzyme, and with B. S. Simic and Hugh's old friend Brian Lloyd, another paper appeared on the activity of ascorbic acid in hypervita-minosis-A in the guinea-pig. But in 1953, Sinclair became seriously interested in the subject of the dietary fat and published a paper on the diet of the Canadian Indians in the *Proceedings of the Nutrition Society*, in which he summarized nine previous publications on this subject and provided an extensive bibliography. Sinclair states in the discussion section:

The Eskimo used to subsist on a carnivorous diet, eating meat raw or lightly cooked in sea water, and consuming head, skin, viscera, marrow, blood and stomach contents. This was an excellent diet adequate in all respects and only failing when the hunting of seals or other animals was prevented. Despite almost complete lack of hygiene they enjoyed a healthy happy life. The Bush Indians fared similarly upon cooked meat with some berries and nuts. The introduction of the white man's diet in return for his exploitation of the furs and extermination of the buffalo changed matters. The Indian has so far been most affected, but wherever the white man contaminates the Eskimos, the diseases of civilization follow. The expectation of life amongst the Eskimo in north Greenland is only 25 years in men and 27 in women … Tuberculosis is responsible for nearly half the deaths of Eskimos and Indians in Alaska … Other diseases that are common amongst civilized persons appear less frequently [among this population] despite the inadequacy of the diet of bannock [white flour, lard and sodium bicarbonate] …

About this same time Hugh wrote an article for the *British Medical Journal* on the assessment and results of obesity, concluding that there is a correlation between obesity and atherosclerosis. This paper, also published in 1953, mentions Ancel Keys:

> Therefore Keys, whom we were fortunate to have working with us for a year in Oxford recently, has been led to suggest that increased dietary fat raises serum cholesterol and produces atherosclerosis: since increased dietary fat will tend to cause obesity it is not surprising that atherosclerosis is commoner in overweight persons.

It is worth noting that essential fatty acids are not mentioned in this publication, and that Sinclair does not challenge Keys' views on the subject of dietary fat and serum cholesterol at this stage; in fact, Sinclair states in his summary: 'Increased dietary fat (but not increased dietary cholesterol) tends to raise the serum cholesterol level and to produce atherosclerosis.'

Although there were now many new faces in Sinclair's laboratory, important people in his life were disappearing. John Ryle died in 1950, and his great experiment in Social Medicine did not long survive him. Two years later Rudolph Peters received a knighthood, and rumours grew about who would replace him when he retired as Head of the Department of Biochemistry. More dramatically, in 1952, Sinclair's good friend Sir Jack Drummond and his family were brutally murdered while on holiday in Provence, France. A simple farmer, Gaston Dominic, was arrested and convicted of the murders, but there was always doubt. William Reymond, in 1997, published an account of the crimes, suggesting that Jack Drummond had lived a double life, involving himself in post-war activities that may have attracted the notice of foreign agents, who subsequently committed the murders.[3] Whether the truth about the murders will ever be known is uncertain, but it is safe to say that Jack Drummond's death silenced a voice that spoke out strongly on Sinclair's behalf.

For a time, it appeared that trace minerals would become Sinclair's principal area of research. The Laboratory already had a considerable reputation in one area of this research: iodine. Largely through work on goitre, conducted by Dr Dagmar Wilson, a close friend of Dame Harriette Chick, Oxford produced significant work on the dietary effects of deficiency in this mineral. The Government had accepted advice from the Medical Research Council that all table salt in the United Kingdom should contain iodine, and in 1951, Oxford University accepted a capital grant and a recurrent grant of £2000 for ten years from the Nitrate Corporation of Chile Ltd. to support Dr Wilson's work in Ceylon, Nigeria, and Sierra Leone by experimental work within the Laboratory of Human Nutrition.

To build on this opportunity, Sinclair sent a proposal concerning trace minerals to the Rockefeller Foundation: its focus was the metabolism of the central nervous system. The proposal harkened back to Sinclair's early days in the Department of Biochemistry. Rudolph Peters' name was not mentioned, but Dr Ritchie Russell's work on disseminated sclerosis and Swayback in sheep was. The primary purpose of the proposed work was to study the part played by deficiency or excess of trace elements in the origin of diseases of the brain and spinal cord. Its secondary purpose was to investigate the fundamental role of trace elements in animal nutrition, including their metabolism in man and lower animals. Hugh Sinclair would direct the work, with advice from Dr Richie Russell and assistance from Brian Lloyd. Sinclair wrote:

> There are many chronic and incurable diseases of the brain and spinal cord in which there is degeneration of certain parts or tissues of the nervous system which is so selective as to make it highly probable that this selectivity depends partly at least on the chemical physiology of the tissue concerned … Where several cases of disseminated sclerosis have occurred in one village, chemical analysis of the soil has been carried out … Dr Ritchie Russell, Dr Bourdillon and I formed a team to prosecute such research. Dr Bourdillon initiated the treatment of certain diseases at present incurable with small quantities of a large number of elements known to occur in

living tissues in small amounts; because most trace elements are absorbed from the intestine with difficulty and apparently in some cases by special mechanisms … and because similar difficulties occur at the blood–brain barrier, Dr Bourdillon has been studying questions of permeability.

The proposal was not funded.

In 1953, Hans Krebs received the Nobel Prize in Medicine for his work on the citric acid cycle, and in 1954, he succeeded Rudolph Peters as Head of the Department of Biochemistry at the University of Oxford. Hugh had a hand in recruiting Krebs:

5 April   PRIVATE   HMS to Krebs
I am sure that a number of people here, like myself, feel very strongly indeed that we must get you as successor to Peters. You will have seen that applications have to be in by 29 April and you probably know that the Electors consist of: The Vice Chancellor, the Waynflete Professor of Physiology, the President of Trinity, the Regius Professor of Medicine … and Sir Edward Mellanby. Mellanby was staying with me a few days ago and he seemed to have all the right ideas. I am sure you would find here all the facilities you would need and it would be wonderful to have a rejuvenation of Oxford Biochemistry.
With kind regards … yours sincerely …

15 April
(Written on letterhead from The Ambassador Hotel, in Atlantic City, New Jersey.)
Dear Sinclair, Your letter reached me during a visit to the United States. I very much appreciated your kind encouragement. A copy of the advertisement was sent to me a few weeks ago and I have been thinking matters over for some time. I am of course very much attracted but my feeling is that I am not really free to apply without having first discussed the fate of my MRC Unit with the MRC. I have long-term responsibilities towards the MRC, and towards the staff of my unit. I feel strongly that I cannot just try to walk out. These problems can hardly be dealt with by correspondence and I therefore do not find it possible to cope with the matter before I am back home on May 11th. I noted in the advertisement that the Electors are not necessarily confined in their choice to those who apply. If they wish to consider me I shall be available immediately after my return. With kindest regards, and many thanks,
Yours sincerely, H. A. Krebs.

Sinclair persisted, and eased the way for Krebs' candidacy, despite the fact that he was not in the country. In his autobiography, Hans Krebs later acknowledged that Hugh Sinclair played an instrumental role in his considering the position at Oxford by advocating his candidacy.

Like Peters, Krebs made every effort to support his scientific staff. His watchwords were said to be: 'Never leave until tomorrow what you can do today,' and 'I do not require more than 24 hrs a day of your time!' He begrudged any effort spent on what he saw as trivialities. Although Krebs and Peters shared an exacting view of scientific investigation, on a personal level they could not have been more different. Unlike Peters, who enjoyed the formalities of academic life, Krebs rarely attended scientific meetings, played little part in Biochemical Society or journal affairs, and refused many invitations to travel abroad. In 1955, the year after Krebs became Head of Biochemistry at Oxford, Sinclair's Laboratory of Human Nutrition was transferred to the Department, and formally terminated shortly thereafter.

Hans Krebs, the son of a Jewish doctor, was born in Germany in 1900. After studying in Göttingen, Freiburg, Munich, Berlin, and Hamburg, he received his MD degree in 1925. For four years, he was an assistant to Otto Warburg, the leading biochemist of the day, at the Kaiser Wilhelm Institute for Biology in Berlin-Dahlem, where Krebs was greatly influenced by Warburg's exacting approach to methodology. In 1930, he took up the position of *Privat-Dozent* in International Medicine in Thannhauser's clinic in Freiburg — a post with considerable clinical responsibility. Nonetheless, he pursued his research on tissue metabolism, heavily funded by the Rockefeller Foundation, using Warburg's methods. Sir Frederick Gowland Hopkins invited Krebs to the University of Cambridge when Hitler was obviously moving towards war and there were rumours of the planned action against the Jews. Krebs could only take ten Marks out of the country, but arrangements were made for him to export his laboratory equipment; with that, and continued help from the Rockefeller Foundation, Krebs quickly established himself at Cambridge. An associate said:

When I first met him, I was disconcerted by the sudden distant smile, and a gaze which looked abstractly at the wall behind my ear. I learned after many such encounters in the corridor and laboratories that it was well worth heeding his apparently diffident suggestions, and that his awareness of one's work, even thought, was thorough.[4]

In the laboratory, Krebs would quietly watch people working, anticipate mistakes they might make, gently intervene, and — without rebuke — demonstrate how a procedure should be done to the high standards he demanded. For most people, this teaching technique built confidence and teamwork. This was very unlike Sinclair's approach to laboratory management, and marked a considerable difference between the two men.

In an interview, Agnes Mayer remembered working in Sinclair's laboratory at this time The two men differed, she felt, but one scientific approach was not superior to the other. Sinclair encouraged a congenial environment in the laboratory, and provided opportunities for each researcher to develop their own style, even at the expense of making repeated mistakes. She recalled that Sinclair followed the highest standards of experimental design, and that with each new experiment he sat down and discussed how variables affecting the outcome could be controlled.

Agnes began work in the Laboratory of Human Nutrition in 1950 as a volunteer and continued working for Hugh until 1957, as his Readership in Human Nutrition was ending. As there had been little work in Germany after the war, she had applied to Oxford to complete her studies and find employment. Her letter of application eventually reached Hugh. Sinclair offered her work, but no salary, and for many months, until Hugh helped her get a Wellcome Fellowship, she lived on meagre personal savings. Agnes had trained in Germany as a microbiologist, and Hugh needed someone to do microbiological assays for vitamins, especially B_6 and B_{12}. At the time, everyone in the Laboratory was working on the development of one or another new assay method for measuring substances in food and tissues. Sinclair believed that the complex link between diet and disease could only be found with improved methods of biochemical analysis, and so encouraged his staff to explore new research techniques wherever possible.

Not surprisingly, Sinclair saw the members of the Laboratory staff, including himself, as subjects for experimentation. On one occasion, they all took antihistamine tablets to observe any physiological effect; the only observable consequence was that Agnes and Hugh developed transient facial redness. Hugh's interest in antihistamines never came to much, but he thought there might be some connection between radiation burns and histamine, and for a time pursued animal experimentation along these lines. His work was inspired by data then available from victims of the bombing of Hiroshima, and like other scientists, Sinclair wanted to understand the mechanism by which radiation damages the skin.

But Agnes's first assignment turned out to be one of Sinclair's little jokes. She was young and innocent, and when he told her to produce good wine out of a two-gallon bottle of soured wine he had standing in the laboratory, she did her best to oblige. Using barium hydroxide, she tried for days to alter the wine's acidity. At last she gave up, and suggested that he should buy new wine. Amused by her perseverance, he shifted her to more suitable tasks. Some months later, with Hugh's support, Agnes was accepted at St Hugh's College and began work towards a BSc, which she received in 1956.

When Hugh had a guest from Germany, he would sometimes invite Agnes to join them for a dinner. He also invited her, along with others from the staff to Lady Place, where his mother, now a thickset matriarch, would see they were properly looked after. On one occasion, a group from the laboratory was invited to join Hugh on an outing to a large estate near Reading, from which he had purchased a number of pieces of garden statuary. Travelling in an old, battered truck, rusting away at every seam, students and technicians helped Sinclair haul the heavy pieces back to Sutton Courtenay. With Hugh at the wheel, the truck, filled with lurching statues and giggling young people, hurtled along the country roads. Once their cargo was unloaded at Lady Place, Rosalie invited them to wash, and then served an elegant tea of cucumber sandwiches and cakes.

In many ways, these were the best years in Sinclair's career. Although he did not have the independence he craved, he had a fine staff. His publications were prolific and varied, including papers on vitamin B_{12}, vitamin deficiency in alcoholism, the effect of combined deficiency of thiamine and of pantothenic acid on the rat nervous system, and chronic thiamine deficiency. As Agnes Mayer [later Agnes Huber] pointed out, if any scientists today produced as much work in as short a time as Sinclair, they would certainly receive promotion, not criticism.

About this time, Hugh wrote a small piece on the teaching of nutrition, in 1954. Sir Wilson Jameson praised it, and upon receiving a copy Sir Robert McCarrison wrote to Hugh:

Dear Hugh

Many thanks for the reprint of your paper … I have read it twice over and find it altogether admirable. Particularly do I applaud — and endorse — your recommendation that 'the teaching of nutrition should not be made a separate subject, but should permeate the curriculum'; and your repeated emphasis upon nutrition as the prime factor in the preservation of health pleases me greatly.

It is my experience that most practitioners of Medicine, whether GPs or consultants, are <u>still</u> more interested in treating the sick than in preserving health; and too many of them look upon 'vitamins' as drugs, rather than as essential food factors. I would wish that this paper of yours were reprinted in such journals as the Lancet, the BMJ, the Practitioner, and the Student Medical Journal, so that it might be widely read by all those whose business it is, or will be, to look after the people's health.

With congratulations and best wishes

Yours ever — Robert McCarrison.

At home, circumstances were strained. John Sinclair had suffered several more heart attacks, and for some months he had lived at home, keeping himself to the ground floor to avoid the stairs, but getting in the way of both Hugh and Rosalie. John died in November 1954.

But Hugh had little time for grief; the financial clouds were beginning to close in over the Laboratory of Human Nutrition. On 15 December 1954, Hugh Sinclair wrote to Agnes Mayer:

I write to confirm what has now been stated verbally by Miss Buxton or me to all of the staff of the Laboratory and students (except two members of staff who are away), namely that the Registrar of the University has informed me that the Hebdomadal Council has regretfully come to the conclusion that it has no alternative to closing the Laboratory of Human Nutrition as soon as the necessary procedure can be put in motion.

I understand that legislation to this effect is likely to come before Congregation of the University in the middle of February and, if passed, will close the Laboratory at that date. In that case you would on that date be given notice that your position would terminate one week later.

I need hardly add that Miss Buxton and I shall be very pleased to give all members of the staff any help or advice we can, and no obstacle will be put in the way of any member of the staff who wishes to seek another post before the date on which the question of closing the Laboratory comes before Congregation. The University always has an obligation to students reading for Advanced Degrees, and Dr Acheson or I, as their Supervisors, will continue to do all we can to help those in our Laboratory.

Yours sincerely, Hugh Sinclair

During 1955, the Laboratory of Human Nutrition was transferred from the Nissen huts on the grounds of the Churchill Hospital to the Department of Biochemistry, where the staff continued their work under ultimate direction of Dr Hans Krebs. Negotiations for the merger of the Laboratory and Department began in January, and went on for several months, but the actual removal of equipment and supplies took place shortly before June.

Hugh invited Krebs to dine in his rooms at Magdalen for the initial discussion of the matter, hoping to retain as much autonomy as possible. The setting was impressive: well furnished rooms situated in New Buildings, a large block of living accommodation standing by itself between the Cloisters and adjoining the Magdalen deer park. The food was excellent: Hugh had arranged the menu with the chef, and his scout arranged the table and saw that the food was properly presented. Over dinner, the two biochemists talked about an amalgamation of their work. The men parted in cordial agreement that something could be worked out. It appeared there would be a reprieve for Hugh's laboratory, but he felt obliged to tell his staff that he could give them no assurance that it would continue.

Sinclair had anticipated the legislation to close the laboratory would be published by the University in January, 1955, but no notice appeared. He felt this was due to opposition on the two Faculty Boards to the closure of the Laboratory. Hoping for the best, in March he again wrote to Hans Krebs:

> Dear Hans,
>
> I want to write this note while awaiting my train to Liverpool to say that I spoke this morning to all the staff of the Laboratory of Human Nutrition, saying that I had reason to believe that the Lab would continue and outlining the type of merger within our Department, which I very much hoped would occur.
>
> I shall of course write more fully later, but should like to say in this indebted and hurried way that the staff joins me in welcoming enormously the possibility of coming within your Department, and that they and I want to convey to you as quickly as possible their gratitude. Forgive this way of doing it,
>
> Yours, HMS

Matters bumbled on. In July 1955 Sinclair and Krebs met to discuss the future of one of the staff and several others matters, including Hugh's future within the Department. The grant from the Wellcome Foundation to Oxford was at the heart of the issue; it was required that the funds be used for human nutrition research, but the Foundation did not require the University to maintain a Reader in Human Nutrition. This meant Sinclair was expendable. Krebs said the Registrar had told him he had heard from 'a highly responsible member of the University' that Sinclair intended to leave Oxford: he would understand if Sinclair wished to do so. Sinclair (anxious to remain at Oxford) countered, saying he had no interest in leaving; but when the University had decided to close the Laboratory, during the previous December, he had naturally considered work elsewhere.

Krebs said he recalled Sinclair telling him that he had been offered positions at other universities. Hugh responded by saying that over the past years he had been offered six chairs, but all of these offers were before the suggested closure of the Laboratory, and were therefore turned down. The matter between the men reached a stalemate, and it was agreed Sinclair would remain at Oxford as Reader in Human Nutrition, at least until his current appointment was completed.

Hans Krebs respected Agnes Mayer's work, and invited her to join his research team. She felt this was to take her away from the nutrition work, and owing to her feeling of loyalty to Sinclair, decided to leave Oxford. Sinclair had been a great teacher, an inspiration, and above all else — kind. No matter what other faults he had, when touched by the needs of others he was capable of genuine benevolence. Agnes contacted her old laboratory in Stuttgart, inquiring about employment there. When she left England to return to an uncertain future in Germany, Hugh took her to the train station. Some months later, she received a letter from Dr Fredrick Stare, who was then Chairman of the Department of Nutrition at Harvard University, inviting her to join his department. She took the opportunity without question, and began work towards her doctorate. Some years later, she learned that Dr Stare had written to Sinclair, telling him he wanted to recruit the best and brightest to his department and asked for thoughts on any outstanding young scientists he could recruit. Hugh recommended Agnes.

A focus had formed in Hugh's research interests. Although he was still keenly interested in skin and the nervous system, as changes in these were fairly easy to monitor, he had become increasingly interested in the biological action of essential fatty acids. He and his staff, now working in the Department of Biochemistry, began collecting and comparing international information on dietary levels of vegetable fat and rates of cardiovascular disease. Was there a lag period between dietary changes and their effects, they wondered? Perhaps five years? How could the necessary data be collected and correlated? Was it possible to relate dietary data from 1934–8, with clinical symptoms in 1950, for example?

On 10 May 1955, Hugh Sinclair wrote to Lawrence W. Kinsell, at the Institute for Metabolic Research, Oakland, California, asking for more reprints of his work on dietary fats and atherosclerosis. Kinsell said he was coming to England, and would drop by to meet Sinclair. What interested Hugh was Kinsell's clinical application of ideas they shared about the importance of essential fatty acids. Sinclair was now convinced that essential fatty acids were important parts of biological structure, possibly as esters of cholesterol. When saturated fats replaced these essential fats, for whatever reason, the resulting structures were biologically wrong. It did not matter which tissues were involved — nerves, kidney, skin, aorta — when a large enough concentration of unsuitable molecules accumulated, the early processes of disease began. What is more, he believed other nutrients played a role in these processes, and early on experimentally questioned the importance of vitamin A.

By early 1956, Hugh Sinclair knew he would not be able to continue his work within the Department of Biochemistry, and began branching out boldly on his own. His interest in dietary fats, and his attendance at two annual International Conferences on Biochemical Problems of Lipids, had put him in contact with an international community of scientists engaged in research dealing with essential fatty acids. The Third International Conference was to be held in Brussels, in July 1956; Hugh wanted to ensure he was part of the debate that was sure to occur. Drawing on the literature, concepts evolving in his own laboratory, and partially completed experiments by scientists elsewhere, he began to write.

While this new endeavour was ongoing, he also began making plans to host at Oxford the Fourth International Conference on Biochemical Problems of Lipids — the first international meeting to focus entirely on the subject of essential fatty acids. The Oxford conference was arranged with Krebs' approval, although he would fail to participate in it. However, Howard Florey did attend.

During this crucial period, Hugh published several key papers. One was a review article on 'Vitamins and the nervous system', that appeared in the *British Medical Bulletin*, in 1956. He began with a quote from Sir Edward Mellanby, written in 1934:

> The field of nutrition has constantly been handicapped because much of the work has been in the hands of narrow specialists — biochemists with little or no biological or medical knowledge, and medical men with little or no fundamental grasp of biochemical or physiological training. By anybody with a wider grasp of physiological, biochemical, pathological and clinical subjects, a rich harvest will be gathered.

Sinclair went on to make a case for combined clinical and laboratory research by again quoting Sir Edward: '... to survey the field, pick out the best lines of attack on any particular problem, [and] keep in touch with the literature ...' Certainly Hugh would have liked to have written those words himself.

In another document, a chapter on nutrition and skin in man, Hugh Sinclair, now Vice-President by succession of Magdalen College, included a section on essential fatty acids in which he reviewed the subject from the original work by Evans and Burr, in 1928, through to the work being done in his own Laboratory of Human Nutrition on the consequences of essential fatty acid deficiency on skin permeability. This chapter, a *rapport* to the 4th International Congress of Nutrition,[5] summarizes his thinking on essential fatty acids at that time:

> ... I have produced ideas to stimulate discussion. I have intentionally departed from the customary dull review of well-established facts to discuss some of the work of my colleagues and myself, and to speculate — some will say frivolous or fantastical — about its implications. Too often dermatologists give different names to essentially similar conditions; too often research workers on lower animals describe in superficial terms the lesions they produce. If I pour new wine into old bottles, a reaction may ensue; if this reaction is discussion or, better, controlled experiment and observation for which dermatology is ideally suited, I shall be content. A unifying hypothesis is needed and I offer one. On other occasions ... I have listed and discussed the signs on the skin that should be recorded in assessing nutriture, and I have given a few of our preliminary correlations between such signs and objective measures as blood analyses. The results of all our surveys (in England during the war, in the acute famine in the Western Netherlands, and in chronic malnutriture in post-war Germany) are on

Hollerith cards, but these are stored awaiting facilities and staff for their analysis. This will undoubtedly shed light on the interpretation of such signs. During the Second World War we started to estimate EFA and pyridoxine in blood because of the prevalence of skin signs and the lay opinion that these were caused by 'lack of fat' — an opinion completely without scientific evidence. But then the difficulties were too great. Now such estimations are possible, and these combined with controlled therapy may advance our knowledge enormously.

We know that a relative deficiency of essential fatty acids is becoming increasingly prevalent in the diets of the more highly civilised countries. This includes an absolute deficiency of linoleic and arachidonic acids, and of vitamin E (tocopherols) which stabilises them, and vitamin B_6 which is required for their metabolism. It includes also a relative abundance of more saturated fat and of the more stable isomers of EFA such as *cis-trans*-linoleic acid or conjugated linoleic acid. … The hypotheses I present are easily tested therapeutically, and I offer them to those who have the inclination and facility to do so.

By the time these words were published, Sinclair had already shocked the scientific community with his famous letter to the *Lancet*, published in 7 April 1956,[6] which became a classic contribution to the scientific debate on the importance of dietary fats. The letter was titled: 'Deficiency of essential fatty acids and atherosclerosis, etcetera'. Peters and Krebs strongly opposed its publication, saying it was too speculative. The letter set about insulting the reader as it makes Sinclair's seemingly outlandish case:

Scant attention seems to be paid by the medical profession and by food administrators to a very important change in the dietaries of the more civilised countries that has been occurring over recent decades with increasing intensity. I refer to a chronic relative deficiency of the polyethenoid essential fatty acids (EFA) … Your readers with stereotyped minds should stop reading at this point … You will now conclude, Sir, if indeed you have not done so long ago, that I have strayed into the realm of fantasy. You will remind me that long ago Sir Thomas Browne wrote: 'Imagination is apt to rove, and conjecture to keep no bounds.' My inclusion of 'etcetera' in the title invites the scorn we so readily pour on vendors or patent cure-alls. I have no wares to sell.

The debate that would dominate the remainder of Hugh Sinclair's scientific career had begun.

CHAPTER 17

Exchanges in the *Lancet*

Hugh Sinclair was concerned that new forms of food processing might influence human nutrition and, when he wrote his letter to the *Lancet*, this was uppermost in his mind. Coincidental with the publication of Hugh's letter, the British Government was debating the possibility of allowing the production of highly processed wheat flour. This troubled many nutritionists because, as bread was a major source of nutrients in the British diet, the change meant that the majority of British people would run the risk of becoming deficient in certain vitamins. During the war and the years that followed, when food had been scarce, the milling and extraction processes used to make flour were minimized as a means of preserving as much of the grain's nutrient value as possible; but the baked goods it produced left much to be desired. Now that Great Britain was free from war-time privation, a public tired of being forced to eat heavy cakes and beige-grey bread was demanding finer and whiter flour. Technology used to produce a more acceptable grade of flour not only caused the extraction of nutrients, but also the introduction of bleaching agents and substances classified as 'improvers'.

Sinclair was among those against improving flour quality. His case was threefold: First, the more heavily processed flour would contain fewer essential fatty acids. Secondly, bread was a major source of pyridoxine, or vitamin B_6, which plays an important role in the metabolism of linoleic acid, and more extensive processing of flour would reduce levels of this nutrient. And finally, bleaches and other chemicals destroyed much of the vitamin E in wheat that protects essential fatty acids from oxidation. But Sinclair was concerned about more than the nutrient value of flour. He also was concerned that the process of hydrogenating vegetable oils in the production of margarine would destroy essential fatty acids in these oils, and cause the formation of unnatural fatty acids called *trans* fats. This, he believed, would lead to further long-term essential fatty acid deficiency, which would be the cause of an increasing number of people suffering from degenerative illnesses, such as coronary heart disease, certain forms of cancer, leukaemia, seborrhaeic eczema, and diseases of the central nervous system.

In his letter to the *Lancet* (see Appendix D), Sinclair acknowledged that the foundation for this theory originated from his own experimental work, which he referred to as 'humble in scope', and to a careful assessment of the scientific literature. [Some experiments, which he did not mention in the letter, were informal, but informative. For example, during the early 1950s, Sinclair and Dr Ramalingaswami conducted an autopsy on an old, and very fat, cat that had enjoyed many years at Lady Place being fed fish and large bowls of cream by Rosalie. Sinclair and Ramalingaswami specifically wanted to examine its coronary arteries and see if they contained a build-up of fatty deposits associated with atherosclerosis. To their surprise, there was hardly a sign of fatty deterioration. Both men thought it plausible that the essential fatty acids found in the oily flesh of fish had in some way protected the cat from the effects of the high levels of saturated fat in the cream.]

In Sinclair's letter, the polyunsaturated fats of greatest interest were linoleic acids and one of its metabolic end products: arachidonic acid. Because of their molecular structure these are classified as omega-6 fatty acids. Note that the fatty acids in fish oil that are now believed to play a key role in the prevention of coronary heart disease have a different molecular configuration, and are called omega-3 fatty acids. Sinclair would later become identified with the 'Eskimo diet', in which fish oil plays a central role. However, at the time he wrote his letter of 1956, fish oil was not mentioned.

Formal experiments conducted in Sinclair's laboratory produced useful results that were mentioned in his letter. In 1954, Sinclair and Dr V. Basnayake found that rats fed a diet entirely free of fats and sterols, but otherwise complete, rapidly formed cholesterol deposits in the epidermis, or outer layer of skin, which increased in size with time. Obviously the rats' bodies were producing cholesterol. This work was reported at the 2nd International Conference on Lipids, in Ghent, in July 1955.

Sinclair began his letter to the *Lancet* with an attack on a fellow nutritionist: Professor J. Yudkin, who had written a letter to the same journal during the previous year, suggesting the possibility that essential fatty acid deficiency in man occurs rarely, if ever. Sinclair completely disagreed, and told friends and associates that he wanted to stimulate scientific debate on the subject in an open forum. He wrote to people interested in both sides of the argument and asked them to write to the *Lancet* expressing their opinions, and he called for new research into the subject. This stimulated an exchange of letters in the *Lancet* that continued for more than a year. Unfortunately, during this time Sinclair's contributions were greatly hampered as he had limited access to laboratory space, and was forced to speculate from work published by other scientists. Some colleagues sent him data and their own ideas on the matter: most important among these supporters was the American, Larry Kinsell.

The first published response to Sinclair's letter was from E. F. St John Lyburn:[1]

> Sir,
>
> With reference to Dr Sinclair's stimulating letter last week, I should like to add that when essential fatty acids are applied to mammalian epidermis there is immediate absorption.
>
> The results in cases of psoriasis and acne, after treatment in wet high thermal conditions combined with applications of fatty acids, are very encouraging. …

This letter supported Hugh, but gave no hard facts to justify the reported observations.

In the second issue of the *Lancet* following Hugh's letter,[2] John Yudkin responded to Hugh's attack. When describing certain fatty acids, Sinclair had used the scientific nomenclature tetraenoic, dienoic, and trans-fats: John Yudkin seemed to think this pretentious:

> Sir,
>
> After the gentle reprimand which, Dr Sinclair has given me (April 7) for daring to comment on matters in which I have not worked experimentally, it is good to see that he himself has not hesitated to refer to the work of others. In building his house, he has used as his cards some twenty-five reports of original work, in which he has himself been associated with four. Of the diseases of which he speaks, his work has been on atherosclerosis, not on carcinoma or leukaemia or peptic ulcer or disseminated sclerosis or any of the other parts of the 'etcetera'.
>
> My purpose, however, is not merely to say *tu quoque* to Dr Sinclair. It is to give one tiny warning to his readers, and one tiny suggestion to Dr Sinclair. The warning is that the decorations on the edifice — terms like tetraenoic

and trans fatty acids — are bright and glittering but they may well dazzle the eyes of those who are viewing it from the non-specialists enclosure. As a result, it may be that the cards look like real bricks and that the stability of the edifice is assumed to be much greater than it really is.

On 28 April, the *Lancet* published an article titled 'Effects of feeding different fats or serum-cholesterol', by B. Bronte-Stewart and others.[3] This focused on heart disease, and basically supported Sinclair's ideas. In the Addendum to their paper, the authors stated:

> Since submitting this paper we have seen Sinclair's (1956) stimulating analysis of the possible rôle of deficiency of essential fatty acids in the aetiology of atheroma. ... We cannot comment critically on his analysis here, but we are intrigued by some of the possibilities which it raises in interpretation of our own experiments with fats of differing degrees of saturation ... [They then commented on data, presented during the Minnesota Arteriosclerosis Symposium, which showed pronounced falls in serum-cholesterol levels in cases of familial hypercholesterolaemia when subjects were placed on a diet rich in corn oil: an excellent source of linoleic acid.] These views are in line with our short-term investigations of the effects of hydrogenated and natural fats and oils which we suggest are related to their unsaturated/saturated fatty acids content.

In the same issue of the journal, Professor Ancel Keys wrote from the Kyushu University Medical School, Fukuoka-shi, Japan, and joined the debate.[4] Only a few years earlier Keys had worked with Sinclair in his Oxford laboratory, and agreed with his views on the importance of essential fatty acids. Now, dismissing issues other than coronary heart disease in a long letter to the Editor, he says, in part:

> The central issue of Sinclair's speculation is atherogenesis and the regulation of the concentration of cholesterol in the blood. His thesis is that our present plight in Britain, the United Sates, and other countries where coronary heart-disease is so alarmingly frequent, is due to dietary deficiency of the 'essential' fatty acids — linoleic and arachidonic — largely ascribable to the evils of hydrogenation and deficiencies of vitamin E and pyridoxine ... [Keys then explores various perspectives of related research, including those involving studies of mode of life, contending that the important issue in the development of heart disease is the total amount of fats consumed.] ... it is clear that a great deal more research is needed on all

aspects of the problem, including food economics and the psychology of food preferences. While this is in course it may be well to ponder whether a diet containing 40% of its calories as fats is really essential for human happiness and, if not, how may it best be reduced.

Also in the 28 April issue of the *Lancet*, an Editorial on Fats and Disease appeared, which broadly discussed differing views of the subject of fats and their roles in biological materials. Commenting on Hugh's letter, the Editor of the *Lancet* stated:

> Most nutritionists have held that human diets are never significantly deficient in these acids [EFAs], but Sinclair thinks that relative deficiencies, associated with high intakes of saturated fatty acids and a low intake of vitamin B_6, may predispose to a variety of degenerative disorders including atheroma and myocardial infarction ... [The Editorial then focuses on cholesterol and blood lipids, and on problems related to the importance assigned to the balance of dietary fats, the palatability of dietary fats, and issues related to the margarine industry.] ... comments on Dr Sinclair's hypothesis from Professor Yudkin in our last issue and from Prof. Ancel Keys this week indicate that systematic knowledge of the ratio of essential to total fatty acids is far from complete for single artificial dietary fats, and it is almost non-existent for the total fats consumed in the diets of any large group of persons ... Sinclair also postulates that the use of flour of low extraction-rate would significantly lower the effective content of the national diet in essential fatty acids and so predispose to atheroma and myocardial thrombosis and increase the prevalence of other diseases of civilisation. Here he is on very uncertain grounds and can muster no direct experimental support. He is right, however, in urging the need for more research on these problems; and in the present state of our ignorance it is surely rash to commit most of our population to consuming bread made from a flour about whose nutritive properties we know so little and have so many genuine doubts ...

Other letters to the Editor poured in. Hugh's friend, Larry Kinsell, published material supporting Sinclair's views with facts and case information. Others sent in views on gallstones and hypercalcaemia in infants. A. C. Frazer, in a letter published in the issue of 28 July 1956, supported the use of cod-liver oil as a source of vitamin D and essential fatty acids, and pleaded that greater support be given to food research as a whole, concluding:

There is no other way of ensuring that a progressive policy involving the application of scientific advances to food technology can be followed with safety and confidence.

As this published debate on dietary fats continued, Sinclair also engaged directly in the argument over the processing of wheat flour. An expert panel, chaired by Sir Henry Cohen, had published a Report on the nutritive value in flour of varying extraction rates.[5] This was preceded by several other papers and reports including one by E. M. Widdowson and R. A. McCance.[6] Sinclair strongly disagreed with the findings of both documents, and on 19 May 1956, little more than a month after the publication of his now famous letter, he wrote to Lord Boyd-Orr, stating:

> My cousin, Viscount Thurso (whom I am aware you know), was staying with me recently and told me that Lord Hankey was to introduce a debate in the House of Lords on bread as soon as the Cohen report was published. I believe this report was held up because it very surprisingly goes against the advice of the Government's medical and scientific advisers and the advice given by a very powerful committee of the Medical Research Council. At the annual meeting of the Nutrition Society next Saturday (26 May) some of us hope to raise the matter and I wonder if there is any chance of your attending this meeting in the afternoon. We should be delighted to see you and might remind you that it is now about a year since you attended the meeting at St Mary's.
>
> It seems tragic that our present excellent flour should be abandoned because of the interests of the millers and apart from the extraction rate many of us have increasing doubts about the advisability of including chlorine dioxide as a so-called improver. I have set forth in some detail my own views upon the bearing of this upon atherosclerosis, coronary thrombosis and certain other disease that are increasing rapidly in civilised countries; the editor of the Lancet was anxious to publish my letter at once (7 April, page 381) because of the imminence of the Cohen report; he has passed encouraging editorial comment upon it in the issue of The Lancet of 28 April which also contains some original work from South Africa indicating the importance of essential fatty acids in preventing atherosclerosis.
>
> Lord Hankey has sent me an extract from his speech but he points out that he is ill-equipped to deal with the nutritional and medical problems involved. You of course are the only person who can do this …

The letter was addressed to Lord Boyd-Orr via the Athenaeum, where the two men sometimes met.

Lord Boyd-Orr wrote at once to Sinclair, saying he was out of touch with the latest research on the subject, but if Hugh would coach him, he would come to London and 'see that your views were put forward.' Could Hugh send him copies of the Cohen report and the *Lancet* letter, as well as 'a note from you on the results of the investigations and your views.' Again, Hugh should send all this to the Athenaeum. Hugh's response was dated 28 May:

> I enclose a copy of the Cohen report and of my letter to The Lancet. You will see from the report (p. 27) that 'The conclusions reached by the Panel differ from those presented in their evidence by the Government's medical and scientific advisers and by the Medical Research Council'. These advisers stressed that bread is an important source of nutrients other than vitamin B_1, nicotinic acid and iron — these being the ones which are to be added to white flour. Some of these other nutrients are known (e.g. pyridoxine), others probably await discovery. The report (p.25) says that 'Flour, or the bread made from it, contributes nearly a third of the total calories and protein of the average British diet; it is therefore, important that the flour should be as nutritious as possible'. If much of the pyridoxine (vitamin B_6) is removed, it is <u>not</u> as nutritious as possible. For instance, the report states (p.11) that national (80%) flour [i.e. flour in which 80% of the wheat berry remains after milling] contains 0.29 mg of pyridoxine per 100 g., whereas white flour has only 0.11 mg. According to a paper given to the Annual General [meeting] of the Nutrition Society on 4 May by Miss Hollingsworth and others of the Scientific Adviser's Division of the Ministry of Food, the average <u>per capita</u> consumption of pyridoxine daily is this country in 1952 was 1.70 mg of which 0.73 mg (i.e. 43%) came from bread and flour. The total would fall from 1.70 mg to 1.25 mg if white flour were substituted. We know that pyridoxine is needed by man: convulsions caused by its deficiency occurred in infants fed a proprietary milk product, ...
> The estimated daily requirement of man is about 1.5 mg daily, so that a fall from 1.7 to 1.25 might be very important.
>
> It is known that monkeys deficient in pyridoxine develop arteriosclerosis, and it is known that pyridoxine is needed in the body to convert linoleic acid (an essential fatty acid) into arachidonic acid ... This I have elaborated in my letter to the Lancet (p. 381) which I enclose. A subsequent issue (April 28) had an editorial on the subject to which Lord Hankey intends to refer, and also contained an original article from South Africa

which gives experimental support for my view that deficiency of essential fatty acids ... is important in arteriosclerosis and coronary thrombosis.

It is true that bread, although an important source of pyridoxine if of high extraction, does not appear to be an important source of essential fatty acids: according to Miss Hollingsworth's calculations bread and flour only contain about 2 g daily out of a possible total of 14 g in the diet. But these essential fatty acids are very unstable and easily oxidised ... This hydroperoxide formed by air (which is the reason for the drying of paints which contain linseed oil) is biologically inactive (indeed, it is actually toxic since it acts as an anti-vitamin), it is a strong oxidising agent and so will oxidise more linoleic acid ... [it] retains the two double bonds and hence estimates as linoleic acid in foods although biologically inactive. Bread of high extraction made without improvers retains its essential fatty acids and also large amount of vitamin E which, being an anti-oxidant, protects essential fatty acids from oxidation. Wheat germ is an excellent source of both essential fatty acids and of vitamin E. But both [?agents] (now fortunately abolished) and chlorine dioxide (the 'improver' now used) are strong oxidising agents and destroy both essential fatty acids and vitamin E.

Surely the excellent policy of the war regarding bread should not be reversed without more research. A great deal is made in the Cohen Report of the McCance experiments in Germany, but this has been strongly criticised: it was of comparatively short duration and the so-called 70% extraction flour used had analyses nearer 80% extraction.

I shall be in the Athenaeum at lunch and tea tomorrow (Tuesday) and hope I may have an opportunity of seeing you. And I look forward to hearing you on Thursday.

A few days later, Hugh wrote to Dame Harriette Chick, congratulating her on her election as President of the Nutrition Society, and reminding her of the Lords' debate concerning bread, to be introduced by Lord Hankey on 7 June. The debate had been postponed until after the publication of the Cohen report. He was sure she would agree that this was a disastrous report — and that the flour used during the war should not be replaced without protest. Sinclair informed Dame Harriette that Lord Boyd-Orr had said he would try to get to the debate and asked her, in her capacity as President of the Society and also as an expert on B vitamins, to drop Boyd-Orr a line encouraging him to take an active part in the debate.

31 May from Harriette Chick to Hugh Sinclair.
I hope indeed that Lord Boyd Orr will stand up for the 'side of righteous-
ness' in the Lord's debate … I imagine that he will need some briefing —
will you be able to do that, for Lord Orr was never a person to bother with
small detail and in this matter there are many important details …

Particulars from relevant and important papers were enclosed with the
letter.

4 June 1956   Lord Boyd-Orr to Hugh Sinclair

If I can get a sleeper and room I will go to London and speak on Thursday,
but I need coaching. The Widdowson and McCance report seems to have
unduly influenced the Cohen Panel.

Within hours of receiving this letter, Hugh received word that Boyd-
Orr had been in a car accident. He was not seriously hurt, but was unable
to attend the debate. Hugh pressed on. On 8 June, he wrote to Boyd-Orr.

My cousin very kindly took me to the debate in the House of Lords yesterday and I think Lord Hankey introduced his motion with a very telling
speech. It was a great pity that you could not be present. Would you not
consider writing to *The Times* on the subject.

In this letter, Sinclair agreed with Boyd-Orr regarding the McCance
and Widdowson report, and made two additional points: At the end of
the experiment the children on the lowest extraction flour started to 'fall
off'. The second point was that this lowest extraction flour, alleged to be
70 per cent, in fact had a composition nearer that of 80 per cent for
national flour … He noted that Dame Harriette planned a meeting of the
Nutrition Society to discuss the report, and Hugh encouraged Boyd-Orr
to attend.

It is not the purpose of this book to explore the history of bread in
Great Britain, although that is a worthy subject. Here, space limits us to
the fact that soon after the exchange of letters above, ordinary bread flour
became whiter and finer. Although a few nutrients were added to replace
those removed during processing, plain bread flour never again contained the same rich blend of nutrients available during and immediately
after World War II.

Amidst all this debate, Hugh was planning to host a conference on
essential fatty acids, to be held during the summer of 1957 at the

University of Oxford. He was also desperately seeking money to support his Institute. On 5 September 1956, Hugh wrote to the London address of Sir Edward deStein, a man who accumulated great wealth as the head of the Gallaher Tobacco Company, saying:

> I wonder if you will allow me, as Vice-president of your former College [Magdalen], to seek your private advice. A mutual friend has advised me to take this course … I have for many years conducted research upon human nutrition. During the last war I started and directed the Oxford Nutrition Survey which carried out work for the Ministries of Health and Food …
>
> For some years I have worked upon a highly revolutionary concept: I have evidence that the dramatic increase in many degenerative diseases which is occurring in civilised countries is caused by the sophistication of our foods, particularly in respect of the processing of fats as in the manufacture of margarine and cooking fat … I have obtained a considerable amount of experimental evidence concerning these extremely important diseases, but we have neither the facilities nor the accommodation for adequately working upon them in our very cramped quarters in the University. Yet such work must be done.

Sinclair's letter then struck a more human note. Mentioning illnesses suffered by Prime Minister Winston Churchill, and American President Dwight D. Eisenhower, he recalled his own brother's death from heart disease during the previous year, and the partial paralysis of his cousin, Viscount Thurso, the former Leader of the British Liberal Party. He quoted statistics: 20 years earlier, in 1936, about 3000 persons over the age of 75 died of coronary thrombosis per hundred thousand, 10 years later the figure had risen to about 6000, and in 1956, he claimed, the figure approached 20 000.

> And it is known that a considerable number of persons below 50 years of age (like my brother) are now seriously affected.
>
> This is a very serious increase in only one of the diseases which are killing more and more persons in civilised countries as processed fat is increasingly eaten. Since we are unable here for lack of accommodation to do the fairly simple experiments required for absolute proof of how to prevent them, I put forward my views in a long and provocative letter to *The Lancet* last April. The response was very gratifying, for the thesis I put forward is now being taken up in various centres; for instance, I have recently heard that an

American business man has given a research group in the US $250,000 year-ly for ten years to work upon this effect of diet upon arteriosclerosis. But over the years we ourselves have had so much more experience of this research and have collected so much data that we feel disinclined to leave this urgent and immediate problem to our colleagues in the US. For a comparatively small capital sum and a small recurrent sum for a few years we could get the required information. I write to seek your advice about a possible benefactor who might be approached along these lines.

It is of course obvious that the relevant Ministries, the Medical Research Council, the existing Foundations and the Food Industry should be inter-ested. They are; and I might mention in private that plans are being made for establishing an Institute of Human Nutrition where these and related problems will be adequately studied. But these will take some years to start, and it would be wrong to wait until then to continue our researches. If we could find at once a far-sighted benefactor who would provide a small sum of money for a short period, we could we believe solve one of the most pressing problems in medicine.

Would Sir Edward be his guest at either the College, or the Athenaeum, Sinclair asked?

In a letter dated 7 September, Sir Edward offered none of his own funds, but suggested that Sinclair contact Sir Charles Dodds at the National Research and Development Corporation, which was under Treasury control, and was charged with the duty of providing funds for inventions, research, and development. Dodds, deStein wrote, could be of assistance to Sinclair, as he was the medical person.

By now, friends at Oxford were worried about Hugh's need for funds to support his research, and attempted to help. Sir Howard Florey dropped Hugh a hand-written note, dated 28 November:

Dear Hugh,
Enclosed is a buchshie tip that might interest Unilever. It is all very tentative but could easily be put on a firmer basis by a few simple experiments. It would not have as much interest to me as the things I am doing; so I would not do the experiments unless they were of interest to somebody. And that is not blackmail, as the experiments would cost nothing or nearly nothing.

The only real interest to me would be to obtain general free soap for Citral sensitive people. Ever …

Florey's letter then described the rudiments of a research proposal for Unilever, involving the toxic action of Citral and Geraniol on the blood vessels of the eye; both these substances were used as scenting agents in the manufacture of soaps and detergents. Howard Florey suggested two simple and one more complex experiments which he would help conduct. In the end, he would send Hugh data and details of the work. Hugh responded by inviting Howard Florey to attend the conference on essential fatty acids he was planning for the following summer.

CHAPTER 18

Plans for Nuneham Park

It was 1957. Elvis Presley rocked and rolled, the Russians launched Sputnik, and Sir Wilson Jameson and Lord Porritt proposed and seconded Hugh Sinclair for the high honour of joining the Court of Assistants of the Worshipful Society of the Apothecaries. It was the year Hugh became a Founder member of that Society's Faculty of the History of Medicine and Pharmacology. This was also the year Hugh reapplied for his Readership in Human Nutrition, and was refused.

At the third International Conference on Biochemical Problems of Lipids, held in Brussels during the previous year, Hugh had arranged to have Oxford University host the first major international conference on essential fatty acids, in July 1957. Invitations were sent to experts throughout the world, both clinicians and chemists, resulting in an impressive list of participants. Unfortunately, the universality of the meeting was marred when scientific politics entered the scheme of things. A respected member of the scientific community wrote to Hugh and suggested that Dr George Burr and his wife, Mildred, from California, should be invited to attend, as it was widely believed that George Burr was the first person to identify the physiological importance of essential fatty acids. Although Hugh had visited the Burrs' laboratories and been cordially received by them during his trips to America, he refused to send them an invitation. As an explanation, he claimed their work was out of date in the context of the planned conference sessions. Hugh was again contacted, and again reminded of the Burrs's pioneering contributions to the study of essential fatty acids. It was unlike Sinclair to be inhospitable, but he responded:

> … the Congress has no funds available for this purpose since the membership fee has been kept as low as possible. In the second place, there are those (and I am one of them) who consider the pioneer work of Herbert Evans really led to the discovery of EFA. I think we might have a brief informal discussion about this perhaps during the conference since I know there are many sides to this

difficult question and personalities come in rather strongly to it. Of course, if Dr Burr cared to come as a member, we should all be delighted to see him but he has not worked in this field for many years now …

Hugh's reason for not inviting the Burrs may have been related to the fact that another speaker, A. E. Hansen, of the University of Texas, was presenting a paper on essential fatty acid deficiency in infants. Hansen's work at the time involved maintaining young human infants, absent from their mothers, for three weeks to six months in a metabolic unit where they were fed various dietary formulae, some of which were fat free. Certain of these were supplemented in linoleic acid. The children deprived of linoleic acid developed skin symptoms that were described by Hansen as 'not unmindful of that observed in young puppies with fat deficiency'. The paper Hansen presented in Oxford concluded that the inclusion of linoleic acid in the diet corrected the clinical evidence of unsaturated fatty acid deficiency. Here was human dietary data in line with Sinclair's own research on animals, and the general thesis of his letter to the *Lancet*. The use of infants in such a way is abhorrent today, but half a century ago it was less shocking. The Burrs, however, had objected to this form of experimentation. The facts are unclear, but it is possible that because Hansen's work added significantly to the content of the conference, Hugh was willing to exclude the Burrs to avoid conflict at the meetings.

Reading the proceedings of the conference, it is striking that although the Burrs were not among the one hundred and six conference participants, their work was repeatedly quoted in the papers presented there, and even formed part of the foundation of Sinclair's own contribution to the meetings. Such are the politics of science. The conference participants included some of the leaders in the area of research, including: E. H. Ahrens, Jr, the Rockefeller Institute for Medical Research; A. Antonis, the South African Institute for Medical Research; R. T. Holman, Hormal Institute; P. A. Isherwood, Unilever Limited; A. T. James (the man responsible for the creation of the gas–liquid chromatograph), National Institute for Medical Research, England; and others from Germany, France, India, Norway, Holland, Poland, Denmark, and Sweden. Larry Kinsell attended not only as a leading expert in the field, but also as Hugh's greatest source of support.[1]

Participants were housed in Magdalen and St Hugh's Colleges. Hans Krebs was invited, but to Sinclair's relief was unable to attend; Howard

Florey attended, however. In his welcoming address, Professor E. R. H. Jones, the Waynflete Professor of Chemistry, reviewed the then current state of knowledge concerning fatty acids, but before reaching the meat of his subject he stated:

> Two essential factors in ensuring the success of any conference are that the subject should be both important and timely. These are certainly true today. The relationship between heart and other diseases and fatty acids in the diet is now a very lively issue and one of paramount importance to mankind.

What better support could Hugh Sinclair have received for the topic he championed?

Until now, the study of essential fatty acids had been restricted by a dearth of scientific technology: that was to change. A. T. James and A. J. Martin had published a paper the previous year concerning the behaviour of common saturated and mono-unsaturated long-chain fatty acids in the gas–liquid chromatograph, and the types of results produced.[2] At the 1957 Oxford conference their paper describing the behaviour of polyunsaturated fatty acids on the same equipment was a highlight of the conference. This was a major breakthrough in the identification and measurement of these delicate molecules, and would allow rapid progress in both clinical and experimental research.

The conference was divided into six sessions, covering topics pertaining to both the chemical aspects and biochemical functions of polyunsaturated fatty acids. Hugh chaired the final general discussion of the group, and, with A. L. Macmillan, also of Magdalen College, presented a paper on the structural function of essential fatty acids. He began by reminding his audience that, in the 1930s, George Burr and his wife were the first to suggest that tissues of fat-deficient rats may have 'abnormal membranes'. [Later, in 1942, Burr wrote: 'undoubtedly the unsaturated fatty acids function primarily as building stones in cell structures and the mobile lipids'.]

Twenty years after Burr's original observation, and some months before the Oxford conference, Macmillan and Sinclair tested the hypothesis that in fatty acids deficiency there will be defective formation of new tissue in rodents. They concluded that increased skin permeability in rats and erythrocyte fragility in mice were manifestations of structural flaws in the tissues of fat-deficient animals. Increased metabolic rate was also observed in fat-deficient animals, and it was believed this was due to a structural defect in the mitochondria within cells, and therefore a disruption of certain key oxidative processes. The paper was well received.

Other papers explored the relationship between blood cholesterol levels and heart disease. One such piece of work was presented by Larry Kinsell, and was entitled 'Essential fatty acids, lipid metabolism and vascular disease'. As his working hypothesis, Kinsell stated the possibility that essential fatty acids lower plasma cholesterol by decreasing cholesterol synthesis 'as the result of increased efficiency of the systems involved'; conversely, in a deficiency of these molecules, cholesterol production would be increased. He said:

> There is no possible substitute for long-term clinical and chemical observation to determine whether relatively or absolutely high EFA diets have major value in the prevention or treatment of modification, much lesser dietary modification should have significant prophylactic value.[3]

Like George Burr, Ancel Keys, a man well known for holding views on cholesterol that contradicted Sinclair's, did not attend the conference. Like Burr, his work was present in the minds of participants, and the thoughts were not always flattering. Unbeknownst to Keys, he was frequently the subject of spiteful jibes contained in letters between Sinclair and Kinsell. Ignoring discretion, during the conference Sinclair made his feelings towards Keys an open joke. On the afternoon of Tuesday 16 July, the conference adjourned for a time so members could travel to Stratford-on-Avon to see a performance of *Julius Caesar*. To remember the occasion, Hugh revised one of Antony's soliloquies, part of which read:

> If you have plaques, prepare to shed them now.
> You all do know this corn-oil; I remember
> The first time Larry Kinsell gave a dose;
> Twas on a summer's evening, in his ward;
> That day we overcame the Keysii.
> [The entire soliloquy can be found in Appendix E.][4]

Overall, the conference was a great success. Sinclair took time from other activities to edit and publish the proceedings, which became a classic text in the study of polyunsaturated fats. Sinclair never wrote a definitive book on his field of study, unlike many other scientists. However, the published Proceedings of the conference approached the magnitude of such a contribution, and became a classic in the study of dietary lipids.

Despite what many might now think was clear evidence of the validity of Sinclair's view that there is a causal relationship between essential fatty acids deficiency and degenerative illnesses, there was little improvement in his relationship with Hans Krebs and others in the Department of Biochemistry. Knowing he would never be allowed to conduct experimental work in the department again, Sinclair set about the task of finding the funds needed to purchase a property large enough to house his Institute of Human Nutrition. He made some contacts with possible donors in England, but looked abroad for help as well. Larry Kinsell introduced him to a number of corporate leaders in the United States, hoping they would help fund his work in England. At the same time, Sinclair explored links with South Africa, and hoped to establish collaboration with the University of Witwatersrand, in Johannesburg. For funds to support the venture, he approached Sir Ernst Oppenheimer, a South African financier who had previously given generously to Oxford University. Sir Ernst was a sick man, having already suffered two heart attacks. Sinclair sought him out and explained the importance of essential fatty acids in the prevention of heart disease. Sir Ernst was personally interested in Hugh's ideas, but his aides hinted to Hugh that the possibilities of a contribution were modest. Nonetheless, Hugh's hopes were bolstered by the fact that there was interest within the University of Witwatersrand for collaboration; the only barrier was money.

Over the years Hugh would repeatedly approach industry and private individuals for funds to build his institute. There is no room here to describe all these occasions, and so the following story is presented as a typical example of his behaviour under this kind of stress.

About this time, Hugh was told that Nuneham Park was again available. There were rumours that the University had found a tenant who would occupy only part of the estate, and other tenants were being sought. Hugh was desperate to secure the remainder and, if possible, block any other use of the Park. He contacted several potential contributors, and hinted that a philanthropist was considering funding him [this must have been Oppenheimer]. He also claimed that senior figures at Oxford were strongly in favour of the property being transformed into an Institute for Human Nutrition. Hans Krebs, Hugh said, also welcomed the possibility of his setting up a scientific programme there. To block anyone else obtaining the property before he could secure funds, he went to see the University Land Agent, and afterwards wrote a confidential note to himself of their discussion:

24 October
I said that I had discussed the following matter privately with the President
of Magdalen and with Professors Krebs and Jones, and I now sought the
Land Agent's advice in confidence.

    A group of persons were interested in setting up a central Institute for the
study of Human Nutrition in this country which would be financed primarily by industry and located near a University but would not be part of a
University. The Scientific Adviser's Division of the Ministry of Food and the
Secretary of the Medical Research Council had shown interest unofficially.
There seemed from discussions so far held to be little difficulty in raising
the necessary recurrent sum, as is indeed done by the Nutrition Foundation
and by the National Vitamin Foundation in the USA; and the representatives of various important firms in this country had welcomed the proposal.
The appeal had not been launched pending obtaining a capital sum to buy
and equip suitable premises … in this letter, which was signed by 18 leading
persons interested in nutrition from 12 different countries, including Sir
Robert Robinson and Sir Robert McCarrison, specific mention was made of
the possibility of placing the Institute near the University of Oxford …

The Land Agent wanted to bring the matter to the attention of the
[Oxford University] Chest the next day. Sinclair asked that his name was not
mentioned at the meeting, but he told the Land Agent he could say that the
potential use for the property would be a large scientific research institute,
financed by industry. He could add that government officials had shown
interest in the project. Much of this, of course, was fantasy on Sinclair's part.

After the meeting, Sinclair was contacted. The University would
consider an offer from him, but he needed to act in a matter of days. Sinclair responded with a rather vague statement of intent, based on his
belief that Sir Ernst would eventually provide the money. The University
wanted more specifics, and gave him a few more days to find the funds.
They also offered Hugh the freehold of the mansion and grounds under
certain conditions, which would give him rights over parts of the
property not rented by the University's tenant. The price was £25 000.
But there were also conditions, the most important of which was that he
must obtain the consent of the appropriate government ministries.

It was now 29 October. To protect the University, it was agreed that the
contract on behalf of Sinclair's Institute must be signed no later than 8
November. The names of the proposed purchasers must be disclosed, and

evidence must be provided that the purchasers were in a financial position to complete the sale.

On 2 November, Hugh wrote to Professor B. S. Platt, now of the Human Nutrition Research Unit, the Medical Research Council Laboratories:

> I have been wanting to write to you for some time about some proposals which were put forward by a number of people over a year ago for trying to raise money from industry for financing nutritional research … The best suggestion seemed to be that money so raised might finance research through the very few existing laboratories for this purpose (such as yours) and also run an Institute that would be near a university but not part of a university. Private approaches to industry and to various people have met with a warm response and it is hoped to launch the appeal fairly soon from a memorandum which has been in a draft form for over a year and is now being tidied up and printed.
>
> I hoped I might have an opportunity to see you … but you were very busy. However, I have been asked by a very powerful person in industry to let him know at once in what way these proposals fit in or conflict with the proposals of London University for a Nutrition Institute under John Yudkin at Queen Elizabeth College. I know that Yudkin has approached certain persons in industry and also the White Fish Board about building an Institute near his present Department … could you tell me if his ideas overlap with mine, and (privately if you wish) if Yudkin has your support.

On 7 November, Sinclair wrote to the Land Agent, thanking him for his letter of 28 October, and asking for more time. He had been in London, he said, at meetings of the Physiological Society, and none of the companies that had promised money could produce it in the time set out in the Agent's letter. In fact, there is little reason to believe Hugh had received anything other that polite responses to his desperate attempt to find finances, and he failed to consider the time required by any large corporation to agree a contribution of the size he sought.

The University gave Hugh the additional time requested. To rally support, he planned a weekend party at Lady Place, and invited, along with other influential people, Hans Krebs and his wife, Howard Florey and his wife, and several key people from major corporations. Thanks to Rosalie's masterful charm and a good wine cellar, the gathering went well, but there was no sign of any rapid resolution to Hugh's financial plight. It was about then that Sir Ernst Oppenheimer had a third and fatal heart attack, leaving Hugh little hope of funds from South Africa.

Platt wrote to Hugh on 2 December. He had been away and just received his letter, he said. He knew nothing of any industry interest in the subject, and knew nothing of any proposals for a nutrition institute to be attached to the University of London. But he promised to look into the matter.

Desperation deepened. Hugh wrote to Morris Motors, reminding them of the early discussions he had had with Lord Nuffield in 1942, during the war. Hugh wrote:

> I believe he would wish to be informed about a very important project for the
> benefit of humanity that is planned for this neighbourhood and which
> indeed he would pass as he travels from his home to his office. This project is
> the establishment of a National Institute of Human Nutrition at Nuneham
> Park …

Hugh then described the problem of degenerative diseases and processed foods in the diet, especially unstable fats. He delineated how this problem results in defects in the structure of cells and of cement-substance that binds cells together. As a result, he claimed the body's susceptibility to stress and injurious or infective agents was increased:

> Hence arise those lesions that are treated with cortisone, such as certain skin
> diseases, ulcerative colitis, nephrosis and rheumatoid arthritis … The
> mucosae of both nose and intestine are defective: the former allows foreign
> proteins to pass through (hence asthma and hay fever) and also the virus of the
> common cold; the latter allows the virus of poliomyelitis to pass (hence 'polio'
> is now 'adult paralysis' rather than 'infantile paralysis') and fails to protect the
> intestine against the stomach's hydrochloric acid, hence duodenal ulcers. The
> defective structure of cells makes them more susceptible to the harmful effect
> of x-rays (hence possibly the increase in leukaemia) and to the cancer-
> producing effect of certain chemical compounds (hence lung cancer if
> cigarettes are inhaled, and cancer of the stomach and bladder). The require-
> ment of these essential fats is about seven times greater in males during the
> reproductive period than in females; therefore the incidence of these diseases
> tends to be much greater in middle-aged men than women. All these diseases
> are almost exclusively confined to the more highly civilised countries but are
> found in less privileged countries if our diets are introduced …
>
> Nutrition is now the most important single environmental factor
> affecting health, and the need for research and education will become yet
> more insistent … To meet this urgent need a number of persons have

proposed the establishment of a central or national Institute of Human Nutrition, where fundamental work would be conducted in close association with medical specialists and University scientists; disorders of the skin and nervous system would receive particular attention. In addition, the Institute would collect and integrate particular existing knowledge, and would disseminate this by education through various channels.

A draft Formation Agreement for his Institute had been prepared, Sinclair added, and Nuneham Park was available to house the Institute. Oxford University had offered to sell this fine property and its grounds for £25 000. If Lord Nuffield would help with the purchase of Nuneham Park, it could be re-named Nuffield Park. A change small in letters but immense in significance — and 'National Institute of Human Nutrition, Nuffield Park' might tempt him to visit the staff and postgraduate students frequently as he passed.

For maximum effect, Hugh used his most prestigious title at that time, and signed the letter 'Hugh Sinclair, Vice-President of Magdalen College'.

Lord Nuffield responded on 12 December. After thanking Hugh for the letter, he said he found the proposed National Institute of Human Nutrition interesting, 'if somewhat ambitious'. Owing to previous promises and commitments, he was unable to help.

It is important to remember that much of what Hugh Sinclair believed about essential fatty acids sounded like gibberish to many scientists: too many clinical consequences were being attributed to their biochemical activity. When these people were asked by potential donors from outside the scientific community to comment on requests for funds to support his work, many were dismissive.

On 21 December 1957 Hugh Sinclair and Hans Krebs had a conversation about Hugh's future within the University. Krebs challenged Sinclair to justify his continuing in the department. Unlike Rudolph Peters, who had forgiven Sinclair both his personal insults and his failure to publish, Krebs spent little time trying to understand Sinclair's more creative nature. On 30 December, Sinclair wrote his response to Krebs' challenge. At Hugh's request, the President of Magdalen College first read it, and then it was hand-delivered to Hans Krebs on 3 January. The rambling communication was six and a half pages of single-spaced text recounting Sinclair's current version of the past. Some of the text is included here to indicate Sinclair's state of mind at the time.

First, Hugh apologized for the letter's length, but said he needed considerable space in which to put his case accurately. Then he tramped over cold ground.

He began by suggesting that the criticisms he suffered in 1948 were because of his difficulties with Professor John Ryle, and 'the Brigadier in charge of Public Health in Germany'. He recounted how he first met Ryle in Cambridge, and admired him personally, and how they eventually had a disagreement over the value of experimental work; Ryle saying that no experimental work — on humans or 'lower animals' would be conducted in any building over which he had jurisdiction, forcing Sinclair to close his small animal house. Senior associates recommended Sinclair leave the University, he said. He told the Regius he was resigning because he could not see how nutrition could be studied without the application of the 'experimental method'. In return, Sinclair claimed, the Regius said a Committee was examining the future of Social Medicine and Human Nutrition: in the end, it was agreed they should be separated. None of this, Sinclair said, indicated a rift between Ryle and himself, who he claimed remained one of his most important supporters

As for the work in Germany, Sinclair wrote:

> In August 1945 I agreed to organise the nutritional work for public health purposes in Germany because I thought it my duty to do so although I wanted to get back to research. The war was still on, and it was essential — amid the protests that the allies were vindictively starving the defeated Germans — that scientific facts should be available to the Control Commission and upon these should be based the rationing policy for the three Western Zones … When I found repeatedly that the temporary Brigadier in charge of Public Health (an ex-IMS Colonel who hated the Germans) was altering the facts to suit political purposes, I told the Minister in charge that I must resign. I was asked to continue because the Minister put the greatest reliance upon knowing the true facts, and (although I also held the rank of Brigadier) I was put in the Senior Officers Mess with the Major-Generals in charge of departments to show I was not subordinate to Brigadier Martin in charge of the Division of Public Health. This was not my wish and it was a mistake, for I was aged 36 and the Brigadier over 60.

Sinclair suggests that Krebs should talk with one of the men who was with him in Germany, and also one of the men who was a Minister for part of the time the events in Germany were taking place. He also pointed out that he had received the Medal of Freedom with Silver Palm from the United States, and was offered a nomination for the Legion of Honour by French officials: an offer which he refused. He quoted the entire Citation for his Medal of Freedom because it specifically mentioned 'mutual scientific co-operation.' He continued '... this charge of unco-operation in Germany is constantly brought against me, particularly by McCance who was working in Germany for a totally different reason from mine ...'

Further on in the letter, Sinclair stated that Peters had been unsure that unsaturated fatty acids were required by man, and claimed that:

> ... on 22 March 1956 I asked you whether you thought essential fatty acids were worth working on in the Laboratory and you said you did not think deficiency occurred in man although you assumed Evans's work on lower animals was all right. Ancel Keys, who worked with me knowing my interests in fats and coronary thrombosis, could not see then the relevance of the work Basnayake and I were doing on the deposition of esterified cholesterol and phospholipid ... I felt this attitude to the unimportance in man of EFA and perhaps B_6 ... to be so strong that it might prejudice the chances of a student for a research degree, and therefore no student I supervised (other than two intelligent Asiatics — Ramalingaswami and Basnayake — and Macmillan very recently) ever worked on them.

Beginning with a recounting of his own struggles to study EFA and degenerative conditions in humans, including some self-experimentation, Sinclair then launched into a discussion of the three physical facilities he needed to conduct research on this topic. First, he needed clinical material: current offers from various clinicians, dental school, a commercial firm, and the Ministry of Health would help fulfil that requirement. Second, he needed a laboratory in which to conduct his 'chemical estimations'. The laboratory he had constructed at Lady Place, his home, in 1934, would be where he would undertake that work. Third, and lastly, he required a place where he could draw together the extensive work on the Hollerith cards, now spread among several locations.

> I have learnt to produce by dietary means in rats and mice deposition of cholesterol and phospholipid analogous I believe to that occurring in atheroma in man, duodenal ulcers, osteoporosis, nephrosis, ulcerative colitis, a condition similar to the basic defect in the collagen diseases, increased sensitivity to x-rays and to chemical carcinogens, increased

sensitivity to vitamin D, and a condition superficially resembling
rheumatoid arthritis. In pigs I have similarly produced conditions
superficially resembling the arthritis of gout and spondylosing arthritis. I do
not like to publish scientific papers until I am sure of the facts, and much of
the histological work has not yet been completed (we have never had a
histological technician) … I believe Human Nutrition forms an excellent
bridge between preclinical and clinical departments, and such bridges
appear to be all too few.

If therefore the Electors thought fit to continue my readership I could
continue to 'engage in advanced study or research' [though I should prefer
to interpret these as not being alternatives], and to 'lecture and give
instruction', but it would be necessary to delete the obligation to 'supervise
research in the Laboratory of Human Nutrition'.

During their earlier conversation, on 21 December, Krebs asked
Sinclair for the names of people of Hugh's generation who might give an
opinion on his work. Sinclair listed Fredrick Stare, at Harvard, a
Professor at the University of Illinois, who had worked in Peters'
department after Sinclair did, and a French clinical nutritionist who took
part in the tripartite meetings in Germany during the days of the
Nutrition Survey. There were also names of people in the Ministry of
Food, and clinicians both in Britain and abroad.

Appended to this long letter was a list of papers, books, and chapters
Sinclair had published since he first applied for the Readership in 1948.
He also listed his major lectures, both in Great Britain and abroad.

27 January 1958   Hugh Sinclair to Hans Krebs:
A very senior and experienced member of the University who saw my letter
to you of 30 December, said that from what he had heard he thought it was
very important that the letter be circulated to the other Electors. He added
that this was quite usual procedure and the Registry would do the
duplication and circulation.

Would you have any objection to this? If not, I am not quite sure what
would be the correct procedure — for you or for me to send a copy to the
Registry, and I should be most grateful for your advice.

I apologise for troubling you further.

There is no record of a response. Hugh's Readership in Human Nutrition
was not renewed. No longer a member of the Department of Biochem-

istry, he would, however, continue enjoying his work as a tutor at Magdalen College and lecturing wherever and whenever he was asked. His reputation as an inspiring and entertaining speaker was already well established, and he rarely refrained from accepting an opportunity to state his views on essential fats in the hope of recruiting others to hold his view of their vital importance to human health. As we will see, other subjects caught his attention, but it was the challenge of explicating the conundrum of dietary fats that fascinated him.

On the same day Hugh handed his long letter to Hans Krebs, he received a letter from the University Chest Estates Office: the proposed renting of Nuneham Park had failed to materialize — and the property was up for sale. Was he interested?

Indeed he was, and in his attempts to find the necessary funds he called on people he hardly knew, tossed about names like confetti at a wedding, and made financial promises he could not keep. He repeatedly crossed the paths of silent enemies. For almost two years he manoeuvred and schemed, and sometimes overlaced facts with wishful thinking — all without result.

Finally, it became obvious that — at least for the foreseeable future — the answer to his search for a place was clear. Lady Place would be ideal. He looked at the mock-Tudor building and saw in it things he had described a thousand times, and would describe a thousand times more: a great library and information bureau; a source of data that scholars might use when seeking new relationships between diet and disease; a laboratory filled with modern equipment and benches for a team of research scientists; and an animal house for experimentation. Here he could build clinical links within the local medical community, and here was a place from which he could continue to entertain and fire the imagination of those he hoped would fund his dominion. Sinclair had entered his scientific 'wilderness', but at least the wilderness was his.

CHAPTER 19

Grand dreams and little money

In late January 1960 Hugh Sinclair received a letter from Hugh Smith warmly acknowledging his Christmas letter, and saying that he had been following the debate in the *Lancet*. Now living in Arizona, Smith had had time to reflect on the nature of modern science. The world had changed, he observed:

> I fear that research workers on this side, at least, are far too concerned with electronic gadgets to ever take time to really think what they are to be used for. When one recalls the simple equipment that men like Prof. Peters used to carry out their studies, one cannot refrain from wondering. I am not, of course, opposed to labour saving devices, but only the façade that goes for competency so often in our labs.

Hugh was pleased to hear from his old friend, but would gladly have purchased some of the electronic gadgets Smith decried if funds had allowed. The financial pit was deepening. Sinclair wrote and edited works about nutrition whenever possible for income, and was pleased when Robert Maxwell, of Pergamon Press, asked him to edit what he planned to be a twenty-volume encyclopaedia of human nutrition; unfortunately the series was truncated. Tutoring at Magdalen College provided a small income, and there were profits from his flower and produce shops, but all this barely covered the running costs at Lady Place and his not inconsiderable personal expenses. Frequent lunches at the Athenaeum were a necessity. The day before his 50th birthday, Hugh was invited by Hans Krebs to lecture on nutrition for 'three guineas' an hour. Hugh gladly accepted, and as a result in May 1961 his name was added to that of members of the Faculty of Physiological Studies.

To breathe new life into his Institute, he changed its name to the National Institute of Human Nutrition and prepared a new proposal with which to solicit funds: this time, he sought money from the food and pharmaceutical industries. Instead of presenting the entire package of ideas to each potential donor, as he had done in the past, Sinclair divided the proposed work into a number of ambitious projects to 'sell' according to the interests of specific companies. Projects focused on the analysis and dissemination of information, clinical studies, and experimental work. He also included plans to purchase new premises for his work, as the familial tension at Lady Place was, at times, unbearable.

An approach to the food industry was appropriate, he believed, because the food industry would become more important as populations increased, making 'processed and synthetic foods' more needed. Sinclair thought that it was not the duty of the food industry to study the long-range effects of such food upon man. 'The extensive problems in human nutrition cannot be watched by industry, by public health authorities or by a university department,' he wrote. 'They can and should be watched by a properly constituted National Institute of Human Nutrition where fundamental research is continually being undertaken. Such basic research has a long-range value to industry and to society.'

Hugh's ambitious plans were badly flawed, however. He overlooked the political problems an independent institute as potentially powerful as the one he proposed would face. With so much at stake, government organizations and private corporations already in existence would surely want to subsume his plans, or discredit them, before they had a chance to breathe air. To make matters more difficult, he had no commitments of support from any member of any group of potential financial contributors. Attempting to raise money without some cash in hand is always difficult. Next, he had no formal agreement of cooperation from any clinical facility. And lastly, he had no evidence of any significant work in progress. This was the most serious flaw in his plans: Sinclair had little technical or administrative capacity even to begin most of the research work described in his proposal.

A more modest and better focused beginning might have flourished. One idea with potential was that of a central library on Human Nutrition. Sinclair owned a fine collection of reprints and reference materials, which he estimated to contain more than 70 000 items. Using this collection as the basis for a comprehensive reference centre for work on human nutrition could have resulted in a valuable national resource. Unfortunately, he had little time for something of such limited scope.

Sinclair decided his Institute required a building with remarkable architecture that would inspire people to support his cause. The search for an appropriate architect led him to seek out Lord Alfred Bossom, an internationally known architect and a friend of the Sinclair family when they were living in Maidstone, Kent. Hugh sent Lord Bossom a copy of the new proposal for the Institute, offered him a leading role in its future, and asked him to help solicit funds. Lord Bossom agreed that careful consideration should be given to the building that would house an undertaking as important as that which Hugh proposed. However, he had little time to give the venture as he had been made a Peer. Sinclair let the matter drop.

Although funds were limited, Sinclair found time and money to attend scientific meetings around the world. While in the United States he visited his old friends, including Larry Kinsell and Fred Stare and his wife. He also met Dr Paul Dudley White, President Eisenhower's cardiologist, and motored to Canada to spend a day with scientists whom Sinclair considered to be authorities on the Eskimos.

Towards the end of 1960 Hugh Sinclair wrote to Sir Wilson Jameson and invited him to lunch at Magdalen. He wanted to know if Jameson felt Lord Bossom was a suitable candidate for the position of Honorary Treasurer and Chairman of the Executive Committee of the proposed National Institute of Human Nutrition. He added that he hoped Lord Bossom would initially sketch a building for the Institute that might attract capital grants from philanthropists. Jameson was unwell and unable to lunch, and subsequently did little to help. In the end, it was left to Sinclair to approach Lord Bossom again. First, however, he set about drawing some architectural plans of his own. Early in 1961, he wrote again to Lord Bossom:

> A little over a year ago I had some correspondence with you about the proposed National Institute of Human Nutrition, and in particular I hoped to have the privilege of discussing with you the type of building we might erect … I now venture to inquire if you might happen to be free to attend this year [the Judge Randolph Dinner, at Magdalen College] on Saturday March 25. I am sure you would enjoy the company and you would I hope stay the night. Yours sincerely …

Bossom accepted the invitation, saying he would be pleased to stay overnight and discuss the kind of building that might be suitable for the proposed Institute.

Hugh responded, naming others whom he thought would attend the dinner, including such notables as the Speaker of the House of Commons, Lord Gladwyn, and the President of the Royal Society, who was then Sir Howard Florey. But Hugh knew he could best impress Lord Bossom by entertaining him at Lady Place. Pleading that the plumbing at Magdalen College was inferior, and accommodation less than comfortable, Sinclair changed the invitation to include an evening at Lady Place, about which he wrote:

> My mother, aged eighty, is living there and would I know like to have some news of Maidstone. My father bought Barming House in 1912 and apart from the war we lived there until his death in 1924 when we moved to Wateringbury. Barming House was an enormous mansion, which was then pulled down to make a housing estate …

Hugh then named many of his father's titled friends. In the text of the letter, he exaggerates on some points: for example, he said Barming House was owned by his father, when in fact, it had been purchased by Sir John Jackson for the family and was never owned by the Colonel. This was part of what was becoming the Sinclair Narrative: the adjusted story of his life devised to meet his professional ends.

Hugh met Lord Bossom in London and drove him to Oxford. With the help of Rosalie's great charm, the entire event was a complete success and on his return to London, Lord Bossom wrote:

> My dear Dr Sinclair, will you please give my sincere thanks to your mother and to your sister for their most gracious hospitality and care on Saturday evening, and may I add to this my warm appreciation of your having me as your guest at the Magdalen dinner. Needless to say, I was exceedingly interested in the information you gave me about your most important research work in connection with the food for the human race. It is one of the most fascinating topics I have encountered for a long time and I do hope to hear much more about it. Incidentally, will you lunch with me in the House of Lords after Easter? I should be so pleased … Personal regards Yours most sincerely, Bossom of Maidstone.
>
> PS. Would you like me to ask a few appropriate questions regarding the increase of certain diseases?

Hugh had found a new champion, but it was weeks before he answered Lord Bossom's letter. Tragedy intervened.

29 April 1961

Dear Lord Bossom, I must apologize for my discourtesy in not replying earlier to your letter of 27 March. Shortly after I received it my sister (whom you met at my house) was killed in a motor car accident when taking some guides to a camp. As a consequence of her wide public activities, particularly with children, I have had five hundred letters which paid tribute to the work she did. Apart from the sympathy these letters expressed, the only consolation that I have is the assurance that there is nothing she could have done to prevent the accident or protect the children in her charge. I have been thinking a great deal about your admirable suggestion that we should get Lord Woolton [Minister of Food during the war] to associate himself with the proposed National Institute of Human Nutrition. His outstanding achievements as Minister during the war make it very desirable his name should be associated permanently with the Institute and the magnificent site we have in mind for our building would link it geographically with the names of Churchill and Nuffield. I wonder what the best approach to him is? I should be delighted to accept your kind offer that I should lunch with you in the House of Lords and I wonder if you might think it appropriate that we could discuss it with him then. … With kind regards, Yours sincerely …

Lord Bossom wrote expressing his condolences and suggesting three dates for a meeting in the House of Lords. All the dates were suitable for Sinclair. As Sinclair's closest contact with the Ministry of Food during the war was Sir Jack Drummond, he had had no significant contact with Woolton. He needed an introduction. A letter was drafted to be sent by Lord Bossom to The Rt Hon. The Earl of Woolton:

Of course everyone knows of your great reputation concerning food for the nation, and Dr H. M. Sinclair of Magdalen College, Oxford, has expressed the real wish to have the opportunity to exchange a few words with you. He is particularly interested in the various stimulants now being used to get better results in agriculture, and I understand a great many people in the University of Oxford have the matter seriously at heart. There is no question that he would ask for money or anything of that sort, but you are looked upon as such a universal authority on matters relating to food that I know he really would appreciate such an opportunity. If you felt able to fit this in with your busy life I would be so pleased if you would join us for lunch at the House of Lords and I am suggesting one of the following three dates …

13 May 1961
My dear Bossom, thank you for your letter of 8th May. I have a great
admiration for Dr. Sinclair and shall be very glad of the opportunity of
meeting him. I shall be in London on Wednesday, 31st May, and delighted
to join you at lunch in the House of Lords. With all good wishes to you.

Yours sincerely, Woolton.

The date was set. The day before the lunch, Hugh wrote to Larry Kinsell
thanking him for a carbon of a letter containing gossip about people they
knew, and bragging about his up-coming lunch with Lord Bossom and
Lord Woolton. He failed to mention the interesting fact that both these
men were of his father's generation, and had had no direct contact with
him during any previous part of his career.

The luncheon went well, but about a fortnight passed before Sinclair
wrote a letter of appreciation to Lord Bossom. Bossom's suggestion of
Lord Woolton as President of the National Institute of Human Nutrition
was 'a brilliant one', Hugh wrote, and that it was clear that owing to Lord
Bossom's advocacy, Lord Woolton was deeply interested in the project.
He also asked that Lord Bossom be Honorary Treasurer for the Institute.
Bossom responded: he was pleased to accept, and would help Sinclair's
cause by raising a question for him in Parliament. After this exchange of
letters, Sinclair failed to contact either Lord Bossom or Lord Woltoon.
Finally on 21 July, Lord Bossom wrote to Sinclair, suggesting that they
meet again soon. Another five months passed before Sinclair responded.

In part, a deep grieving over Catherine's death caused this rude inattention. Both Hugh and his mother were desolate. He had lost a close friend
and she a daughter and companion. Such suffering should have bound
them together, but instead it drove them further apart. At this time, Hugh
occupied New Buildings III, 4, in which he had been resident since at least
1945, and where he now retreated to his rooms away from Lady Place.
Rosalie was left alone with servants for company. She wrote to Hugh, asking him to give her more space in the house so she could more easily
watch the comings and goings around the property; specifically, she
wanted to claim the morning room. As an excuse, she played on her
advancing age. This was the final straw for Hugh, and years of bitterness
poured out. Despite the fact that he now spent little time at Lady Place, he
begrudged his mother the space she felt she needed. On 29 May, he wrote
to her:

… You are over 80 and I am over 51. Half your children have died and there is
not the slightest reason why I should not die before you: Granny lived to be
over 90 and John died at 48. I cannot be expected to live here on the
assumption that you will die soon and then I can do what I want to regarding
the house. Those people I've consulted are surprised you should want two sitting rooms to yourself, one just so that you can watch the front door better …
One possibility is for you to live here with a companion and I'll make a home
for myself somewhere, because at the age of 51 it is not unreasonable to want
to have a home where I can entertain my friends as I would like … I don't
think, Mother, you always weigh up the pros and cons quite right. To give two
examples: At Gransden you could not let me have a bedroom to myself
because — you told me — if I did you would not be able to put up Granny,
Aunt Dorothy and the two children. In consequence, as I couldn't sleep with
John (since he kept the windows closed and the room smelt of cigarette
smoke), I had each night to wait until he was asleep and then go and sleep in
my dressing gown on the ottoman in the big spare room, returning before he
woke in the morning. As I was working very hard I had a nervous breakdown
through lack of sleep and received treatment at the Charing [Cross] Hospital
(I said I was studying Anatomy there). It could be argued from this that Aunt
Dorothy's son was vastly more important to you than your own. Secondly,
John knew perfectly well (as I did) that, having had 3 coronary thromboses,
he would have to give up gardening and going for walks, and might soon have
difficulty in going up stairs. He told me the way he would probably have to
spend the day was reading and playing billiards if we had a table … So
naturally I got a table for him. But you complained that this would mean that
occasional bridge-parties occupying both the drawing rooms … and the
billiard room could not occur … I thought that … [on these] occasions extra
tables might have been willingly sacrificed for John's daily exercise … Hugh

In September 1961, Rosalie suffered a blackout, fell on the stairs, and was
admitted to the Radcliffe Hospital. She wrote to Hugh telling him of the
accident and asking him for a receipt for a £2000 loan she had made to him.
This, it was said, was needed to cover work on Lady Place, but it also
covered much of his international travel, although this was a matter rarely
mentioned between them. The loan from his mother seemed to lighten
Hugh's financial burden for only a time, but he soon asked if she could see
clear to increasing the sum. Her answer was negative: Catherine's estate
would soon be settled and Hugh would benefit from that.

Hugh was desperate for laboratory space in which to carry on his experimental work. He wrote to the Administrator of the United Oxford Hospitals, asking if he could use the old washhouses on the Churchill Hospital campus as a biochemistry research laboratory. The letter, dated 12 December, was full of exaggerations. Sinclair wrote in confidence that he hoped to build accommodation for the National Institute of Human Nutrition very shortly, and announced that the Earl of Woolton was now President of the Institute and Lord Bossom was Honorary Treasurer.

There was news of a charitable trust that might be a source of funds for the Institute, and Hugh informally approached Sir Charles Dodds, newly appointed Chairman of the Executive Committee of Hugh's Institute, on the matter. Dodds told Hugh that he should hire a professional fund-raiser to help with an appeal to industry. What would be the design of the Institute, he asked? Hugh described a twenty-storey contemporary building, unlike anything then existing in the City of Oxford, with two wings, leading from the south. The possible site was a field of seven acres situated north of the Churchill Hospital, he said. Early in January 1962 another letter arrived from Lord Bossom, whose name Hugh was now using widely as an active supporter of his planned Institute. Bossom had received word that Hugh now contemplated building an edifice twenty stories high in the City of Oxford. The letter read, in part:

> I do congratulate you on designing your project as successfully and working out such a comprehensive solution, but may I call your attention to a very definite risk you are running of having your entire project bitterly condemned by many lovers of Oxford who will think you have carried the idea of a tall centre unit too far. When I suggested you should have a higher construction in the middle, I am afraid I was not envisaging anything very much taller than now exists in the town. I should think ten storeys might be taken and respected, but twenty will dominate everything else in sight ... I do think you have got a magnificent idea and a good plan, and I think you want to be awfully careful not to have some rather unfortunate comparatively minor arrangement that would cause condemnation of the entire project, for there is no doubt that a twenty-storey building in the middle of Oxford would have to be seen no matter in which direction you approached the town ... Do let me know some time when you come to London as I would like to talk it all over with you again some time.

Hugh responded to Bossom a month later, defending tall buildings, and reporting that new plans for hospitals to be built in the area called for 'tall' buildings. He was willing to reconsider, however, and base the building on a ten-floor arrangement. Should they engage an appeal organizer, he asked Lord Bossom? And, could Lord Bossom meet him for lunch at the Athenaeum?

Dropping Lord Bossom and Lord Woolton's names where necessary, and being liberal with his exaggerations, Hugh began recruiting people for an Appeal Committee.

> We have a magnificent site at Headington overlooking the city of Oxford on which to have our permanent building …' he wrote. 'Lord Bossom (who is of course a very distinguished architect) has proposed that we should have a main building ten stories in height which will carry the name of a single benefactor if we are fortunate in securing one, and in front of this will be the six-storey Woolton Building for education of the public in food practices throughout the world …

When all this is read, it should be kept in mind that there was no written agreement between Hugh and anyone as to their formal role in the structure of the planned Institute.

War continued at Lady Place. Rosalie again wrote to Hugh, asking for his permission to reallocate the rooms in Lady Place so she could better store some of her furniture, now that both John and Catherine were gone. He responded a few weeks later, threatening to leave Oxford:

> Mother, I do not see from your note how we can go on living here together. … I'm quite prepared to go to America and start again, but I'm not prepared at the age of 52 to go on wasting my time … So will you please write down for me which rooms you want for your things and then I'll decide what I'll do about the future?
>
> H.

Rosalie responded with a long letter, which she ends: 'Mother/feeling very unwanted and misunderstood'.

Later in May 1962 Rosalie wrote to Hugh regarding his threat to move to the United States. In her eighties, but still writing in a fine hand, Rosalie Sinclair addressed her son, saying that she understood he had a life of his own; she said she had always hoped he would marry, and that she would have gladly fallen in with whatever plans he then made. But she did not understand what he wanted now. She and Catherine had often talked of moving to a smaller house together, she said. With Catherine gone, a smaller house on her own might still be the right solution. Certainly, if Hugh went to America she could manage for a time by herself, but a permanent relocation would be a different matter. She had moved to Lady Place to help him in his work; did Lady Place continue to be necessary for his work? Perhaps there was more scope for research on nutrition abroad — he must make his plans and not let her stand in the way. Hugh's response was to begin a legal process that eventually led to the breaking up of Lady Place into smaller rental units to supply badly needed income.

The National Institute of Health, in the United States, was granting money for projects on ageing. On hearing of this, Hugh designed a research study, wrote a brief proposal, and sent it to Washington DC. The United Oxford Hospitals [UOH] wanted to participate in the work, and had agreed to support the clinical aspects of Hugh's project as he had developed the plan with a member of its staff. Obviously under the impression that there was a very good chance the funds would be forthcoming, the UOH Administrator wrote to Hugh, suggesting that a formal application be made to the Hospital's Board for Sinclair to make use of the huts on the hospital grounds. Hugh agreed at once, and asked that the Board accept any grant made.

In July, Hugh received word from Washington that his grant application was not funded, and he so informed the United Oxford Hospitals. Unaware that this grant application might fail, the UOH were about to distribute Hugh's proposal for research into ageing to people at the University, including Hans Krebs, who was then the Head of the Department of Biochemistry and Chairman of the Research SubCommittee. On learning this — and not wanting his proposal scrutinized by Krebs — Hugh responded in haste:

> To save the trouble of Sir Hans, your Sub-Committee [sic] and the Board I have decided I do not wish to continue with the project.

He was, however, still interested in the offer of help regarding accommodation in two disused huts on the Churchill Hospital campus:

> You will recall that these were the latrine and washhouse of a former prisoner-of-war camp. When I was permitted to use them in 1946 I removed the latrines … myself and did all the plumbing to convert one into an animal

house and the other a laboratory. They have been empty for the eight years since I vacated them in the summer of 1954. Of course I recognise the force of the argument that if, for instance, the Nuffield Professor of Obstetrics were to die tomorrow his successor might want these erstwhile latrines for laboratories, but they might not be ideal for his purpose being far from any beds, having no gas or electricity, no heating or other facilities. (I have my own laboratory benches and equipment in store.) Nevertheless, I would (if allowed to use them temporarily) vacate them at a month's notice whether required in that unfortunate event or for any other purpose.

The Administrator of the United Oxford Hospitals wrote in return, expressing regret Hugh had decided to discontinue his research plans: the first part of Hugh's letter had been circulated to the research SubCommittee (and therefore Hans Krebs), and the second part, concerning the huts, had been sent to the Establishment Committee. Along with a formal letter, the Administrator penned Hugh a private and personal note, saying how sorry he was to hear the project had been turned down in the United States. He also wrote:

> My opinion of a number of people had gone down since all this happened but I had better not write more even privately!

Obviously, Hugh's feeling of being undermined by members of the Oxford Community was received with sympathy by at least one of its members.

In early December 1962, a letter arrived from Lord Bossom: he had not heard from Hugh since February. He had received a letter from Lord Woolton, and a copy was sent to Hugh. Had they communicated, Bossom asked? The following is from Woolton's letter, as found in Sinclair's files:

> 3 December 1962
>
> My dear Bossom, You will probably remember inviting me to lunch with you and Dr Sinclair of Magdalen some months ago. From my point of view it was of great interest because, as I look back, I realise that I did not make the public realise the outstanding part that nutrition played during the war — through the operations of the Ministry of Food — and the part, almost as important, that it might play in the normal life of the nation. I ought to have done more to make the nation realise this, but at the time I was very much preoccupied with the more direct issue of trying to keep the country alive on the meagre food resources that were at my disposal. Dr Sinclair set my mind working on this subject and I have thought a great deal about it.

Last week I was the guest of the General Medical Council at its Seasonal Dinner, at which there are no formal speeches, but the members are, as I'm sure you know, very distinguished members of the profession. I took the opportunity of talking to them on this subject. During, and immediately after, the war, my name was associated with stringency, unpalatable food, and the like, but I did succeed in educating women as to the value of nutrition ... Now the atmosphere has changed. The General Medical Council sat down to a 6-course dinner, excluding the trivialities of nuts and fruit, etc. I reminded them of the vast fortunes that were being made by chemists and specialists in slimming and of the fact that according to the women's papers, a woman's 'vital statistics' — and I probably misquote them — are something like 26, 36, 38. I always associated 'vital statistics', with infant mortality and duration of life — not waist measurement! Now is the time when so much could be done to educate the public to a proper appreciation of food values and I have been wondering if you and your friends in Oxford have made any further progress. At any rate the attitude of these eminent members of the medical profession demonstrated a deep interest in the subject.

Ever yours — Fred Woolton.

On 4 December 1962, Hugh Sinclair responded to Lord Bossom. He was delighted to have received Lord Bossom's letter, he wrote, and had been meaning to write for some time. Over the summer, at the International Congress of Food Science and Technology, held in London, an eloquent appeal was made for the establishment of an international nutrition institute in Great Britain 'at once'. Also, Hugh reported, he and Miss Dorothy Hollingsworth, both permanent members of the Nutritional Congress of Representatives of European Countries, heard that nutritional institutes had been set up in France and Holland, and were being planned for Belgium and Scandinavia.

'Whether we join the Common Market or not, we must not lag behind and rely for our information on central Institutes in other Common Market countries,' Hugh wrote to Lord Bossom.

Hugh briefly sketched his expanding plans for the Institute. The library and museum must be called the Woolton Building, he noted. This building will not only house the facilities to counter 'food fads and fallacies', but also overcome ignorance:

Only last night on the radio Dr Cicely Williams (who wishes to join the National Institute's staff) said that the main World nutritional problem was ignorance; and from her experience in 40 different countries that comprise part of the two-thirds of the world's population that is undernourished, better use of food through education would bring great dividends in health and productivity. It is suggested that some of the leaders in different fields might be assembled at a luncheon to discuss the Institute. I enclose a list of some obvious names. I should greatly value your advice about this especially since late January might be the best time ... As Lord Woolton says in his letter to you, 'now is the time when so much could be done to educate the public to the proper appreciation of food values'.

Sinclair attached a list of 34 names he suggested might attend a meeting at the end of the following month, including: Lord Boyd-Orr, Lord Cohen of Birkenhead, Sir Edward deStein, Sir Charles Dodds, Viscount Hankey, Lord Rank, Lord Sainsbury, and Sir Solly Zuckerman. Although the list included the President of the British Food Manufacturing Industries Research Association, the Secretary of the Medical Research Council, the Chairman of the 1st World Congress of Food Science and Technology, the Chairman of the Food Manufacturers' Federation, the Chairman of the Milk Marketing Board and a Vice-Chairman of Unilever, the only Oxford name was that of the Chairman of the Oxford Regional Hospitals Board. The academic scientific community was ignored.

On 12 December, Lord Bossom responded with sound advice:

Dear Dr Sinclair. Thank you for your letter of the 13th. If you could produce a good programme with expert speakers who would not take up too much time and have the luncheon under the auspices of some recognised highly placed authority, I should think it would be a most valuable procedure. To do this, however, would be a considerable expense, and probably largely wasted unless it was backed by the most reputable authorities that exist. If you are coming to London for any purpose in the near future let me know, and we could talk over the matter. I think your list is an exceedingly good one; though a lot of them might be quite willing to lend their names in some way but would not be able to give much money if that would be your main object. If it is influence you want, the group you have outlined could get it if they are convinced; if it is money, I think the list ought to be considerably revised.

Yours most sincerely. Bossom of Maidstone.

From that point onward, Hugh largely ignored both Bossom and Woolton.

During much of 1962 Hugh Sinclair's energies were drained off into a defamation case, in which the manufacturer of a product containing essential fatty acids, claiming to protect against the effects of mild radioactive fallout, used Hugh's name in one of their marketing brochures. From the way the text was written, it could be construed that Hugh Sinclair was in some way connected with research conducted by the company. This was not the case, but Hugh had met with officers of the company, and had shared his views on the protective nature of essential fats. Based on material in Sinclair's files, it is probable that he pursued the matter as much for money, which he desperately needed, as for a public apology. The case was to drag on into 1964. The company involved had limited resources, and Sinclair had trouble meeting legal costs. In the end, it all came to nothing. What would have happened if the time and resources poured into this pursuit could have been used publicly to capitalize on the help offered by Lord Bossom and Lord Woolton can only be guessed at.

Rosalie Sinclair now suffered giddy spells when getting up from a chair or the sofa. She no longer drove her car, and her doctor suggested she use taxis: she said they were too dear.

On 13 February 1963 Lord Bossom and Lord Woolton lunched in the House of Lords and discussed Sinclair's plans before contacting him further. They believed any contributions to his Institute would come from companies, but not individuals. The previous day, Sinclair had written to Lord Bossom excusing his lack of communication by commenting on the recent scientific meetings he had attended. Then, he made a fatal mistake. He asked privately for Lord Bossom's thoughts on Sir Charles Dodds as a possible candidate for the position of Chairman of the Executive Committee of the National Institute of Human Nutrition. This was asked despite the fact he had previously asked Lord Woolton to serve as President of the Institute, and had already discussed the matter with Dodds. As it turned out, Lord Bossom wrote to Hugh Sinclair on both 13 and 14 February 1963. The first note, written before he received Sinclair's letter concerning the Chairmanship, was warm and supportive. The second note was more direct. He was afraid Woolton and Dodds might clash, and that Sinclair should address his question to Lord Woolton, as he had previously asked him to be President of the Institute.

Sinclair answered Bossom on 5 March, saying that he had been introduced to someone with experience of organizing appeals — in other words, he did not enjoy the advice of Bossom and Woolton on the matter of funds.

Sir Edward deStein, of the Gallahers Tobacco Company, was another man targeted by Sinclair as a potential contributor to the Institute. While Hugh was Vice-President of Magdalen College (1956–57) Sir Edward had been his guest at the Judge Randolph Dinner. Now, some five years later, Hugh invited him to attend the same event; deStein was delighted. Using a now standard technique, Hugh wrote to his guest explaining that, owing to the condition of the plumbing arrangements at the College, he would be pleased to put Sir Edward up for the night at Lady Place: there would be room for Sir Edward's chauffeur as well. He included a list of important people invited to the dinner; neither Lord Bossom nor Lord Woolton was mentioned. On a copy of a letter containing directions to Lady Place, Hugh made a number of hand-written notes about deStein, including:

> He is having another operation … in a few days. Voice husky — recurrence of carcinoma. He is interested in how fish recognise food … He built up one thing at a time: deStein's bank … Gallahers. G. has spent [much money] on its research into tobacco and cancer …

According to his letter of thanks for Hugh's kind hospitality, Sir Edward said he enjoyed the occasion. He was off on travels, but hoped to entertain Hugh after his return. There was no promise of funds.

Lord Bossom wrote to Sinclair on 18 March suggesting they meet with Lord Woolton some time after Easter to discuss further plans for the Institute over lunch in the House of Lords. The very next day, 19 March, Lord Woolton wrote to Sinclair in a less than supportive tone:

> Dear Dr Sinclair … I am … and perhaps due to increasing age — failing to see where there is a point of contact. Everybody who is interested in this problem is interested in so many other things, and whilst I am sure that knowledge is growing all the time we are not getting together an organisation that might centralise such knowledge. I am sure that the scientific interest is there, just waiting to be aroused …

He said that he had and would continue to speak in support of a national nutrition institute, but many of the people Hugh listed in his earlier letter were friends of his and Lord Bossom's, and Woolton felt he could not make a personal appeal to them. More directly, he had been, for some time, a member of the appeal committee set up to create a National Heart Foundation.

Hugh ignored the implications in Woolton's letter, and wrote to Lord Bossom suggesting that he, Sinclair, should host a luncheon for seven, rather than three {Bossom, Woolton and Sinclair], at the Athenaeum. Lord Bossom agreed, but said Sinclair should be in contact with Lord Woolton, as it would be he who would invite one particular guest, who Sinclair called 'one of the greatest figures in the food industry', to the meeting. Dates for the luncheon were agreed and changed several times owing to engagements on the part of the three key participants. 29 May was the date finally agreed. The luncheon was held, and — according to a letter from Hugh Sinclair — was a success. Lord Woolton was anxious to have something concrete come from the meeting, and suggested a more substantial proposal from Sinclair for the Institute.

Over the decades to come, Hugh Sinclair frequently repeated that Lord Woolton was so anxious to see a National Nutrition Institute founded that he persuaded Hugh not to take a professorship in America, but stay in England and build the organization they both imagined. A comment to such effect may have been made over one of the two lunches Sinclair shared with Woolton, but there is no evidence that it was a serious plea, nor that Hugh had any interest in working closely with Lord Woolton on such an endeavour. However, these ideas were now an indelible part of the Sinclair Narrative.

This was the chance in Hugh Sinclair's life-time to see his ideas made substance: what better champion than Lord Woolton? And yet, the records show that Hugh treated both Lord Bossom and Lord Woolton with casual indifference. Why? Certainly both men wanted to participate as active players: but that probably did not suit Hugh, who trusted no one. He wanted and took no advice, and he certainly could not accept even a hint of criticism, no matter how warm the spirit in which it was offered.

After returning to England from his travels, Sir Edward deStein invited Hugh to visit his home and see his garden. Hugh did not respond. Weeks passed. Sir Edward was so distressed at not hearing from Hugh he wrote to him, in tortured script, inquiring as to his health. Hugh wrote a long letter in response, telling deStein about the meeting in the House of Lords with Lord Woolton, and claiming:

> [Lord Woolton] is now retiring from his Directorships and Committees so that he can devote his time to this very important project for which we have a superb site available in Oxford.

Hugh said he could not meet deStein until the very end of June, as he had so many other commitments. Sir Edward wrote again. Hugh did not respond. In late June, Sir Edward wrote a card to Sinclair, suggesting that Hugh's answer to deStein's letter must have gone astray. Would Hugh like to come and stay in his home for all or part of the week? There is no record of a response.

The pros and cons of fluoridation were being hotly debated in the general press, and Hugh was increasingly interested in the subject. The author Barbara Cartland (Mrs McCorquodale, later Dame Barbara), then a County Councillor for Hertfordshire, was concerned about the implication of artificially introducing fluoride into the water supply, and sent Lord Bossom material to review on the matter. In turn, Bossom sent this on to Sinclair. Hugh wrote back commenting on recent findings on the subject, then stated:

> The effect of fluoride on our health, as in the case of the effect of agricultural residues and food additives, is just the sort of urgent problem which the National Institute of Human Nutrition would study impartially and objectively. We must get it started urgently. I wonder if Lord Woolton had any success in following up the important line he suggested at the luncheon you so kindly arranged. …

There is no evidence of a meaningful reply.

CHAPTER 20

Life without mother

Rosalie now suffered giddiness and problems with a knee that forced her to walk with the help of two sticks. Nonetheless, she continued to attend bridge parties and the theatre in London, and planned family parties for a hundred and more each summer. She also entertained for her son.

> Hugh — I presume your pupils come to tea this coming Sunday. I have to get things ready by degrees ... How many will be coming, not many children I hope, they do not give your pupils a chance to get near the table ... Mother. Summer, 1962

Once a year, and sometimes more frequently, a hardy group of Oxford students were transported from the University by car — or rusting and noisy lorry — to Lady Place for tea. The uninitiated were surprised when, immediately upon arrival, they were presented with all manner of garden implements and invited to clean up the flower beds and mow the lawns before tea. It was good humoured hard work. When Hugh felt sufficient sweat had been shed, he called for the cricket bats. Those not interested in such sport were pointed in the direction of the boathouse or, later, a large water tank, where they could picnic and swim. After a time, Rosalie — in suitable attire for an Edwardian hostess — greeted each guest, welcoming him or her to enjoy a 'proper tea' of assorted sandwiches, tea, and fruit cake.

People visiting Lady Place during the summer months were always invited to swim or 'bathe'; bathing was featured on invitations to the Sinclair's annual garden party. When purchased, the property included a boathouse with a mooring. Forced by the need for cash, the little house and its land were eventually sold, leaving no place to take even a quick swim on a hot afternoon. The addition of a proper swimming pool would have brought additional taxes, but the installation of a large water tank that qualified as a small reservoir was free from government levy. Hugh had searched out such a tank and had it placed in an open area a little distance from the house. Eventually, water plants, frogs, and scum took hold, but courteous invitations to bathe were still offered with great enthusiasm.

One frequent guest who remembers these times with fondness is Dr David Horrobin, now well known for his work on the clinical aspects of essential fatty acids. He first encountered Hugh Sinclair in 1963, while being examined by members of the Magdalen College Body of Fellows as part of his application for a prize Fellowship. Hugh was impressed by the research paper Horrobin submitted prior to the examination, and was sympathetic towards his candidacy. The younger man knew of Hugh from his Winchester College friends, but had little interest in the essential fatty acids which consumed Sinclair's thinking. David was working on neuroendrocrinology at the time, and continued to do so while he was at Magdalen.

In later years, after David Horrobin launched a biotechnology company using essential fatty acids as its main therapeutic focus, Hugh Sinclair proudly told people David was his most brilliant student, that it was he who interested David in essential fatty acids while at Magdalen. Also, he claimed most of David's ideas originated from him. This part of the Sinclair Narrative requires some exploration.

In the classical sense, everyone who worked closely with Sinclair was his student. But, while at Magdalen, Horrobin's main involvement with Hugh was as part of the tutorial team teaching chemistry and physiology to medical students: the team included Brian Lloyd, who was now a teaching Fellow of the College. Horrobin's interest in essential fatty acids was fired some years later, in the early 1970s, while he was Professor and Chairman of Medical Physiology at Nairobi University, in Kenya. A visiting American scientist, Howard Berne, gave a lecture on prolactin at the Nairobi medical school. Berne was an expert on osmo-regulation in fish and birds, and talked about the hormone 'prolactin' as the chemical regulator allowing eels and salmon to move back and forth between fresh and salt water. In humans, prolactin is secreted by the anterior lobe of the pituitary gland, and is best known for its importance in female reproduction and lactation. As Horrobin was interested in the brain, he was aware of prolactin but had done no work on its function. After hearing Berne's lecture, he began working on the role of prolactin: males and females secrete about equal amounts of the hormone, so its role in the body most go beyond female reproduction. It turned out that prolactin is an important renal hormone.

The next series of questions Horrobin and his colleagues raised pertained to the messenger mechanisms of prolactin, and after a time Horrobin was convinced the hormone worked through another molecule, prostaglandin E_1, which contains an essential fatty acid [dihomogamalinolenic acid] in its molecular structure. As prostaglandin E_1 has a short half-life, Horrobin went back one step further and looked at its molecular precursors. It was then that he stepped into the world of essential fatty acids. This was in the late 1970s, when interest in essential fatty acids was increasing. In the early 1980s, at a conference organized to celebrate the 50th anniversary of the discovery of essential fatty acids, Horrobin again encountered Hugh Sinclair; but that part of the story comes later.

By 1965, Hugh was in serious financial difficulty and decided to sell his collection of rare medical books. Thirty-five years of careful and scholarly selection showed in its quality. Collections of writings by Boerhaave, Jean François Fernel, and Thomas Willis formed the base of the library, with significant works by William Beaumont, Charles Darwin, William Harvey, William Osler, Isaac Newton, and Edward Jenner adding considerable weight to the significance of the sale. The collection was sold for £90 000 to the University of British Columbia, in Canada. Hugh celebrated by buying a new lorry and two cars for his household: a small car for the servants and an Aston Martin for himself. For practical reasons, he chose a model that turned the famous 007 two-door design into a sleek estate car, ideal for transporting produce to the Sinclair garden shops nearby. For the moment, the sale of his books made him feel wealthy, and he took pleasure in living up to that fact.

Lord Bossom died on 5 September 1965 and his personal secretary wrote to Hugh on 8 October, asking whether or not the files pertaining to the National Institute of Human Nutrition could now be destroyed. Lord Woolton had died during the previous year and there appeared to be no other source of interest in the matter. Hugh contacted Lord Woolton's family at once, in hopes of obtaining financial support from his estate for the Institute. It took him longer to respond to Lord Bossom's secretary.

We certainly intend to go ahead with the National Institute of Human Nutrition, of which he [Lord Bossom] agreed to be Honorary Treasurer and had shown a considerable interest. Lord Woolton's death also deprived this important project of a most valuable supporter, and I recall that Lord Woolton said at a luncheon in the House of Lords not very long before his death, which Lord Bossom attended, that he intended to give up all his various commitments so that he could devote as much time as possible to launching the National Institute. To some extent the National Institute of Human Nutrition will I hope be a lasting memorial to him since when Lord Bossom visited me here to inspect the site we have for our building he planned to have a Woolton building associated particularly with education.

I was recently told that Lord Bossom had set up a foundation but I do not recall having seen anything about this and I do not know if the National Institute of Human Nutrition would come within its terms of reference. Yours sincerely …

Hugh was courteously told he had been misinformed concerning a foundation.

About this time, Hans Krebs contacted Sinclair about meeting a certain food industrialist interested in a national centre for the study of nutrition. Hugh sent copies of plans for the Institute and a statement regarding Lord Woolton's and Lord Bossom's planned participation in the venture. The group lunched at the Athenaeum. The food industrialist was interested in Sinclair's ideas, and requested a statement of his thoughts on a 'specialist centre for the study of nutrition'. Nothing came of the contact, largely because Sinclair had little to offer other than ideas. He also had his mind on other matters: as Master of the Worshipful Society of the Apothecaries, he was to preside at a dinner at which the Queen Mother would be the guest of honour.

15 August 1967, My darling Mother. This is just a line before leaving [for Japan] … I hope you will go out in the Aston Martin … In Tokyo I am being received by Princess Chichibu who I think is the aunt of the present Emperor. Prince Chichibu, now dead, was at Magdalen in the early 1920s … I shall meet at her house the Emperor's nephew who wants to come to Magdalen next year … The ducks seem to be growing quickly. It doesn't matter if they get on the top pond but best to feed them by the seat on the lower one. When I get back I want to get on with the herb garden. It is nice about the Apothecaries. In the first week of December we have a week of celebrations for our 350th anniversary (1617). There will be a big Livery Dinner on Wednesday Dec. 6 at which I preside (it is for members of the Livery only, no guests) and the Queen Mother is coming. The Lord Mayor will be on my other side. There are two Soirées for ladies on the 4th and 5th or the 7th (I am not sure which of the two). I think I can take one guest to each, and I wondered if you would like to

go to one and if you don't want to go to the other as well I'd ask Marigold [Viscount Thurso's widow]. There is a service in St Pauls on Sunday 3rd with a reception at the Apothecaries afterwards … Best love from Hugh.

My darling Hugh
… [I] am so glad you have been appointed Master of the Society of Apothecaries and in such a memorable year! I wish I were a bit younger and more active but I do want you to understand that I shall quite understand if you would sooner take someone else anytime instead of me — If I am OK I would like to go to St Pauls on Dec.3 and lunch at the Athenaeum — I do hope you enjoy Tokyo …

On 3 December Hugh's chauffeur packed a cooler and carriage umbrella in the Aston Martin and drove Hugh and Rosalie from Sutton Courtenay to St Paul's Cathedral for the Sunday Service of Matins. Before his mother and the Lord Mayor, and the Sheriffs with their ladies, before Judges and Commissioners, and representatives of the Royal Society and Royal Colleges and Deans of several medical schools, Hugh read the Lesson. (Rosalie later wrote a note saying how well he read.) That evening, Lady Thurso was his guest at the Oriental Club.

On the 5th a symposium was held at St Bartholomew's Hospital on Medicine and the City of London, 1617–1967: Sinclair presented a paper in the morning and chaired the afternoon session before attending the Soirée, with Miss Dorothy Hollingsworth as his guest.

On 6 December, a photograph was taken of the Queen Mother standing beside her host, the Lord Mayor, and the Master of the Apothecaries, Dr Hugh Macdonald Sinclair. She is majestic in ermine and satin, with kid gloves rising above her elbows. Hugh, buoyant in the fur-trimmed gowns of his office as Master, is caught grinning broadly and nervously twisting the signet ring on the little finger of his left hand. His long oval head is now almost bald, and the ice blue eyes (which benumbed many ill-prepared Magdalen students during tutorials) twinkle under slightly raised eyebrows behind dark-rimmed spectacles. It is the face of Puck. The photograph took pride of place as the cover of Hugh's Christmas card in 1967.

The following morning, Sinclair left Sutton Courtenay by train, dressed at the London home of the Worshipful Society of the Apothecaries, and received for the Faculty Dinner. The Archbishop of Canterbury was the honoured guest. The next day he attended meetings, received for the

Sinclair's 1967 Christmas card. The caption inside reads: 'H. M. Queen Elizabeth the Queen Mother, Accompanied by the Lord Mayor and Sheriffs, visits the Worshipful Society of Apothecaries on the occasion of the 350th Anniversary of the granting of its Charter, 6 December 1967'.

Yeomanry Dinner, attended by the Minster of Health, the Right Honourable Kenneth Robinson, and again spent the night in London. Hugh Macdonald Sinclair had come to the position of Master of the Apothecaries by succession through the ranks of the organization in 1967, but there could have been no more knowledgeable, charming, and enthusiastic host for the occasion.

Lady Place rang with the laughter and noise of Hugh's guests during the weekend following the Anniversary celebrations. Rosalie could not have been more proud of her younger son, and there is no doubt that these few halcyon days were the zenith of Hugh Sinclair's life.

Although he had no laboratory facilities, other than at Lady Place, he was not idle. Since 1963, Hugh had given eight Advanced Lectures on Nutrition in each Trinity Term; contributed chapters and sections to five published books, and awaited publication of four similar works; he had planned, and was now Editor-in Chief of Robert Maxwell's *International encyclopedia of nutrition*; working with Dorothy Hollingsworth, he was revising Sir Robert Hutchison's *Food and the principles of nutrition*; he had written, with Dr F. C. Rodger, a monograph on nutrition and eye diseases; and as Fellow by Special Election and Lecturer in Physiology and Biochemistry at Magdalen College continued to teach medical students and biochemists, and 'supervise other biochemists'. As for his research, he continued his work on the relation of nutrition (especially essential fatty acids) to chronic diseases by correlating the dietary pattern and disease prevalence in various parts of the world.

In March 1968, Hugh was reappointed a University Lecturer by the Board of Biological Sciences; a post which he held until his retirement. The duties of such a post include carrying out advanced study and delivering intercollegiate lectures, or holding classes announced in the Lecture List. No formal limitations were placed on his teaching, holding college offices, examining, or similar work, as long as time was available for his private study. His stipend was £1420 per annum.

After a few weeks, Hugh wrote to Larry Kinsell inquiring about his plans to attend the Biochemical Problems of Lipids, to be held in Loughborough during the coming September. A conference in Milan preceded this meeting, as did an international Physiological Congress in Washington DC. If Kinsell would be coming to Europe, would he like to spend some time at Lady Place? The response was not what he expected. Larry Kinsell had fallen and broken a vertebra, which was taking considerable time to heal. To aid recuperation, he and his wife had boarded a cruise ship for Australia, and did not plan to return to the United States for some weeks. On 28 June, Larry Kinsell answered Hugh's letter: he would not be coming to Europe that year, but invited Hugh to stay with his family when he was next in America. Eleven days later, on 9 July, Larry Kinsell and his wife were found together at their home, both having died of cyanide poisoning. On that morning, Larry Kinsell had rung his laboratory saying that his wife was suffering from food poisoning, and that he would be delayed for an hour. He never arrived. The deaths perplexed everyone, as the Kinsells had appeared to be in good spirits. Not since his sister's death had an event so shaken Sinclair.

During the month in which the Kinsells died, Hugh Sinclair gave the commencement address at Baldwin-Wallace College, in Berea, Ohio,

USA, and received an Honorary Doctorate of Science. His topic was the Biology of Student Revolt.

On a Friday evening in January 1969 Rosalie lost consciousness and fell backwards down the stairs. The Sinclairs were entertaining at the time, and guests helped Rosalie to her bed, where she quickly regained her senses. As Hugh was about to leave for scientific meetings on the continent, he hired a nurse to live in and look after his mother. Rosalie lingered on. Leaving her with nurses, Hugh escaped whenever possible to Magdalen College, or to the United States. His travels and contacts began to pay dividends.

> 12 August 1969   private and confidential
> My darling mother …
> I have a friend who is interested in my research … He is Mr Jack Pye …
> Mr Pye is thinking of building an Institute (Which would probably be called the National Institute of Human Nutrition) in the grounds of Lady Place. The area we have in mind is the flat part beyond the hard tennis courts …
> He is thinking of buying Courtenay Nurseries and also Lady Place, with of course the right of both you and me to live there as at present for our lives. It would be a wonderful scheme if it comes off … [1]

Hugh went to Europe, as planned, at the end of the summer. Rosalie wrote to her son on the first of September, but he had moved hotels, and did not receive her letter until some days later. Rosalie had a final stroke on 15 September and died at Lady Place on 20 September 1969. Hugh made a note on her last letter to him that her strong, clear handwriting remained unchanged despite age and illness.

Rosalie left Hugh an inheritance that would help secure his lifestyle for a few more years. More importantly, her death meant that he no longer needed to worry about her designs on Lady Place.

CHAPTER 21

Plans for an institute of human nutrition

A grand plan was taking form. If Hugh sold Lady Place to Mr Pye, he could then lease back some portion of the property for his Institute, and as its Director, could live there and oversee the property as part of his responsibilities. From Sinclair's point of view, little would change, other than the amount of money in his pocket. To put this idea on a more secure legal footing, and facilitate donations that may be made to the Institute, he decided to set it up as a registered charity, to be known as The Association for the Study of Human Nutrition. To help finance the Charity during its early stages, Christie's, the well known auction house, handled the sale of some of Rosalie's jewellery, including a diamond bracelet and tiara.

Eventually, Mr Pye withdrew his offer to purchase Lady Place, but by this time Hugh had committed most of his personal wealth and belongings to his Charity.

By 1971, Lady Place was not the only property Hugh wanted to develop. As a former Vice-President of Magdalen College, he wanted to make some structural changes there. To help him in this task, he tried enlisting the aid of John Fulton's widow, Lucia, suggesting she join him for dinner in the College, as he wanted to show her the plans he was developing for the College. Hugh and John Fulton had been friends since Hugh's early days at Oxford, sharing a deep interest in the history of medicine and rare medical books, and Luicia had entertained him many times in their home in Princeton. He wrote to her:

> In 1946 we had the plans to replace the old Botany Department across the High Street from Magdalen with a really fine building for graduate students. It was to stretch in two wings on either side of the beautiful Danby Gateway which was built when the Botany Garden was founded in 1626. Our plans then had to be shelved but have now been revised … this is probably the finest site

available for building in Europe and we should put up a fine building to match the College buildings opposite with the Great Tower. It is in fact at the entrance to one of the most important streets in the world — the High Street of Oxford. This building will have flats and other accommodation for graduate students both married and unmarried; as you know, we receive the most distinguished of such graduates particularly from overseas and especially Rhodes scholars such as John and Howard Florey. Since these are two of the most distinguished members this College has ever had, I have in mind that we might call the whole graduate building 'Magdalen Hall', and the two wings on either side of the Danby Gateway the 'Fulton Building' and the 'Florey Building'. Mrs Lasker (of The Albert and Mary Lasker Foundation in New York) gave us during his lifetime the rose garden in front of this building in honour of Howard Florey's discovery of penicillin, and I privately hope that if we make an approach to her the Foundation might give us the necessary money (about $750 000) for the Florey Building. What I am asking you privately is whether you think it is possible that we might raise an equal sum ($750 000) from the friends of John Fulton for the Fulton Building on this magnificent site …

Lucia visited Oxford and was well entertained: 'It was just like old times and I felt as if I were in a dream world …' she wrote in her letter of thanks. However, she did little to support Hugh's ambitious and expensive plans.

From time to time other benefactors appeared; for example, on 11 October 1971 The Princess Helena Moutafian wrote to Hugh:

> … I am one of your most devoted disciples in the work that you do … I have always been interested in the land and the produce of the land and … shudder with horror when I see what sort of food people eat nowadays … [Hugh should visit her in London, she suggested:] In the meantime let us keep in touch until our next meeting.

For some months Sinclair pursued the Princess for funds, but (as in the case of the Lords Bossom and Woolton) discouraged her generosity and help through inattention and rudeness. In the end he achieved nothing, other than a few opportunities to enjoy some good food and wine, and to spend time in the presence of a gracious woman.

Hugh took every opportunity he could to travel, giving several lectures in Nigeria, and then on to California, where he addressed the Hollywood Academy of Medicine. He was described as: Hugh Macdonald Sinclair MD, D Sc, FRCP 'Born in Scotland in 1910 (descendant of Prince Henry

Sinclair who sailed from Scotland to Nova Scotia and back in 1398 … .)' The MD, of course, was incorrect, but was Hugh's way of describing his equivalent British qualification, the DM, which he held.

During the 1971 Christmas period, Hugh hired furniture removers to transfer his books, book-cases, filing cabinets, and journals from Magdalen College to Sutton Courtenay. Soon after, he wrote to the Executrix of Sir Robert McCarrison's estate, asking if the remainder of Sir Robert's library could be moved to Lady Place. He would pay for the bookcases to house them. Two flats were now ready as offices for the Institute, and he wanted to begin hiring new staff at once.

In 1971 a stained glass panel showing the Sinclair family coat of arms was erected in the Hall at the Worshipful Society of the Apothecaries to commemorate Hugh's time as Master of the Society. Hugh was delighted by the piece, and wanted to have a very much larger version created and installed at Lady Place, just above the main entrance, where all who came could admire this symbol of his Scottish heritage. Getting the name of the company responsible for the work at the Apothecaries, he asked for an estimate. Pleased with the response, he then commissioned this fragile monument to his self-image. The bill for the window at the Apothecaries was received in August 1972, and was paid in June 1973. Payment for the larger piece was to be equally slow.

On 17 August 1972, the Memorandum and Articles of Association of The Association for the Study of Human Nutrition Limited were at last signed. The Subscribers were 'Professor Hugh Macdonald Sinclair, DM, DSc, FRCP; Professor Francis Aylward, DSc; Professor Sir William Richard Shaboe Doll, OBE, MD, DSc, FRCP, FRS; Mr Arthur Elliot-Smith, FRCS [a close friend of Sir Robert McCarrison, and former senior surgeon at the Radcliffe Infirmary]; Sir David Paton Cuthbertson, CBE, MD, DSc, FRCPE, FRSE [Hon. Consultant in Pathological Biochemistry]; Professor Herbert Davenport Kay, CBE, DSc; PhD, FRS; and John Clough Talbot, Solicitor'. On 17 November 1972, the Association was registered under the Charities Act 1960 as a charity [number 264748]. Hugh would be its Director, and in time its name would become The International Institute of Human Nutrition. Whenever possible, Hugh would include representatives of the University of Reading, where he hoped to establish both teaching and research activities.

In December 1973 Hugh reported to the Executive Committee of the Council of Management of The Association on his activities since its

Photograph of a model of the planned buildings for the Institute of Nutrition. Lady Place mansion is in the background.

inception. He reviewed the history of plans to set up a Central Institute of Human Nutrition in Great Britain, and reported that contributions were being made through the Robert Moss Charitable Trust, the Citizens for Health Information Inc., of Washington, DC, and private individuals. According to a draft of the Committee report, Lansing-Bagnall Ltd. agreed to provide £20 000 annually for five years from 1 May 1973 'for a specified programme of research into the causation and prevention of coronary heart disease'.

Plans for the physical layout of the Institute had been drawn up which called for eight separate departments. The Committee was reminded that the purpose of the Institute was to collect, integrate, and disseminate 'existing knowledge of human nutrition', and to conduct fundamental research 'to increase our knowledge of human nutrition and evaluate nutritional changes (including those due to foods ingested and air breathed) as foods, water-supplies and air are altered and polluted.' There is no record that anyone questioned whether this might be too ambitious a plan for so young an organization.

It was about now that Dr Mary Gale joined Hugh Sinclair to manage his laboratory investigations. She brought the scientific skills his work had been missing, and was a worthy second-in-command who proved to be both understanding and long suffering. Beginning in 1972, and extending through 1989, their primary work was to study the epidemiology of Western diseases in relation to the dietary ratio of essential fatty acids to non-essential fats. [Although she did not mind also helping with household matters, nor feeding the peafowl that wandered through the gardens.] Approaching the research in this manner was far less expensive than laboratory investigations, and so suited Sinclair well at the time. He had little money for modern equipment. Using only published papers and communications from other scientists, Sinclair and Mary Gale probed and pondered in abstraction the possibilities of underlying metabolic systems altered by the presence or absence of essential fatty acids.

In October 1974, the heraldic stained glass window for Lady Place was completed; arrangements were made to erect scaffolding, and soon another piece of the Sinclair story was in place. By now, Hugh Sinclair had created a comfortable world of his own history. The story he loved telling was one of bagpipes and heroic Scots. One day, when cash was low and Mary Gale was worried about funds to pay staff wages, Hugh drove to London intent on selling some expensive glassware. On his return he produced a far smaller sum than was anticipated, and a fine suit of armour. The armour was placed where it could be admired by visitors to Lady Place, in the pool of coloured light falling from the heraldic window, and Hugh dissuaded no one from thinking that it was part of his grand heritage.

In March 1975, Hugh's first major medical problem appeared, and he had surgery on his lower bowel. But that did little to slow him down. He entered hospital in April 1975 for surgery, and wrote to a friend: 'I am planning an expedition to study Eskimos again soon …' And, so he was. (His papers show that he now also added colon cancer to the group of specific conditions his Institute would study.) During the following month Sinclair sought a new champion:

April 1975 to The Rt Hon. Lord Porritt:
This Institute was set up as a result of the initiation of Sir Henry Tizard, Lord Woolton and others nearly twenty years ago … After various delays … the late Arthur Elliot Smith … and I got it started two years ago. I am taking steps to make over my 20-acre property for it, and we have plans to build the main Institute in the grounds. In the meantime we have converted some existing buildings into excellent laboratories, and we are using most of my

house for the office work and library. The Institute is registered as a charity … The Association for the Study of Human Nutrition Limited, but at its last meeting its Council of Management … decided to apply to change the name to International Institute of Human Nutrition. Much more important, the Council decided unanimously to invite you to become a member of the Council. The Council only meets about once a year, having an Executive Committee to conduct routine business …

Porritt wrote back: 'What a chap you are — and what a saga this is!' he stated, referring to the 1974 report to the Committee of Management.

Now in his sixties, Sinclair was a difficult and eccentric man. Evidence in the files substantiates Sinclair's interest in collecting work of an erotic nature, including flagellation. Since Rosalie's death mischief was afoot at Lady Place, and at about this time Hugh began advertising for staff in a Gay newspaper. There were signs of his growing paranoia: notes written, torn up, and thrown in the waste basket by disgruntled staff were retrieved, pieced together, and their contents written out in long-hand on another piece of paper by Hugh so that he could read what had been said. Both the retrieved scraps and his written document were then placed in a file for future reference.

Highly qualified staff came and went, unable to work with him. Over the years he had a pattern of promising a competent person a specific role within the Institute, and then restricting their activities and badgering them until they felt forced to leave. In one case, he dismissed a staff member by waiting until the man and his wife went for a walk, and then pushing a dismissal letter under the door of the accommodation they rented at Lady Place. Scientists employed by Hugh found themselves gardening for their salary. People failing to return a reprint to its exact original location in Hugh's library, or returning card files out of order, were humiliated. One took action through the Industrial Tribunal, but received only emotional pain and no money for his efforts. The pressure from Sinclair was relentless, and the case was eventually withdrawn. Only Mary Gale was able to survive with dignity. Sinclair's abrasive behaviour was not limited to his staff, but sometimes poured over onto those professional friends willing to help him. Always, he was jealous of the leadership of the Institute, and worried that his control over it might be diminished.

In 1976, and at the age of 66, Hugh Sinclair was planning to depart for Greenland with H. O. Bang and J. Dyerberg of Denmark. Hugh was financed by the British Cancer Campaign through the British Digestive Foundation. He wrote a letter to Unilever, from whom he also sought research funds.

'I leave at the end of the week to fly to Copenhagen,' he said. Then he would go on to Greenland where, by dog-sled, he and others would travel to what was believed to be the last group of Eskimos living on traditional diets.

> My share of the initial expenses (travel, equipment) is estimated at £1500, which is covered by a grant from the Cancer Research Campaign (since I shall be sending faeces to Professor Robert Williams, FRS, in connection with cancer of the colon) and the British Digestive Foundation.

During discussion with people at Unilever, Hugh suggested they should fly two Eskimos to the Unilever Research Laboratories in Vlaardingen, Holland, for dietary investigations. He hoped to return to England in late May, accompanied by two Eskimos 'and a seal for them to eat!' After a night at the Institute, he planned to accompany them to Holland for a week.

Letter to Professor James L. Gowans, CBE, FRS, Sir William Dunn School of Pathology, Oxford:

> There is one small group of Eskimo left who are still on their traditional diet of fish, seal and whale, extremely high in fat and protein, with negligible carbohydrate. They are in the Arctic Circle in Northwest Greenland and I am going with a couple of Danish doctors to study them for a month in April. They are profoundly interesting because of their very high fat intake, very low plasma cholesterol and no coronary heart disease.

'A further point of very great interest is that their polyunsaturated fats are of the linolenic class and they have exceedingly low values in plasma for linoleic acid …' he continued, and asked 'whether there is anything practicable that we might usefully study with regard to the activity of Eskimo lymphocytes.'

In fact, the group stayed for only two weeks because the weather broke, forcing Hugh to abandon some of the detailed physical examinations he had planned, and because a lack of electricity at their base camp made it impossible to carry out the anticipated work on platelet function. He spent time reading medical records in the hospital at Umanak, however, and from data collected there he concluded there was an absence of chronic degenerative disease among the population.

Sinclair's sketch of the surroundings of Umanak Hospital, Greenland (May 1976).

The Institute was receiving funds. Mr E. W. Borrow, founder of the Borrow Dental Milk Foundation, offered £40 000 to alter and equip one of the existing buildings at Lady Place for use as a laboratory, to be called the Borrow Research Laboratory. According to Sinclair's files, he also agreed to pay £10 000 per year for seven years to finance research. The work soon fell behind schedule, in part due to staffing problems, and in part due to the overly ambitious promises made by Sinclair. Hugh explained that it takes time to set up new equipment, but Borrow lost confidence in the Institute, and refused to keep the financial bargain. Another company agreed to support the Institute with £20 000 a year for five years. On 31 December 1976 a quarterly cheque for £5000 arrived with a letter stating that it was the last the Institute was to receive. Hugh explained to his Management Council that this was due to internal company matters.

The company was experiencing a recession, and working a four-day week, but more funds would be forthcoming: Hugh expected another £20 000 before long from another source. At this time funds were badly needed. The Institute's bank overdraft (secured against Lady Place) was approximately £51 000. This, Hugh assured everyone, would be made up by the money owed by the two organizations mentioned above, plus £1500 Hugh expected from the Robert Moss Charity. Further, Hugh had lent the Institute £14 000 of his personal money to keep the overdraft below £40 000, which was the limit arranged. New funding sources were needed.

By now, Lady Place was completely consumed by the Institute. Rooms had been transformed into the library, offices, and laboratory facilities. Staff lived in various parts of the house or outbuildings that had been converted into flats at low rents, as part of their remuneration. Sinclair's own rooms were used for entertaining visitors to the Institute, and spare bedrooms housed guests. His life had become indistinguishable from that of the Institute. One member of the Institute Management Council noted that no one was willing to question Hugh seriously about his activities and project plans because most of the money involved was his own. This became increasingly true.

> 27 May — letter to Lord Porritt
> I had a most interesting time in the Arctic investigating the last remaining Eskimos who are on their traditional diet of seal and fish, which they (like myself) ate raw. They have no heart disease or cancer; their diet is of course totally devoid of fibre but they have none of the diseases (such as carcinoma of the colon) that are attributed to dietary deficiencies of fibre. Your former colleague at St. Mary's … is studying the bacteria and steroids in faeces I collected from the Eskimos — with some difficulty — since they discarded these more or less at random and they are immediately devoured by the Husky dogs for a reason that is not understood. So I had to compete with the dogs — some of which are alleged to have rabies. We are bringing back by ship as soon as the ice is thawed deep frozen samples of diets, blood, teeth and faeces that will keep my Danish colleagues and myself busy for about two years …

Hugh prepared a report on 'Fluoridated milk in the prevention of dental caries', beginning with an observation he claimed to have made while in Northern Canada in 1943, where he said he 'confirmed the well-known fact that there was no dental caries although teeth ground down through chewing bones.' The report contained the protocol for a proposed study involving children. A copy was sent to Lord Porritt, who suggested Hugh contact Borrow again. Hugh explained that the Borrow Foundation owed the Institute a considerable amount of money, and asked if Lord Porritt might help encourage the release of these funds. Hugh's request was fulfilled.

By the end of 1976, Sinclair estimated he needed about one and a half million pounds to build his Institute, and sought donors whose names could be attached to the building: James Goldsmith and the Sainsbury family were targeted and pursued, but without result. By this time Hugh had a reputation as a maverick who frequently failed to complete work promised, and little support was forthcoming. Another possible source of funds opened when Dr Oliver Gillie wrote an article published in the *Sunday Times*, 13 June 1976, suggesting that high levels of red meat in the diet may be responsible for increased levels of cancer. Hugh publicly refuted this, and used it as an excuse to solicit funds from the Court of the Worshipful Company of Butchers for Institute research. He tied the question of red meat together with his research on Eskimos and bowel cancer: their lack of cancer and heart disease, he wrote, '… makes nonsense of the claim that "man is not adapted to eat large quantities of meat".' Hugh asked the Worshipful Company for a considerable donation. '… the meat industry has been challenged, and I believe our Eskimo studies will show what, if any, truth lies in those allegations.' No research support resulted from this, but Hugh had taken another controversial public stand.

Hugh also sought funds from the Government by first approaching the Ministry of Agriculture, Fisheries and Food [MAFF]. By this time, considerable work had been published on essential fatty acids, and it was increasingly apparent that Hugh's original letter to the *Lancet* was a landmark document containing significant scientific insight. Reviewers of Hugh's proposal raised questions about the structure of the project, and requested justification for various items in the budget. Eventually, he received a letter from MAFF:

> … We feel that your work is well worthy of support and the main problem as you know has been that it falls at the edge of the various departmental responsibilities. However, we have been in contact with Professor … and Dr … and they feel that an application to MRC has a very fair chance of success and is certainly likely to get a quick answer. If this should fail DHSS can make smaller grants for up to three years and Dr … will be contacting you

about this shortly. I so hope that the building funds for the institute will materialise and I am quite sure that this Ministry would be interested in commissioning work in future so that we would be very keen to keep in touch with you.

The DHSS grants were restricted to £20 000 over 3 years. That was too small a sum to meet Hugh's growing financial needs. This, in part, resulted in his selling his beloved Aston Martin, resorting to using Rosalie's old Daimler. In 1978 he wrote to a friend in Canada:

> I greatly regret to say that I sold the Aston Martin. It had a distinguished existence since when I was Master of the Apothecaries in our 350th year and we had a week of celebrations, I had a driver and on occasion we preceded the Lord Mayor's car. I entertained on occasion the Queen Mother, the Archbishop of Canterbury and the Minister of Health …

In his search for funds, Hugh appealed to the Secretary of the Medical Research Council:

> Dear Secretary, over 30 years ago I studied Eskimos in northern Canada in view of their diet extremely high in protein and fat (but very rich in essential fatty acids) and the almost total absence amongst them of 'Western' diseases such as ischaemic heart disease and cancer. Unfortunately mortality statistics were hampered by uncertainty of ages and a relatively short life-span because of accidents to male hunters.

He went on to describe his recent work in Greenland:

> '… but one small settlement of about 150 in Northwest Greenland are still mainly on their traditional diet of seal and fish, and have a similar expectation of life as ours (they are Christians, and have a cemetery from which I have obtained ages.) … They are the only people in the world who eat no plant food (except for occasional berries in summer), and therefore their diet is free of fibre; but they have none of the diseases attributed by Dennis Burkitt, Hugh Trowell and others to dietary deficiency of fibre. Their diet is exceedingly high in meat and fat, but they have no carcinoma of colon or breast … They probably have no ascorbic acid in their diet but scurvy is unknown; their very high fat diets produce no ketosis probably because γ-oxidation occurs (as I suggested years ago). I want immediately to appoint two chemists and two technicians to analyse the samples I have stored [from Greenland — diets, plasma, platelets, erythrocytes, faeces, urine, milk, skin, lipids, and teeth].

Then I plan to visit this summer by contrast the small colony of Welshmen in Patagonia who are subsisting mainly on mutton (very low in EFA); the proposal is that I will make the clinical observations and Professor Brenner of La Plata (a distinguished lipid chemist) will do the biochemical analyses. Then in the autumn I plan to go myself onto a diet rigidly limited to marine animal food and water to follow serially everything reasonable, particularly platelet function in relation to change in body fatty acids and prostaglandins ... [He briefly describes more planned work] ... Later, I propose with the aid of WHO and the collaboration of Professor Ramalingaswami (Director, All India Institute of Medical Sciences, New Delhi) to compare the fish-eating and coconut-eating islanders of the Pacific ...'

In May 1978 Sinclair submitted a full application for long-term support for this work from the Medical Research Council. The work was to take five years, and would have included a series of exploratory studies in several international locations to compare nutritional factors with chronic degenerative disease, including ischaemic heart disease. Hugh proposed returning to Greenland in the summer, hiring a small ship and, with Danish colleagues, transport about 25 young male hunters south to the hospital at Umanak, where electricity was available and they could do a variety of tests of platelet function. They could also perform needle-biopsy of adipose tissue, and collect semen for the study of prostaglandins. His explained his idea:

It is possible that carnivorous Eskimos, like carnivorous cats and lions ... may have lost the ability to desaturate and elongate fatty acids and hence convert linoleic to arachidonic acid. Analyses by Drs Bang and Dyerberg of the plasma lipids of Eskimos living in Denmark support this possibility, for plasma cholesteryl arachidonate is zero. If this is so, Eskimos relying mainly on 'Western' diets ... might suffer unless they take marine oils. Since Eskimos even in Greenland are now changing to a 'Western' diet, it is obviously very important to test this possibility: cats on a diet rich in linoleic acid and low in arachidonic (as in the 'Western' diet) become severely ill. Therefore if possible the ability to desaturate and elongate linoleic acid should be tested ...

Sinclair also planned to conduct work on himself. For three months, he wrote, he would consume only seal, fish, molluscs, crustaceans, and water. Such a diet would be high in omega-3 fatty acids. Unilever Research would pay for food not provided free by the fishing industry,

Sinclair said, and someone at the Radcliffe Infirmary had agreed to assist with venipuncture and tests of platelet function.

Sinclair's grant application was refused. Advisors to the Medical Research Council were particularly concerned about the 3-month dietary experiment Hugh planned to conduct on himself. Also, there were serious questions about the lack of detail in the experimental design. Although the project was seen to have some intrinsic personal interest, it was felt that a properly controlled study using a group of volunteers might better produce useful results, and could show whether a radical dietary change, such as that which Hugh proposed, could change parameters as fundamental as platelet function.

Early in 1979, with the help of Mary Gale, Hugh Sinclair engaged in a dangerous self-experiment that would last 100 days. Specifically, he wanted to know more about the effect of omega-3 essential fatty acids on blood-clotting times. Data collected would include somatometric measures, blood studies, and tissue biopsies. Friends provided clinical and analytical support. Frozen seal meat was obtained from Canada, and the menus at Lady Place and Magdalen College were adapted to meet his very strict dietary regime of water, seal meat, and other seafood. Medical colleagues warned him that what he was about to undergo was dangerous, as his blood-clotting times were sure to be affected and rise. Other sources recount the scientific details of Hugh's self-experiment. Here, it is worth noting some of the more personal details of events.

Since his mother's death, Hugh took most of his evening meals at Magdalen. Frozen seal carcasses arrived about once a week, and were unwrapped and hung in the kitchen refrigerators before cooking. The meat stank. At that time, the kitchen ventilation was poor, and wretched smells of cooking seal rose and hung about for hours in the Common Room, much to the disgust of others at Magdalen. Fellows would call and ask if Sinclair was dining in; if he was, they stayed away. A Gaudy dinner fell during the months of Sinclair's seal diet, and between 150 and 200 old members of the College were invited. Sinclair ate stinking seal while the other guests dined on fillet steak. He was not always so particular, however. Dr Ralph C. S. Walker, a friend of Hugh's at Magdalen, reported that on an occasion when the Fellows brought their wives into dinner Sinclair enjoyed a fine meal like the rest, as he said it would be ungentlemanly for him to eat seal in the presence of ladies.

During the experiment, friends worried that Sinclair's blood-clotting times were becoming dangerously long, as anticipated, and Mary Gale struggled to keep him true to the experimental protocol they had agreed. Friends worried because his skin lost colour and developed the translucency of a much older man. Physical observations, blood, and tissue samples were collected, but it would fail to produce any significant scientific results.

In 1980 Sinclair described his experience on the diet in a chapter, titled 'Advantages and disadvantages of an Eskimo diet', which he contributed to in *Drugs affecting lipid metablism*, edited by R. Fumagalli, D. Kritchevsky, and R. Paoletti, and published by Elsevier/North Holland Press [pages 363–370]. In this paper he recognized the importance of both fish and vegetable oils, and recommended a diet containing both:

> Despite the diet being very high in fat and cholesterol, LDL and plasma cholesterol were low; VLDL were extremely low and HDL high … The experiment shows that high fat diets can be accompanied by low plasma cholesterol and 'unstickiness' of platelets provided they are relatively high in EFA of the linolenic class. But platelet phospholipids retained arachidonic acid, and both classes of EFA are desirable, the linoleic class being more important … consumption of fatty fish should be encouraged in addition to vegetable oils.

Beyond these words it was some time before more was written about this extraordinary undertaking.

CHAPTER 22

Game to the last

The International Congress on Essential Fatty Acids, Prostaglandins and Leukotrienes held in 1980 celebrated the 50th anniversary of two apparently unrelated events: the discovery of essential fatty acids by George Burr, and the discovery of powerful hormone-like substances, called prostaglandins, by the Swedish scientist, Ulf Svante Hansson Von Euler. By the time of the conference, there was increasing evidence that essential fatty acids were precursors of prostaglandins, giving these delicate fats new scientific importance.

About this time Hugh Sinclair met David Horrobin again, and an old friendship was renewed. Fifteen years earlier, when they had worked together as tutors at Magdalen College, Horrobin had not been interested in Hugh's views on essential fatty acids. However, while working in Kenya, the younger scientist had investigated the regulation of renal function by essential fatty acids, via prostaglandins. Now he was seeking clinical applications for his findings, and Sinclair was an ideal partner with whom he could hypothesize causal effects. Not since the death of Larry Kinsell had Sinclair found a colleague so interested in the biological functions of essential fatty acids. Their main point of disagreement concerned the relative importance of the omega-3 and omega-6 fatty acids. By now, Hugh was convinced that the omega-3 varieties, like those found in fish oil, were the more vital. David Horrobin's primary interest was in the omega-6 forms, and he would eventually convince Hugh that both are of major clinical significance.

In 1982, David Horrobin edited his first major book on the subject of essential fatty acids: *Clinical uses for essential fatty acids* (Eden Press Inc., Canada). Its content and the debate it stimulated were reminiscent of Hugh Sinclair's 1956 letter to the *Lancet*. By bringing together a combination of his own research on this subject, controlled research conducted by others, and hypothesis, Horrobin built the case for essential fatty acids which Hugh Sinclair had failed fully to elucidate earlier. His starting point echoed Hugh's letter: a deficiency in essential fatty acids causes a progression of metabolic mistakes, cellular flaws, tissue damage, and structural weakness that could result in many of the degenerative illnesses common in Western cultures.

Horrobin's book went further, however, and drew together evidence that the key to the progressive nature of many degenerative conditions rests in the decreased activity of a single enzyme, $\Delta 6$ desaturase. This enzyme is the first in a series of enzymes, or natural catalysts, that transform the molecular structures of both omega-6 molecule and omega-3 fatty acids into more biologically active forms. Under the most extreme deficiency of this enzyme, both linoleic and linolenic acid [the primary forms of omega-6 and omega-3 fatty acids obtained in the diet] would be useless as anything other than sources of calories. Any decrease in the activity or availability of this enzyme hinders the normal transformation of these basic fats into molecules needed for normal cellular structure, function, and communication, and prepares the ground for many non-infectious chronic illnesses. Therefore, degenerative diseases can result either from the deficiency of essential fatty acids, or from a failure in the enzyme system that converts them into more active molecular forms. As $\Delta 6$ desaturase production by the body can be obstructed by many factors, including viruses, cigarette smoke, and certain medicines, these same factors simultaneously obstruct the metabolism of essential fatty acids.

At the time Horrobin's book was published, the biological significance of essential fatty acids was hotly debated. Many scientists found it implausible that a deficiency in the metabolites of a fat as ubiquitous in the diet as linoleic acid, or as common as linolenic acid, could result in as many disease processes as Sinclair — and now Horrobin — suggested. The Horrobin book was criticized for containing too much speculation and too few controlled studies. However, for scientists who had spent years investigating essential fatty acids, the book was a triumph because it set out to explain the role of essential fatty acids in normal tissue, the causes and possible effects of deficiency, and to suggest therapeutic uses of these delicate nutrients.

If nothing else, the Horrobin book gave new credibility to Hugh Sinclair's ideas and offered a scientific foundation that might explain them. The 1985 International Congress on Essential Fatty Acids concluded with a banquet honouring his 75th birthday, where he addressed the group. As usual, Sinclair was an inspiration to those who heard him speak. He said that the first evidence of essential fatty acid deficiency was in 5000 BC, when Moses killed the fish and, in consequence, the hearts of the Pharaoh and the Egyptians were hardened. He then sketched his early observation on the Canadian Indians and Eskimos, and discussed his controversy with Ancel Keyes. A friend remembers that when Hugh finished speaking his audience knew they had just heard a great teacher and a great scholar. He returned to his seat through a standing ovation.

In his seventies, Hugh Sinclair was now fêted and filmed and invited to speak at prestigious scientific events around the world. He had at last achieved the acclaim for work on essential fatty acids that he had sought for so long. To all who would listen, he told the story of his life as he remembered it: of how he bought Lady Place as a student at the age of twenty-four, and how he had 'lived amongst the Eskimos'. He linked his name with the great and good, even to the embarrassment, and sometimes great boredom, of his friends. Unfortunately, the personal story he told was remembered and repeated, including his memories of the circumstances surrounding his leaving the Department of Biochemistry at the University of Oxford. His recitations were corrupted by time and his own insecure need for recognition, but it all made good listening, and he enjoyed the telling.

Sinclair continued to teach at the University of Reading as a Visiting Lecturer. Students were treated to Hugh's view of the assessment of *nutriture* of the population, methods of research in human nutrition, and the nutrition of specific human groups. Well supported with an extensive collection of slides, historical documents, and boxes of bones — many probably inherited from Robert McCarrison — Sinclair held the interest of these students with his every word. Before his death, in 1990, Dr Ann Walker, Lecturer in Human Nutrition at Reading, taped a series of these lectures and included them as chapters in her book *Applied human nutrition — for food scientists and home economists* (Ellis Horwood, 1990). From all accounts, the Reading students often went out of their way to question him personally on various aspects of science, and he frequently questioned those from abroad about their diet and family medical histories. To the end, the Teacher was always the Student.

But it was not all hard work. Although he was past handling a bat himself, Sinclair continued to arrange cricket matches between the Lawyers and Doctors. *The game of cricket,* by H. M. Sinclair, describes one such event:[1]

> The practice of medicine is much older that the game of cricket. It took man some time to discover that it might be fun to defend a couple (initially) of

sticks with a third across the top by wielding a fourth against a piece of rock thrown along the ground. This discovery may have been made in the Weald of Kent, and the first probable reference to cricket comes from the wardrobe accounts of Edward I in 1300 in that county …

He went on to describe a match between the Doctors and Lawyers, which the Doctors won, and then wrote:

Any intelligent person who has read as far as this may wonder why I reported last year's game. Well, this year we lost by five wickets, but no matter: we enjoyed our encounter with the Law.

The financial situation of the International Nutrition Foundation had become critical; overdrawn to the extent of £200 000 and on the verge of liquidation. Sinclair planned to transfer his interests in all the properties he owned, plus other assets, to the Foundation by way of a deed of gift. Thereafter, the Foundation would dispose of property to meet its debts without loss due to tax. In an attempt to help raise funds, several members of the Council of Management drafted a document requesting financial support from The Wolfson Foundation, in London. (Hugh had previously been in contact with the Foundation, but with little success.) The Council members worked alone. Funds were sought to establish a Chair in Human Nutrition in the Faculty of Clinical Medicine of the University of Oxford. The benefits of nutrition research and the academic justification for such a position were defined. In preliminary correspondence, the Board of the Faculty of Clinical Medicine expressed willingness to consider accepting the grant, if given. A draft of the proposal was then given to Hugh, who had been attending a lipid symposium in the United States. Probably seeing this document as infringing, and possibly endangering, his authorty within his Institute, he responded with outrage, and in part said:

… I regret to say I think your letter to the Wolfson Foundation with appendices would be disastrous. First, Foundations lay down carefully to whom applications should be made … Secondly we discussed at length … the best method of approach and agreed that a memorandum should go with supporting letters from those who personally know Trustees … I am personally opposed to a 'Chair of Clinical Nutrition' and if consulted … would say so. Obviously such a person would be primarily responsible for treating patients and would normally … be expected to have beds … I have firmly believed all my academic life that there should be appointments in 'Human

Nutrition' and that the great future in the subject lies in <u>preventing</u> disease and promoting health. But this I believe can only reasonably be pursued by someone who is medically qualified and has access on occasion to patients by co-operation of his colleagues ... I note that the application in your draft letter is for 'a Chair in Nutrition in the Faculty of Medicine of the University of Oxford' and that you state our Council of Management has 'given its strong support for the proposed new Chair.' I can only repeat that I am personally opposed to 'a Chair in Nutrition' since I firmly believe there should be Chairs in Human Nutrition and the reason why Woolton wanted to see established a central institute for this subject was that he was convinced ... that universities might continue to regard <u>Human Nutrition</u> as applied science and rat nutrition as pure science and therefore the latter being highly respectable for a university unlike the former. I am sorry I write so critically and at such length, but I have spent most of my academic life in the wilderness fighting perhaps a losing battle for <u>Human Nutrition</u> which I personally believe to be the most important environmental factor affecting health. Yours ever Hugh.

In the mid-1980s Sinclair was also pursuing money from the British Heart Foundation and from David Horrobin for work with Bates and van Dam requiring the collection, shipping, and analysis of samples from fish-eating people in Alaska and Kenya. Mary Gale prepared a detailed grant application for the projects. A decision by the British Heart Foundation was dependent on a site visit, but Horrobin promptly invested in his old friend's plans.

As usual, Hugh remained available to address various scientific meetings, mixing chemistry with humour. At a weekend meeting of the McCarrison Society in 1985, he began:

> Sinclair's eleventh commandment reads, five days shalt thou labour and do all that thou has to do and on the sixth and seventh thou shalt grow thine own vegetables and make thine own bread and yoghurt.

He continued with a long statement about his personal contacts with great men, and then went on to review the facts of the form and origins of essential fatty acids. For almost an hour, he did what he did best: he made the subject of essential fatty acids come to life. For non-scientists, he explained the structure of fats in terms of eggs and mutton, oxidized linseed scum on the top of a can of oil-based paint, and the lard from free-range pigs. For the scientists, he drew together the current knowledge of prostaglandins and platelet aggregate. Along the way he discussed work by others, and work based on samples taken from his own body during self-experimentation. In the end, his message was simple: it is not the amount of fat one eats, but the amount of essential fat one consumes that is the key to the prevention of degenerative diseases.

A major international health project, known as INTERSALT, attracted Sinclair's attention, and when a friend in British Columbia contacted him about joining the investigation, Hugh was delighted. The pair planned to investigate patterns of non-infectious diseases in populations in Indian villages that share a specific genetic pattern, and who continued to eat a traditional Indian diet. This was a prospective study, following a group of people for a long time. The hope was to get essential fatty acid profiles and blood-pressure information on 200 people and watch for signs of cardiovascular disease, cancer, and other conditions that might be caused by essential fatty acid deficiency.

Here was the research opportunity Hugh had waited for all his life. He would have colleagues who respected his views and experience. David Horrobin, through his new company, the Efamol Research Laboratory, agreed to cover transportation costs both to Canada and to Kenya the following year, and also to finance the cost of necessary field equipment. Most importantly, Horrobin agreed to have his scientists carry out laboratory investigations on Hugh's behalf, free of charge: this meant the samples would be analysed under controlled scientific conditions by trained technicians.

Hugh wrote to his friend in British Columbia with bad news: the British Heart Foundation was unable to fund the work, but he had obtained financial support for about half the costs of the expedition, and the costs of lipid estimations were covered. Hugh hurriedly solicited the help of other friends to obtain Canadian money to support the work. The advice was encouraging. One friend told him not to forget the Americans in Alaska — he should contact President Reagan '… Don't waste time on bureaucrats, but go straight to the politicians.'

In the end, the project was never begun owing to a lack of interest from governmental sources.

But there was a possible new avenue down which Hugh might be able to obtain funds. The tobacco industry was spending millions on research. Some of this potentially involved the protective effect of essential fatty acids. For example, a London team of research scientists, working with

rat urinary bladders as experimental models, showed that several components of cigarette smoke, although not nicotine, inhibited prostacyclin
(PGI$_2$) synthesis.[2] As PGI$_2$ is known to have a cytoprotective effect on
certain tissues, and omega-3 essential fatty acids are precursors in the
formation of this substance in the body, Sinclair questioned whether or
not there was a link between the dietary level of certain essential fatty
acids and susceptibility to the carcinogenic effects of cigarette smoke.

With help from friends, Hugh began seeking funds from the tobacco
industry. He pointed out that highly active free radicals are produced in the
body while one smokes a cigarette. If this were true, highly unsaturated
essential fatty acids could be damaged by free radical action. Was it possible
to slow this damage by protecting the essential fatty acids, or increasing the
dietary amount? Sinclair's evidence was varied. Part was epidemiological:
the evidence at the time suggested that the Japanese smoked heavily, and
right up to the end of their cigarette ('sometimes using a pin', Hugh wrote),
yet lung cancer was rare. The situation was similar when, in 1957, Hugh visited Madrid, and asked about smoking patterns there: lung cancer was
rarely seen in the big hospitals, he claimed. These countries shared two
other characteristics: they both experienced low levels of coronary thrombosis, and in both countries the dietary ratio of essential fatty acid to
saturated fat was high. In a letter to a friend, in 1986, Hugh wrote:

> In 1944 I had noted the extensive smoking amongst Canadian Indians and
> Eskimos and the absence of lung cancer, but there were difficulties of
> interpretation as I did not know how long such smoking had occurred, and
> life expectancy was short (mainly accidents). However, when I was recently
> with Eskimos in [Northwest] Greenland I noted the extraordinarily high
> amount of smoking and collected figures for the sale of cigarettes in past
> years; lung cancer virtually unknown and life expectancy high.

Some of his other evidence was more circuitous. Deaths from lung
cancer were increasing at a faster pace than the rate of smoking, he
claimed: even allowing for a 20-year lag period before the first clinical
signs of the disease appeared, Sinclair felt this indicated that something
more was involved than exposure to cigarette smoke. Was it possible
the declining intake of essential fatty acids was a co-factor in the
development of the disease? He wrote:

> Then there is the difficulty in producing the disease in lower animals, which
> are always kept on an excellent diet. And I studied some cases ... of the

association of carcinoma of the lung with polyneuritis (which I thought could be explained by EFA-deficiency). In the 1960s I had a smoking machine for lower animals and intended to use EFA-deficient ones and controls exposed to the same smoke. I twice had to stay with Sir Edward de Stein who built up Gallaher Ltd …; Sir Edward was very interested in my views … But I had not then (unlike now) the necessary laboratory support …

Then came a most interesting suggestion. For the first time directly identifiable in his writing, Hugh drew together the masses of data collected by the Oxford Nutrition Surveys during the World War II and his current research interests. He continued in his letter:

Another approach on which we have done a little work concerns the prospective study of my wartime nutritional surveys for the Ministry of Health. Apart from EFA, there is increasing evidence of a protective effect of vitamin A and carotenoid (present in green vegetables) on the relation of smoking to lung cancer. In some wartime surveys we recorded the amount of money spent on cigarettes, and have detailed information about dietary EFA and vitamin A (and carotenoid), a clinical examination, a test of night-blindness (the earliest evidence of deficiency of vitamin A) and levels of vitamin A and carotenoids in blood. Many of those examined during the war are now dead. All our results are on Hollerith cards which can now be transferred to a computer, so we can do the necessary correlations. This would be very interesting.

It is notable that many investigators were now interested in Hugh's war-time data, as they provided useful data for correlation between current medical findings and earlier dietary and social circumstances.

Copies of Hugh's letter were sent on to the appropriate people within the tobacco industry, but his ideas fell foul of his reputation as an eccentric.

Although he was condemned as a scientific maverick, he was even more in demand as a speaker, his two topics of choice being The Physiology of Laughter, and the Food and Diseases of Eskimos. Both were popular, although he delighted broader audiences with his adroit mingling of history, medicine, and mischief in his masterful dissection of humour. He described the lecture on laughter as an examination of: 'Who laughs? Why and when do we laugh? How do we laugh?'

The Citizen Ambassador Program, a Program of People to People International, which Dwight D. Eisenhower founded in 1956, and of which President Ronald Reagan was Honorary Chairman in 1987, invited Hugh to serve as a 'citizen ambassador' at a meeting in China pertaining to edible oils. His health was beginning to fail, however, and he was able to accept only a limited number of these generous tributes to his knowledge. Upper abdominal pain, delayed gastric emptying, belching, and vomiting foretold days ahead when the travel would be even more restricted. Cancer was diagnosed, and surgery prescribed. Sound wisdom accompanied the upsetting prognosis, however. A close friend wrote: 'My advice is back to normal living accepting the limitations from a partial gastrectomy. Keeping up your moral …' Hugh took the advice. To a friend he wrote:

> I'm terribly busy! Last night 150 merry makers were at Lady Place 8 p.m. to
> 2.30 am dancing reels on my lawn (an annual event; I'm President of the
> [Oxford University] Caledonian Society). Tomorrow and the next day in
> London at the [Royal College of Physicians] for a meeting … but I'll be
> dining at the Apothecaries in the evening. Today was the Doctors
> Procession and Vice-Chancellors Garden Party. To my garden Party 600 or
> so are invited! The day after our Council meeting I have my annual cricket
> match of the Medicinal Profession against the Legal Profession; I think the
> [British Medical Association] are thinking of presenting a cup for this
> highly popular annual game. On Saturday Magdalen has a Development
> Trust Garden Party: maybe I shall see you there or at the Apothecaries on
> Thursday. Anyhow I hope we meet again soon.

Hugh continued driving his rusting Daimler to Magdalen College for dinner. He worked hard at enlarging his reputation as an eccentric. It is recalled, by those who knew him at Magdalen at the time, that Hugh regularly wore a grey tweed jacket that became so threadbare it lost a sleeve. Nonetheless, he continued wearing it to dinner. When friends objected he simply waved a hand and said, 'Plenty of wear in it'.

Terry Newport, Senior Butler at Magdalen, helped when Hugh entertained guests at Lady Place, as did Hugh's old friend and Deputy Chef at Magdalen, William Jarvis. What was to be Sinclair's last luncheon meeting for the Council of Management of the International Nutrition Foundation was prepared with their assistance. Twenty were expected for lunch; on the day, twelve attended and lunched on smoked salmon mousse, *chaudfroid* chicken, cornet of ham, salad, new potatoes, brandied peaches, crème caramel, and coffee. During the meeting, Hugh reported that some income had been received, but that the deficit for the year was £25 750. The total indebtedness to Hugh Sinclair had risen to £171 843, which was in the form of an interest-free loan.

About this time, the McCarrison Society published a biography of Sinclair, sponsored by the Caroline Walker Trust, for his 80th birthday. Geoffrey Cannon, a distinguished health writer, was asked to prepare the text. In the end, he taped a long conversation with Hugh, prepared a document on the basis of what was told, and turned the task of preparing the final work over to Mary Gale and Brian Lloyd. They selected a group of old friends and students, asking each to make a contribution to the text of the proposed book. Most of those contacted later received letters from Hugh suggesting information they might include. The little book, *Sinclair*, was the third of an occasional series published by the Society, and soon found its way to his friends and acquaintances around the world. Its many contributors included Maurice Partridge, Brian Lloyd, V. Ramalingaswami, and David Horrobin.

Hugh Sinclair had been encouraged to write his view of the history of nutrition, but he never got around to it. During his last years he began a manuscript of his autobiography, which he kept in a leather satchel tucked behind his tattered desk chair. It is said that his chosen title for the book included the word 'Failure'. That word may have reflected self-appraisal, but certainly not a consensus of his associates.

In readiness for the last meeting of the Council of Management of the International Nutrition Foundation, Hugh Macdonald Sinclair prepared a report of his current work:

Preliminary results of a 5-year trial of fish-oil fatty acids on 312 patients suffering multiple sclerosis had been published in the March issue of the *Journal of Neurology, Neurosurgeon, and Psychiatry*; a more detailed paper was to follow. He had written the introduction to E. F. Field's book, *Multiple sclerosis*, in which the author proposes a test of the disease based on Hugh's suggestions regarding the role of essential fatty acids in the origins of the disease. Work on information from his Eskimo diet experiment, and on data from the Oxford Nutrition Survey, was progressing. The Leverhulme Trust had given him a research Fellowship for two years to finance a part-time secretary and support other aspects of this work. With the help of Dr Mary Gale and colleagues scattered around the world, he was involved in studies on volunteers consuming meat from chickens raised on diets containing varying levels of essential fatty acids. In this work, he and colleagues were photographing the *arcus senilis* in patients as an indicator of atheroma. Many of his ideas, he claimed, were drawn from work undertaken during and just after the war. Hugh also reported on his lectures and television appearances during the past year in the United Kingdom, in Italy, and in the United States, where he had been entertained at the Annual General Meeting of the American Institute of Nutrition and elected a Fellow of the Institute: a distinction, Hugh told the Council, unique among scientists in the United Kingdom.

Hugh reported to his Council that he had presented the opening and closing papers at an international conference David Horrobin had organized, in Brighton, England, during January 1990, pertaining to essential fatty acids. An international group of more than 90 scientists participated. The topics covered in their scientific papers read like an update of Sinclair's 1956 letter to the *Lancet*: atopic eczema, acne vulgaris, psoriasis, skin lesions in cats and dogs, coronary heart disease, hypertension, rheumatoid arthritis, Sjögren's syndrome, virus disorders, alcoholism, schizophrenia, tardive dyskinesia, child hyperactivity, certain forms of cancer (in mice), peptic ulcers, irritable bowel syndrome, biliary

In April 1989 Sinclair was elected a Fellow of the American Institute of Nutrition recognizing 'a distinguished career in the Science of Nutrition'.

cirrhosis, cystic fibrosis, diabetic neuropathy, premenstrual syndrome, mastalgia, and breast cysts.

As a result of that meeting, a second book concerning essential fatty acids was edited by David Horrobin.[3] In the Introduction, Horrobin wrote:

> Within the last ten years, and especially within the last five, clinical research on EFAs has accelerated rapidly. It now appears that many of the ideas put forward by Sinclair were broadly correct. This book is, in a sense, a tribute to Hugh Sinclair, for it describes just how far we have progressed in relating clinical medicine to EFA abnormalities.

Knowing that the scientific world now took him and his early ideas about the clinical importance of essential fatty acids seriously, Hugh Macdonald Sinclair celebrated his 80th birthday, on 4 February 1990, with a sense of satisfaction. The next day he travelled to the University of Reading to present his annual lecture on Methods of Research in Human Nutrition. As a birthday surprise, Dr Ann Walker, of the Department of Food Science and Technology, and the Final Year and MSc Food Science students there, presented him with a shield of the University coat of arms, and a kite in the form of a giant green fish. Sinclair was delighted. Little did he know that a research programme in human nutrition would one day be inaugurated on that campus and named in his honour.

On 3 May 1990 Hugh wrote to his family and close friends telling them he was terminally ill. His letter was as much a report as a personal farewell. 'I regret to have to inform you,' he wrote to the members of the Council of Management, 'that I have inoperable cancer.' He recounted the symptoms and circumstances leading to a partial gastrectomy in November, 1987. Most of his friends and relatives never knew the surgery had taken place. His letter continued:

> All went well until March this year when I was attending by invitation a large International Conference on Omega-3 (fish-oil) Essential Fatty Acids in Washington and I developed nausea and vomiting and was unable to complete the after-dinner speech at the Banquet which I had to give, though it will be published. I returned from the USA on 26 March last and immediately attended the International Conference on Essential Fatty Acids in Bath, arranged by the Biochemical Society in honour of my 80th birthday. Because of illness I had some difficulty with my own paper and before the end of the Conference David Horrobin kindly motored me home. I was immediately

admitted to the [John Radcliffe Hospital] ... I assume I might have one
month or two to live and would hope to attend the Council of Management
on 30 June. I have had over eighty years of a largely wasted but very happy life
(apart from the death of two fiancées, relatives and friends) and I have no
regret whatever — except that, having spent much of my life trying to get
teaching and researching human nutrition properly recognised, I am deeply
concerned about the future of the Foundation ...

During his final days, thin and lionesque from his advancing cancer,
Hugh once again indulged in self-experimentation. He asked his medical
friends to try treating his cancer with a lithium salt of arachidonic acid, a
metabolite of the omega-6 linoleic acid. Under the pressure of his
insistence, they agreed that no harm could be done at that point, and
Hugh had his way. During the hours following the administration of the
formulation he produced vomit and black urine. A friend protested: he
must stop this nonsense. 'No', said Sinclair, who was monitoring and
recording his body's reactions to the treatment, 'tomorrow, just halve the
dose'.

On 22 June, Hugh Macdonald Sinclair died. His body was cremated
and the ashes placed in the family grave, in Kent, where he joined Rosalie,
Catherine, John, and the Colonel. It was for those he left to decide
whether this intelligent, driven, and eccentric man had made the most of
his life. Had the boy, who had bowed to the King so deeply that his hat
touched the ground, ever succeeded in pleasing his father, his mother, or
himself? If he had failed to achieve all that was expected of him, were the
elements of his character at fault, or had the circumstances of his life
worked against him? Having attempted to present as fair and complete a
summary of Hugh Sinclair's eventful years as possible, this book leaves it
to the reader to decide.

The matter of Hugh Sinclair's success in his chosen field of science is
less subjective. His life's work did not result in any great discovery, nor
did it lead to the methodical and irrefutable explanation of any important scientific question. Instead, Sinclair did something very different
and — many would agree — more important: he championed unpopu-
lar ideas, inspiring others to test and argue their veracity. From his early
school days until the end, Hugh Macdonald Sinclair's life was a celebration of the mind, and he invested all that was his in making the enjoyment
of knowledge both accessible and acceptable.

Epilogue: Legacy of a phoenix

After Hugh Sinclair's death the peafowl continued their stately saunter among the ramble of weeds and wild nettles that spilled over the paths and beyond the walled garden of Lady Place. Remnants of Sinclair's International Nutrition Foundation persisted, and an assortment of renters provided income to pay salaries and help maintain the property. Dr Mary Gale stayed on for a few years to analyse and publish Sinclair's research findings, and Dr Brian Lloyd, now the retired Director of the Oxford Polytechnic (which later became Oxford Brookes University), was appointed Director of the Foundation. Most of his work pertained to concluding the business of the Foundation, tending to its existing facilities, and computerizing the data from the Oxford Nutrition Survey. Money was in short supply, and it was decided to auction the contents of the house. The auctioneers, Phillips, in Oxford, handled the sale. Many of the finer pieces had once belonged to the Jackson family. But there also were items Hugh and his mother had collected over the years from house sales and London auctions. They chose well. The Sinclair auction was held on 10 October 1990: three days before Hugh's memorial service. Objects of beauty, history, and curiosity went under the gavel. As examples, a mid-Victorian oak chair said to have been made from the wood of HMS *Victory*, or another ship of the period, was sold with a guide price of £150–200. A desk bearing the initials and arms of the Baroness Burdett-Coutts, the great Victorian philanthropist, had a guide price of £4000–6000. A mid-Victorian long wall settee purchased at a sale of the contents of Fawley Court, Henley-on-Thames, in 1952 for £9 was given a guide price of £150–200. A satinwood and ebony strung bedroom suite, by Gillow of Lancaster, which had been purchased at a sale of contents of Friar Park, Henley-on-Thames, for £85 in 1951, was offered in 1990 with a guide price of £2500–3500.

The sale was extensive. Family portraits, pipe racks, carpentry tools, comic lantern slides, fishing reels, and a Massey Ferguson tractor with front

digger, all left Lady Place in the hands of new owners. Even a stuffed croco-
dile Hugh scavenged from the Worshipful Society of the Apothecaries
fetched a price. Personal items were not spared the indignity of the ham-
mer: Colonel Hugh Sinclair's military medals were sold, as were those of his
son, John. A collection of watercolours by the Colonel, a fine and personal
record of his times in Jerusalem, Cyprus, Mesopotamia, Egypt, and the
foothills of Afghanistan, found a new home. One hundred and five lots of
garden furniture were auctioned: along with urns and carved stone seats
were a bronze phoenix, a Japanese toad, a family of cranes, and four
gnome-like musicians whose stone eyes had watched frolicking on the
lawn at Lady Place for decades. The sale was a great success, but perhaps
some items should have been kept back. Most poignant of these was a white
metal presentation casket adorned with a relief view of the Nutrition
Research Laboratory at Coonoor, Tamil Nadu, India, which had belonged
to Sir Robert McCarrison. After Hugh's death, Sir Robert's papers were
removed from the International Nutrition Foundation files and given to
the Wellcome Foundation; perhaps this commemorative casket should
have gone with those papers. Or, perhaps it should have been saved and
handed over to the academic organization that would eventually assume
the Sinclair name and legacy.

But the real stuff of a man is betrayed by his library. On 13 February 1992
Phillips held a second auction of the best of Hugh's remarkable collection
of books. According to the 6 March 1992 issue of the *Oxford Times*, the
occasion was one of the biggest and most historic book sales in Phillips'
history. A director of the company personally took the gavel and conducted
business. The catalogue had caused great excitement, and the assembled
crowd was as much curious as it was anxious to buy. There were rare and
fine medical books, including works by H. Boerhaave (similar to the work
Sinclair gave Leiden University after the war), and by the early French
physiologist, Jean Fernel. Eight copies of Florence Nightingale's *Notes on
nursing, what it is and what it is not* were auctioned, as were eleven volumes
written by Sir William Osler, the Oxford Regius Professor of Medicine
during the early 1900s, and one of the greatest clinicians of the twentieth
century. There was also a small group of books on agriculture and gardening, and collections of literary and miscellaneous works that suggested the
range of Sinclair's interests. Finally, and a great surprise to many who knew
Hugh only through his scientific work, there was an extensive collection of
erotica. This was the stuff that brought in the cash. Bidding on these items

was international, brisk, and punctuated by raucous laughter. Prices reached impressive heights. *The rod*, a poem in three Cantos by H. Laying, had a guide price of £150–200; it sold for £360. Most impressive, however, was the bidding for a collection of pamphlets by J. Bonefonius, including *Cupid's beehive or the sting of love* (1721) and *The second part of whipping Tom* or *A rod for a proud lady* (1722). That item had a guide price of £90–120. Its new owner paid £4400 after some furious bidding. In total, the sale realized more than £85 000.

During the next few years following the auctions, Mary Gale retired and Brian Lloyd moved most of the administrative work from Lady Place to his home in Oxford. The old mock-Tudor home of the Sinclairs had suffered from neglect before Hugh died: now the situation was dire. A few tenants occupied flats in various buildings on the estate. A family occupied the front of the main house, using it as both their home and a music school, keeping up what they could of both the house and its grounds. No one else seemed to care much about the state of things. The grounds slowly fell under the spell of pernicious weeds. Ivy found its way over walls and wooden sills, inviting in the damp. The roof leaked and water pipes broke, flooding rooms and leaving moisture that caused the remaining piles of Sinclair's papers and files to moulder in rooms that were once his office and laboratory. Some yards from the main house, however, the contents of the ugly breezeblock building Hugh had called his information centre fared better. Large storage heaters were installed to help protect its store of memorabilia, reprints, Hollerith cards, and books from the weather.

During the August Bank Holiday weekend, in 1998, an arsonist lit the flame that burnt the mansion at Lady Place to the ground. There were no injuries. The roof collapsed in the intense blaze and, sadly, the large stained glass window over the stairs in the entry hall, which bore the Sinclair family's coat of arms, was broken and smashed into the main hall. There was nothing to salvage. No other buildings were damaged. Fortunately, most of Hugh's papers, books, and personal items that remained in the mansion after the auctions were safe, because they had been moved either to the University of Reading, now the recipients of Hugh's estate, or to Sterling, Scotland, where they were under the protection of David Horrobin.

Not long after the fire, Lady Place was sold and all its buildings bulldozed to make way for an expensive and gated housing estate. The spell was broken: the ghost of Sinclair that once roamed the tumbled-down gardens and inspected the rooms of his beloved house had disappeared.

Lady Place after the arson attack of August 1998.

But that is not the end of the Sinclair story. Following his death, the Inter-national Nutrition Foundation and all the Sinclair assets had been placed in Trust. The Council of Management agreed these considerable resources should be granted to a university that would used them to further Hugh Sinclair's aspirations by endowing a Professorship in Human Nutrition. An international panel of nutrition experts was appointed to review proposals from interested universities. There were three contenders: the University of Oxford, the University of Reading, and Oxford Brookes University. After considerable discussion, the award was made to the University of Reading, where Hugh had spent many pleasurable hours teaching during the last years of his life.

In October 1995, Dr Christine Williams, an internationally respected nutritionist studying the metabolism of essential fatty acids, was appointed the first Hugh Sinclair Professor of Human Nutrition at the University of Reading, and under her leadership, the Hugh Sinclair Unit of Human Nutrition was officially opened in 1999. Work within the unit is diverse, although the study of dietary fats remains a major focus for many projects. The work would have pleased Hugh.

But he would have been most pleased by seeing the extent to which an interest in the metabolic and clinical significance of essential fatty acids now permeates medical research around the world. For example, in 1996,

exactly forty years after Hugh Sinclair published his notorious letter to the
Lancet, an international conference on 'Highly Unsaturated Fatty Acids in
Nutrition and Disease Prevention' was held in Barcelona, Spain. The meeting was co-sponsored by F. Hoffmann-La Roche Ltd, Basel, Switzerland,
the International Society for the Study of Fatty Acids and Lipids, USA, the
Institute National de la Recherché Agronomique, France, and the Japanese
Society of Nutrition and Food Science. Scientists, agronomists, corporate
representatives and product developers, healers, and journalists from 38
countries attended. Some papers and demonstrations pertained to the
most exacting aspects of cellular physiology. Others considered broad
topics, such as the availability of polyunsaturated fats in the human food
chain. Hugh Sinclair was not forgotten. Dr Michael Burr, from the University of Wales College of Medicine, gave the address at the Conference
Dinner, and recalled for the group Hugh Sinclair and his outspoken claims
about the importance of omega-3 fatty acids.

Hugh Sinclair was a man who believed in an idea and was not afraid to
pursue it to the end, even at great personal expense. He was brilliant, but
deeply flawed, and the roots of most of his problems sprang from his own
history. One can argue that, had he been able to accept leadership instead of
combating it, and if he could have shared his ideas instead of anxiously
pulling away from possible collaborators, his story would have been one of
far greater public and scientific success.

Today, it is rare to hear Hugh Sinclair's name linked with any aspect of
human nutrition other than the study of essential fatty acids. But he left us
much more. He left us a blueprint of how human nutrition should be
investigated. Sinclair's blueprint is comprehensive, integrating every aspect
of human physical and social well being. He championed the belief that a
fundamental bond exists between human well-being and the foods we eat.
He argued that this bond must be investigated simultaneously through
clinical and biochemical means, and that the results of these investigations
should be used to influence not only medicine, but also agricultural and
social policy. Perhaps of even greater importance, facts about food and
nutrition should be made clear and widely available to all.

While we in the twenty-first century struggle with problems of environmental pollution, contaminated food, over-processed food ingredients,
and rapidly increasing rates of degenerative illnesses, we should pause and
ask how Hugh Sinclair's plea for a comprehensive view of the relationship
between food and health would profit us. A careful consideration of his

ideas should encourage us to pursue the truth about nutrition, and thus assuage the damage we cause ourselves by ignoring it.

Chapter notes and sources

Chapter 1

Documents quoted on pages 5–6, 8, 9, and 12–13, were used with the permission of the Hugh Sinclair Trust of the University of Reading, England.

Chapter 2

Documents quoted on pages 17, and 25–6, were used with the permission of the Hugh Sinclair Trust of the University of Reading, England.
1. The author is grateful to Dr Brian B. Lloyd for the use of the notebook from which this quotation was taken.
2. Sinclair, J. G. (1931). *Portrait of Oxford.* p. 81.

Chapter 3

Documents quoted on pages 27–8, 38, 39, and, and 40 were used with the permission of the Hugh Sinclair Trust of the University of Reading, England.
Notes
1. McCarrison, Sir Robert. (1953). *Nutrition and health.* p. 15.
2. Passmore, R., Peters, R. A., and Sinclair, H. M. (1933). On Catatorulin: A method of comparing the oxidative factor in vitamin B_1 concentrates. *Biochemistry*, 27, 842–50.
3. Sinclair, H. M. (1990). Methods of research in human nutrition. In: *Applied human nutrition for food scientists and home economists* (Walker, A. F. ed.), pp. 45–6.
4. McCarrison, Sir Robert (1953). *Nutrition and health.* p. 57.
5. McCarrison, Sir Robert. *Nutrition and health.* pp. 67–8.
6. (page 56) Lawes, J. B. and Gilbert, J. H. (1877). On the formation of fat in animals. *Journal of Anatomical Physiology*, 11, 577–88.

Chapter 4

Documents quoted on pages 42, 45, 46, 47–8, and 49, were used with the permission of the Hugh Sinclair Trust of the University of Reading, England.
Notes
1. Sherrington, Sir Charles (1946). *The endeavour of Jean Fernel.* Cambridge University Press, Cambridge.
2. Drummond, J. C. and Wilbraham, A. (1939). *The Englishman's food: A history of five centuries of English diet,* p. 543.

Chapter 5

Documents quoted on pages 55–6, 62–3, 64, 65, 67–8, 69, 70, 71–2, and were used with the permission of the Hugh Sinclair Trust of the University of Reading, England.
1. Harrison, B. (1994). *The history of the University of Oxford*, Vol. III, pp. 167–88.

Chapter 6

Documents quoted on pages 81–2, 83, 84–5, 86, and 87, were used with the permission of the Hugh Sinclair Trust of the University of Reading, England.

Chapter 7

Documents quoted on pages 93, 94–5, 96–7, and 101–2, were used with the permission of the Hugh Sinclair Trust of the University of Reading, England.
Notes
1. Lewis, C. S. (1950). *The problem of pain*. p. 117 (18th impression). Geoffrey Bles, Whitefriars Press Ltd, London.
2. Sinclair, H. M. *The Oxford Nutrition Survey preliminary report*, page 4, 11 May 1942, folder 79, box 10, series 2.1, (Post-war Correspondence), record group 6.1 (Field Office — Paris), Rockefeller Archives, Rockefeller Archive Center, North Tarrytown, New York (hereafter designated RAC).

Chapter 8

Documents quoted on pages 106, 111–12, and 113, were used with the permission of the Hugh Sinclair Trust of the University of Reading, England.
Notes
1. Hugh H. Smith to W. A. Sawyer, 1 October, 1942, folder 61, box 10, series 700 (Europe), Health Commission, England, 1942–1945, Record Group 1.1 (Projects), Rockefeller Foundation Archives, RAC.
2. Hugh H. Smith to W. A. Sawyer, 08 October, 1942, folder 61, box 10, series 700 (Europe), Health Commission, England, 194201945, Record Group 1.1, (Projects), Rockefeller Foundation Archives, RAC.
3. Youmans to Sinclair, 28 December 1942, folder 62, box 10, series 700 (Europe), Health Commission, England, 1942–1945, Record Group 1.1 (Projects), Rockefeller Foundation Archives, RAC.

Chapter 9

Documents quoted on pages 116–8, 121, 122, and 123 were used with the permission of the Hugh Sinclair Trust of the University of Reading, England.
Notes
1. Sawyer's diary note on meeting with Sinclair, 12 August 1943, folder 62, box 10, series 700 (Europe), Health Commission, England, Record Group 1.1 (Projects), Rockefeller Foundation Archives, RAC.

2. *Sinclair.* pp 48–9.
3. *Oxford Nutrition Survey: Army Nutritional Intelligence Units (Preliminary Suggestions,* 30 May 1944, folder 79, box 10, series 2.1 (postwar correspondence), sub-series: Oxford Nutrition Survey, record group 6.1 (Field Office — Paris), Rockefeller Foundation, RAC.
4. Seventh meeting of the Advisory Committee of the Oxford Nutrition Survey. 31 May 1944, page 6, folder 79, box 10, series 2.1 (Postwar correspondence), sub-series: Oxford Nutrition Survey, record group 6.1 (Field Office — Paris), Rockefeller Foundation, RAC.

Chapter 10

Documents quoted on pages 133–4, 135, and 136–42,were used with the permission of the Hugh Sinclair Trust of the University of Reading, England.

Chapter 11

Documents quoted on pages 144, 149, and 150–2, were used with the permission of the Hugh Sinclair Trust of the University of Reading, England.
Note
1. Drummond, J. C., Report to the Advisory Committee appointed by SHAFE (Impressions of the Towns of W. and N. W. Netherlands) 20 May 1945, page 1, file: Drummond, J. C., Ministry of Agriculture, Fisheries and Food Archive, London.

Chapter 12

Documents quoted on pages 156–8, 159, 161–2, 164, 165, 166–8, 170, and 171, were used with the permission of the Hugh Sinclair Trust of the University of Reading, England.
The author is greatly indebted to Dr Brian B. Lloyd for the use of the Hugh Sinclair diaries from the period covered in this chapter.

Chapter 13

Documents quoted on pages 172–4, 175, and 176, were used with the permission of the Hugh Sinclair Trust of the University of Reading, England.
1. *Sinclair.* The McCarrison Society, 1990. pp. 49–51.

Chapter 14

Documents quoted on pages 183–4, were used with the permission of the Hugh Sinclair Trust of the University of Reading, England.
Note
1. RRS–Diary entry, Monday 8 December 1947, folder 62, box 10, series 700 (Europe), record group 1.1 (projects), Health Commission, England, 1942–5, Rockefeller Foundation, RAC.

Chapter 15

Documents quoted on pages 186–7, were used with the permission of the Hugh Sinclair Trust of the University of Reading.

1. Personal communication, 13 September 2000.
2. *Malnutrition and starvation in the western Netherlands.* volume 1, introduction, pp. 2–3.
3. *Malnutrition and starvation in the western Netherlands.* volume 2, appendix 2, pp. 4–6.
4. Pyke, Magnus (1947) *Fat in the human diet*, report to the Ministry of Food, London.
5. Chevreul, Michel Eugène (1815). Recherche chimiques sur plusieurs corps gras, et particulièrement sur leurs combinations avec les calcuis. *Ann. Chim.* (Paris, **95**, 5–50.
6. Chevreul, Michel Eugène. (1823). *Recherches chimiques sur les corps gras d'origine animale.* Paris.
7. *Sinclair*, op. cit. pp. 18–19.

Chapter 16

Documents quoted on pages 201–2, 205–6, 207, and 210–11, were used with the permission of the Hugh Sinclair Trust of the University of Reading, England.

Notes
1. *The Practitioner*, **168**, 133–9.
2. Keys, A., and Sinclair, H. M. (1952). Real nutritional deficiency. *British Medical Bulletin*, **8**, 262–4.
3. *Dominici non coupable les assassins retrouvés.* Flammarion, 1997.
4. Edelman, J. (1982). Obituary. *Biochemical Society Bulletin*, 2, June 4, s
5. Sinclair, H. M. (1957). Nutrition and the skin in man. *Annals de la Nutrition et de L'alimentation* (Paris), **XI**, 147–76.
6. Sinclair, H. M. (1956). [Letter]. *Lancet*, 7 April, 381–3.

Chapter 17

Documents quoted on pages 217, 218–9, 220, and 221–2, were used with the permission of the Hugh Sinclair Trust of the University of Reading, England.

Notes
1. St. John Lyburn, E. F. (1956). [Letter]. *Lancet*, 14 April, 449.
2. Yudkin, J. (1956). [Letter]. *Lancet*, 21 April, 506.
3. Bronte-Stewart, B., Antonis, A., Eales, L., and Brock, J. F. (1956). Effects of feeding different fats on serum-cholesterol level. *Lancet*, 28 April, 521–6.
4. Keys, A. (1956). [Letter]. *Lancet*, 28 April, 576–7.
5. Ministry of Food. (1956). *Report of the panel on composition and nutritive value of flour.* (Chairman: Sir Henry Cohen). HMSO, London.
6. McCance, R. A. and Widdowson, E. M. (1956). *Bread, white and brown: their place in thought and social history.* Pitman, London.

Chapter 18

Documents quoted on 224–5, 229, 230, 231–2, 233, and 234–5, were used with the permission of the Hugh Sinclair Trust of the University of Reading, England.

Notes

1. The papers and discussions from the conference can be found in *Essential fatty acids: Proceedings of the fourth international conference of biochemical problems of lipids.* Oxford 1957, edited by H. M. Sinclair. Butterworths Scientific Publications, London. 1958. Comments by the Waynfleet Professor of Chemistry are found on page xvii.
2. James, A. T. and Martin, A. J. P. (1956). *Biochemical Journal*, **63**, pp. 144–9.
3. Sinclair, H. M. (ed.) (1957). *Essential fatty acids.* op. cit., pp. 142–3.
4. Sinclair, H. M. (ed.) (1957). *Essential fatty acids.* op. cit., p. ix.

Chapter 19

Documents quoted on pages 237, 238, 239, 240, 241, 242, 243, 244, 245, 246–7, 248–9, 251, 252, and 253 were used with the permission of the Hugh Sinclair Trust of the University of Reading, England.

Chapter 20

Documents quoted on pages 254, 256–8, and 261, were used with the permission of the Hugh Sinclair Trust of the University of Reading, England.

Chapter 21

Documents quoted on pages 262–3, 266–7, 268, 270, 271–3, and 275, were used with the permission of the Hugh Sinclair Trust of the University of Reading, England.

Chapter 22

Documents quoted on pages 279–80, 282–3, 284, and 287–8, were used with the permission of the Hugh Sinclair Trust of the University of Reading, England.

Notes

1. Sinclair, H. M. (1982). *British Medical Journal*, **285**, p. 1785.
2. Jeremy, J. Y., Mikhailidis, D. P., and Dandona, P. (1985). Cigarette smoke extracts inhibit prostacyclin synthesis by the rat urinary bladder. *British Journal of Cancer*, **51**, 832–42.
3. Horrobin, D. F. (1990) *Omega-6 essential fatty acids: pathophysiology and roles in clinical medicine*, p. XV.

Reference books

Ashwell, M. (ed.) (1993). *McCance and Widdowson: a scientific partnership of 60 years.* British Nutrition Foundation, London.

Burger, G. C. E., Sandstead, H. R., and Drummond, J. C. (1948). *Malnutrition and starvation in the western Netherlands.* The Hague.

Drummond, J. C., and Wilbraham, A. (1939). *The Englishman's food.* Jonathan Cape, London.

Gale, M. and Lloyd, B. B. (eds) (1990). *Sinclair.* The Founders of Modern Nutrition series, published by The McCarrison Society, London.

Harrison, B. (ed.) (1994). *The history of the University of Oxford: the twentieth century.* Vol. 8. Oxford University Press, Oxford.

Horrobin, D. F. (ed.) (1990). *Omega-6 essential fatty acids: pathophysiology and roles in clinical medicine.* Wiley-Liss, New York.

Humphreys, N. A. (ed.) (1885). *Vital statistics: A memorial volume of selections from the reports and writings of William Farr,* (1885). Offices of the Sanitary Institute, London.

Krebs, H., with Martin, A. (1981). *Reminiscences and reflections.* Oxford University Press, Oxford.

Krebs, H., with Schmid, R. (1981). *Otto Warburgs cell physiologist, biochemist and eccentric.* Clarendon Press, Oxford.

Lewis, C. S. (1950). *The problem of pain,* (18th impression). Geoffrey Bles, Whitefriars Press, London.

McCarrison, R. (1961). *Nutrition and health,* being the Cantor Lectures delivered before The Royal Society of Arts, 1936, together with two earlier essays. (Postscript by Sinclair, H. M.) Faber & Faber, London.

Mitchison, R. (1962). *Agricultural Sir John: the life of Sir John Sinclair of Ulbster 1754–1835.* Geoffrey Bles, London.

Reymond, W. (1997). *Dominici non coupable: les assassins retrouvés.* Flammarion.

Sinclair, A. (1908). *A young man's life.* Andrew Melrose, London.

Sinclair, H. Macdonald (ed.) (1958). *Essential fatty acids.* Butterworths Scientific Publications, London.

Sinclair, H. Montgomerie (1926). *Camp and society.* Chapman & Hall Ltd., London.

Sinclair, J. (1833). *Sinclair's code of health and longevity.* (5th edn.). Sherwood, Gilbert, and Piper, London.

Sinclair, J. G. (1931). *Portrait of Oxford.* Veracity Press, Surrey, England.

Smith, H. H. (1978). *Life's a pleasant institution: the peregrinations of a Rockefeller doctor.* Tucson, Arizona. [Private printing.]

Walker, A. F. (ed.) (1990). *Applied human nutrition: for food scientists and home economists.* Ellis Horwood, London.

Selected publications by Hugh Macdonald Sinclair

The following is a representative selection of more than 300 articles, chapters, letters and other published items attributable to Hugh Sinclair.

Sinclair, H. S. and Brown, W. R. (1924–1927). The butterflies of Winchester. *Report of the Winchester College, Natural History Society,* pp. 18–20.

Passmore, R., Peters, R. A., and Sinclair H. M. (1933). On catatorulin; a new method of comparing oxidative factor in vitamin B_1 concentrates. *Biochemical Journal,* 27, 842–50.

Peters, R. A. and Sinclair, H. M. (1933). Studies in avian carbohydrate metabolism; factors influencing maintenance of respiration in surviving brain tissue of the normal pigeon. *Biochemical Journal,* 27, 1677–86.

Sinclair, H. M. (1933). The metabolism of brain tissue with special reference to vitamin B_1. Thesis for B.Sc. (unpublished).

Peters, R. A. and Sinclair, H. M. (1933). Studies in avian carbohydrate metabolism, IV. Further studies upon action of catatorulin in brain. *Biochemical Journal,* 27, 1910–26.

Sinclair, H. M. (1938). The estimation of vitamin B_1 in blood. *Biochemical Journal,* 32, 2185–99.

Peters, R. A., Sinclair, H. M., and Wood, P. (1939). Discussion on clinical aspects of vitamin-B complex. *Proceedings of the Royal Society of Medicine,* 32, 807–22.

Goodhart, R. S. and Sinclair, H. M. (1939). The estimation of cocarboxylase (vitamin B_1 diphosphate ester) in blood. *Biochemical Journal,* 33, 1099–1108.

Sinclair, H. M. (1939). Estimation of Vitamin B_1 in cerebrospinal fluid, *Biochemistry Journal,* 33, 1816–21.

Goodhart, R. and Sinclair, H. M. (1940). Deficiency of vitamin B_1 in man as determined by blood cocarboxylase, *Journal of Biological Chemistry,* 132, 11–21.

Konstam, G. and Sinclair H. M. (1940). Cardiovascular disturbances caused by deficiency of B_1, *British Heart Journal,* 2, 231–40.

Sinclair, H. M. (1941). Modern therapeutics: the use of vitamins in treatment. *The Practitioner,* 146, 109–18.

Sinclair, H. M. (1943). Malnutrition and peripheral neuropathies. *Proceedings of the Royal Society of Medicine,* 36, 172–4.

Sinclair, H. M. (1944). Wartime nutrition in Britain as a public health problem. *American Journal of Public Health,* 34, 828–32.

Sinclair, H. M. (1944). The evidence for nutritional deficiencies in pregnancy. *Proceedings of the Nutrition Society,* 1, 233–8.

Lloyd, B. B., Sinclair, H. M., and Webster, G. R. (1945). The estimation of ascorbic acid for clinical purposes by the hydrazine method. *Biochemical Journal,* 39, xvii–xviii.

Peters, R. A, Sinclair, H. M., and Thompson, R. H. S. (1946). Analysis of inhibition pyruvate oxidation by arsenicals in relation to enzyme theory of vesication. *Biochemical Journal*, **40**, 516–24.

Sinclair, H. M. (1948). Nutritional oedema. *Proceedings of the Royal Society of Medicine*, **41**, 541–4.

Investigations of the Oxford Nutrition Survey Team (1948). In: *Malnutrition and starvation in the western Netherlands*, **1**, 140–62, and **2**, 262–99.

Sinclair, H. M. (1949). Current therapeutics; vitamin B complex. *The Practitioner*, **162**, 235–46.

Sinclair, H. M. (1949). The Atwater — a nutritional unit. *British Journal of Nutrition*, **3**, x–xi.

Sinclair, H. M. (1950). Nutrition. *Annual Review of Biochemistry* **19**, 339–70.

Sinclair, H. M. and Robb-Smith, A. H. (1950). *A short history of anatomical training in Oxford*. Oxford University Press, Oxford.

Sinclair, H. M. (1951). Nutritional surveys of population groups. *New England Journal of Medicine*, **245**, 39–47.

Keys, A. and Sinclair, H. M. (1952). Real nutritional deficiency. *British Medical Bulletin*, **8**, 262–4.

Sinclair, H. M. (1952). Oral manifestation of deficiency diseases. *The Practitioner*, **168**, 133–9.

Ramalingaswami, V. and Sinclair, H. M. (1953). The relation of deficiencies of vitamin A and essential fatty acids to follicular hyperkeratosis in rats, *British Journal of Dermatology*, **65**, 1–22.

Ramalingaswami, V. and Sinclair, H. M. (1953). Lesions of mucocutaneous junctions in the rat in deficiency of pyridoxine. *Journal of Investigative Dermatology*, **20**, 81–92.

Sinclair, H. M. (1953). The diet of Canadian Indians and Eskimos. *Proceedings of the Nutrition Society*, **12**, 69–82.

Simic, B. S., Sinclair, H. M., and Lloyd, B. B. (1953). The activity of ascorbic acid in hypervitaminosis-A in guinea-pig. *Internat. Ztschr. Vitaminforsch*, **25**, 7–20.

Sinclair, H. M. (1953). Nutrition and health: recent advances. In: *Nutrition and health*, McCarrison (ed.), R., Faber & Faber, London. pp. 107–17.

Ramalingaswami, V. and Sinclair, H. M. (1954). Polycythaemia in pyridoxine deficiency in the rat. *British Journal of Nutrition*, **8**, 386–92.

Basnayake, V. and Sinclair, H. M. (1954). Skin permeability in deficiency of essential fatty acids. *Journal of Physiology*, **126**, 55–6.

Sinclair, H. M. (1955). Vitamin deficiencies in alcoholism. *Proceedings of the Nutrition Society*, **14**, 107–15.

Sinclair, H. M. (1956). Vitamins and the nervous system. *British Medical Bulletin*, **12**, 18–23.

Sinclair, H. M. and Hollingsworth, D. (eds) (1969). *Hutchison's food and the principles of nutrition*. Edward Arnold, London.

North, J. D. K. and Sinclair H. M. (1956). Nutritional neuropathy: chronic thiamine deficiency in the rat. *AMA Archives of Pathology*, **62**, 341–53.

Sinclair, H. M. (1956). Deficiency of essential fatty acids and atherosclerosis, etcetra. [letter] *Lancet*, **i**, 7 April, 381–3.

Kinsell, L. W. and Sinclair H. M. (1957). Fats and disease. [letter]. *Lancet*, **272** (6974), 27 April, 883–84.

Sinclair, H. M. (1957). Food and health. *British Medical Journal*, **2**, 1424–6. (Hugh Sinclair's Work Health Day Oration)

Sinclair, H. M. (1958). Nutritional aspects of high-extraction flour. *Proceedings of the Nutrition Society*, 17, 28–37.

Sinclair, H. M. (1958). Essential fatty acids and the skin. *British Medical Bulletin*, 14, 258–62.

Sinclair, H. M. (1958). Advances in nutrition and dietetics. *The Practitioner*, 181, 468–74.

Sinclair, H. M. (ed.) (1958). *Essential fatty acids: proceedings of the fourth international conference on biochemical problems of lipids.* Butterworths Scientific Publications, London.

Macmillan A. L. and Sinclair H. M. (1958). The structural function of essential fatty acids. Ibid. pp 208–15.

Sinclair H. M. (1958). Deficiency of essential fatty acids in lower animals. Ibid, 249–56.

Sinclair, H. M. (1958). Essential fatty acid and idiopathic hypercalcaemia of infancy. [letter] *Lancet*, 1 (7021), 22 March, 639–40.

Sinclair, H. M. and Jelliffe, D. B. (eds) (1961). *Tropical Nutrition and Dietetics*, (4th ed.). Baillière, Tindall & Cox, London.

Sinclair, H. M. (1961). Some ups and downs of Oxford Science. *Oxford Medical School Gazette*, 14, 6–24.

Sinclair H. M. (1962). Historical aspects of inborn errors of metabolism. *Proceedings of the Nutrition Society*, 21, 1–12.

Sinclair H. M. (1964). Fluoridation falsified. [letter]. *Brititish Medical Journal*, 5382, 554–85.

Sinclair H. M. (1968). Essential fatty acids and atherona. [letter]. *Lancet*, 2 (565), 24 August, 459.

Sinclair H. (1971). Sugar intake and myocardial infarction [letter] *Lancet*, 1 (688), 2 January, 43–44.

Sinclair, H. M. (1972). Nutritional aspects of alcohol consumption. *Proceedings of the Nutrition Society*, 31, 117–23.

Sinclair, H. M. (1973). Deficiency of vitamin E and myocardial infarction. [letter]. *Lancet*, 2 (827), 01 September, 500–1.

Sinclair H. (1973). Fluoride toxicity. [letter]. *Lancet*, 2 (835), 27 October, 962–3.

Sinclair, H. (1977). Polyunsaturated fatty acids in multiple sclerosis. [letter]. *British Medical Journal*, 2 (60696), 1217.

Sinclair H. (1979). Prescription for a better British diet [letter] *British Medical Journal*, 1 (6168), 952–3.

Sinclair H. (1980). Is severe restriction of food really a very-high fat diet? [letter], *British Medical Journal*, 280, 1121.

Bang, H. O., Dyerberg, J., and Sinclair, H. M. (1980). The composition of the Eskimo food in north western Greenland. *American Journal of Clinical Nutrition*, 33, 2657–61.

Sinclair, H. M. (1980). Prevention of coronary heart disease: the role of essential fatty acids, *Postgraduate Medical Journal*, 56, 579–84.

Sinclair H. M. (1980). Advantages and disadvantages of an Eskimo diet. In: *Drugs affecting lipid metabolism*, by Fumagalli, R., Kritchevsky, D., and Paoletti, R. (eds). Elsevier/North-Holland Biomedical Press. pp. 363–70.

Sinclair, H. M. (1980). C. S. Lewis and experimental medicine. [letter]. *British Medical Journal*, 281, 1353–4.

Sinclair, H. (1981). Erucic acid and the Spanish oil epidemic. [letter]. *Lancet*, 2 (8258), 5 December, 1293.

Sinclair, H. (1982). Icosapentaenoic acid and ischemic heart disease [letter]. *Lancet*, 2 (8294), 14 August, 393.

Sinclair, H. (1982). The relative importance of essential fatty acids of the linoleic and linolenic families: studies with an eskimo diet. *Progress in Lipids Research*, 20, 897–9.

Sinclair, H. M. (1982). The game of cricket. *British Medical Journal*, 285, 1785.

Sinclair, H. M. (1983). Magdalen and medicine. *Magdalen College Record*, pp. 24–9.

Sinclair, H. M. (1983). Dietary fibre. [letter]. *Lancet*, 2 (8345), 6 August, 340.

Sinclair, H. M. (1984) Essential fatty acids in perspective: review article. *Human Nutrition: Clinical Nutrition*, 38: 245–60.

Sinclair, H. M. (1987). Food fats, good and bad. *British Journal of Clinical Practice*, 41, 1033–6.

Sinclair, H., Bates D., and Gale, M., *et al.* (1988). A double-blind controlled trial of oils containing n-3 polyunsaturated fatty acids in the treatment of multiple sclerosis. *Journal of Neurology, Neurosurgery and Psychiatry*, 52, 18–22.

Sinclair, H. (1989). Latitude and ischaemic heart disease. [letter]. *Lancet*, 1 (8643), 22 April, 895.

Nightingale, S., Woo, E., Smith, A. D., French, J. M., Gale, M. M., Sinclair, H. M., *et al.* (1990). Red blood cell and adipose tissue fatty acids in mild inactive multiple sclerosis. *Acta Neurol. Scand.*, 82, 43–50.

Sinclair, H. M. (1990). Assessment of the nuriture of the population. In: *Applied nutrition: for food scientists and home economists* (Walker, A. F. ed.) Ellis Horwood, London. pp. 19–35.

Sinclair, H. M. (1990). Methods of research in human nutrition. Ibid. pp 36–54.

Sinclair, H. M. (1990). Nutrition of specific human groups. Ibid. pp 74–96.

Sinclair, H. M. (1990). History of essential fatty acids. In: *Omega-6 essential fatty acids: pathophysiology and roles in clinical medicine* (Horrobin, D. F. ed.). Wiley-Liss, New York. pp. 1–20.

General plan of research: diseases of the nervous system

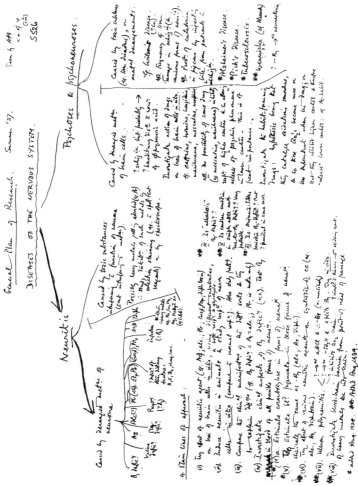

APPENDIX C

Nutritional surveys of population groups *

Hugh M. Sinclair, M.D.†

[Reprinted with permission from the *New England Journal of Medicine*, vol. 245, July 12 (1951), pp. 39–47.
Copyright 1951, Massachusetts Medical Society: All rights reserved.]

IT IS SAID that people who live in glass houses should not throw stones. Having perforce spent much time in recent years carrying out nutritional surveys, I might be chary of criticizing them. But since I have now ceased to be actively concerned in their conduct, I can safely attempt to assess their value and define their limitations. John Clarence Cutter was, we are told, "energetic and possessed of forcible character. He had the courage of his convictions, and his fearlessness counted much. . . ." If in giving this lecture one hundred years after his birth I appear to speak more forthrightly than is customary on these august occasions, I would plead that he would have wished a lecturer to do so.

The lecture is on preventive medicine, and we might do well to recall that this subject is not per-

haps a branch of medicine, which by derivation is the discipline of healing; and healing or medicine does not arise if prevention of disease is perfect. Most of us who study and teach the normal functions of the human body find such activities to be sterile unless we encroach on pathology. Conversely, those who are skilled in prevention of disease must know the normal state of the body and how it may be altered in different environmental circumstances to allow disease. Unfortunately man is a neglected animal, so far as nutrition is concerned; at the end of the last peace there was no Institute of Human Nutrition in Great Britain, although there were some excellent ones for the study of the nutrition of lower animals. The reason, I think, is to be seen from a quick survey of the rise of nutritional science, for which I crave your indulgence.

For convenience let us look at the history of

*The Cutter Lecture on Preventive Medicine, delivered at the Harvard School of Public Health on February 9, 1951.
†Formerly, director, Oxford Nutrition Survey; director, Laboratory of

years ago (more correctly antilog. 5), that is, about 100,000 B.C., paleolithic man was a hunter, eating the animals around him and his family and being eaten by them. Four log. years ago, about 10,000 B.C., the neolithic tribe was pastoral, having tamed animals and learned to take their flocks with them on the nomadic search for new pastures. Nutritional

problems were probably not common; a hostile tribe would, if victorious, kill you and annex your flocks. Three log. years ago saw at its distal end the rise of agriculture; communities became stationary and famines appeared if crops failed; the Tosorthus stele of about 3000 B.C. described a disastrous famine caused by the failure, seven years in succession, of the Nile to flood. At the near end of our three log. years we have the beginnings of nutritional science: About 250 B.C. Erasistratus did excellent metabolic work on chickens (Dr. Stare here is doing the same on ducks, 2200 years later), and about 200 A.D. Galen was studying the principles of dietetics in gladiators (with the same aim as the Harvard Fatigue Laboratory, before it was recently closed). Between two and two-and-a-half log. years ago there was a period of very great importance, starting with the Renaissance and ending with Darwin. Let us examine briefly two instants in this period of two hundred and one hundred years ago.

In 1753 James Lind published his classical work, "A Treatise of The Scurvy." He first reviewed the

literature:

But as it is no easy matter to root out old prejudices, or to overturn opinions which have acquired an establishment by time, custom, and great authorities; it became therefore requisite for this purpose, to exhibit a full and impartial view of what has hitherto been published on the scurvy. . . . Indeed before this subject could be set in a clear and proper light, it was necessary to remove a great deal of rubbish.

(These words might well be applied to nutritional surveys; before they can be set in a clear and proper light it is necessary to remove a great deal of rubbish.) He then described the disease of scurvy; its symptoms, signs and pathology; its causes, prevention and cure. He based his statements upon accurate observation and well designed experiment (as when on May 20, 1747, he took 12 patients with scurvy, divided them into six groups, and gave each group daily one of the following: a quart of vinegar, half a pint of sea-water, a barley-water electary, or two oranges and a lemon). A few years later a young doctor, William Stark, did some excellent experimental work upon himself. Stark was thirteen years old when Lind published his "Treatise"; he graduated at Glasgow and Leiden, and studied in London under John Hunter. He became interested in the function of foods, and on June 12, 1769, he started to study on himself the effects of experimental diets.[1] He did four main experiments in which he produced scurvy in himself and cured it, once with half-a-pint of black currants (containing some 600 mg. of vitamin C). In his final and fatal relevance of such research to medicine was forgotten until Darwin's work a century later. From then onwards the use of the experimental method upon man and when necessary upon lower animals, together with the advance of chemistry, brought rapid progress in nutritional science.

Finally I come to the last period, one log. year ago, namely, the beginning of 1941. This is when the Oxford Nutrition Survey started — an event of no historical importance whatever but one of some slight personal interest to me, since I was rudely cast from the relatively straightforward discipline of experimental nutritional science into the complicated field of nutritional epidemiology — a subject recently adorned by an interesting review by our chairman, Dr. Gordon.[2] Before describing our own work I must ask you to consider the origin of nutritional surveys, and their purpose and conduct. I shall consider at the end of this lecture their relation to nutritional science.

ORIGIN OF NUTRITIONAL SURVEYS

Before the present century there were many attempts to define what food was needed under various conditions, and some were made to determine this experimentally; we have, for instance, already noted the work of William Stark. For experiment he intended to try the effects of fresh fruit and vegetables; however, he made his one false step, for he consulted Sir John Pringle, shortly to become president of the Royal Society, who advised him to omit salt from his diet; Stark followed the "rubbish" of "great authorities," castigated by Lind seventeen years earlier, and died at the age of twenty-nine. Had Stark been less unlucky the whole history of nutrition and indeed of medicine might have been changed. There was need for experiments such as his, although they were handicapped by lack of chemical knowledge, — and here again there was bad luck. Some Dutch merchants, who were losing trade in the East Indies through scurvy, had taken the enlightened step of founding a chair of chemistry at Leipzig for the study of antiscorbutics. The person appointed, Michael, had to send his preparations to the Dutch East Indies to be assayed on scorbutic sailors and to await the results months later. He found an oil extract of the seeds of scurvy-grass to be the most effective remedy. With any other vitamin than the unstable vitamin C his approach would probably have been crowned with brilliant success. The distinguished advances in medicine in the past hundred years that have resulted from the use of chemical experimentation might have occurred much earlier f a little luck had helped Michael and Stark.

One may wonder why experiments upon lower animals were not used by these scientists. The reasons we need not pause now to consider there were almost no attempts to investigate the epidemiology of malnutriture, — except in times of famine.

There were sporadic attempts to measure the food consumed by individuals. One of the most interesting was carried out by John Dalton of atomic-theory fame, who in 1831 briefly described certain experiments he had performed as a young man nearly forty years earlier in which at three different seasons of the year he weighed the individual solids and liquids taken at each meal. It is obvious that there was little interest in what food people ate until one could state the composition of that food and compare the data with what they should eat. From the middle of the last century, when food analyses became available, a number of persons (Moleschott in Italy, Playfair in England, Voit following Liebig in Germany, Payen in France, and Atwater in the United States) evolved dietary standards in terms of aliments, and attempts were made to compare the amounts consumed with the standards. For instance, in 1894 Atwater[3] found that, in the United States, daily calorie consumption could vary between the 1490 kilocalories of five Javanese at the World's Fair, Chicago, to the 8850 kilocalories of 237 brickmakers living in a boarding-house in this city of Boston. In Great Britain pioneer work on dietary surveys was done by Charles Booth (1889–1902) and Seebohm Rowntree (1901), later by Orr and by Cathcart and more recently by Widdowson and many others. In the United States the work of Atwater (1886 and onwards) includes some of the best studies ever made in human nutrition; later dietary surveys were made by Sherman and Gillett (1914–1915), Stiebeling and Phipard (1939), Dr. Stare and his colleagues and many others, and a large number of dietary studies have now been made in different countries.

Observations about height and weight, fatness, and clinical signs were soon added. Koppe (1905)[4] rated the amount of subcutaneous fat in five grades as one of the eight components of constitution. To subcutaneous fat Gastpar (1908)[5] added pallor of skin and mucous membranes, and muscle bulk. Hogarth (1909)[6] rated the nutriture of school children according to muscle bulk and subcutaneous fat; Howarth (1910)[7] used three grades according to color of skin and mucous membranes, condition of hair, muscle tone and alertness. Then came the Dunfermline system (Mackenzie, 1912–1913)[8] in England and the Sacratama index of von Pirquet in Austria. The discovery of vitamins brought supposedly more specific methods. Thus we find that at the end of 1932 the Health Committee of the League of Nations recommended von Pirquet's Sacratama index together with certain clinical signs of vitamin deficiencies, x-ray films of epiphyseal junctions and determinations of protein in serum and hemoglobin and inorganic phosphate in blood. The methods available were summarized for the League by Bigwood (1939)[9] and were put into practice in various countries at about the time the war started; the most comprehensive ones in this country were the Tennessee survey of Youmans, the North Carolina survey of Milam and the Pennsylvania survey of Mack. Let us pass to consider the methods used in these and other surveys.

THE METHODS OF NUTRITIONAL SURVEYS

We are discussing nutritional surveys of population groups. The purpose of such surveys is to define the cause, severity and distribution of malnutriture; this is achieved by assessing the nutriture of individuals within selected groups. No mass method is available. We must not delude ourselves into believing that we are told anything worth-while about the nutriture of populations by the type of study in which the food consumption of a large group, such as a country, is assessed by estimating food imported and exported, guessing food home produced and dividing by the number of inhabitants suitably adjusted through the use of some arbitrary scale of man-values. Such statistics no doubt have their uses, but not for the assessment of nutriture. We may also pass by minimum-cost studies, in which the amount of money believed to be spent on food is divided by the minimum amount believed to be necessary for purchasing an adequate

diet. These studies may interest economists but they do not assess nutriture.

In considering what nutriture is and how it can be assessed I shall follow what I have written elsewhere.[10] I do not think we can usefully consider the methods of nutritional surveys without first considering their purpose.

Nutriture, which is the state of the body or cells, resulting from the nutrition of the body or cells, is one of the components of health, and health is a component of the physical state of the body.[11]

$$
\text{Physical state}
\begin{cases}
\text{Health}
\begin{cases}
\text{Nutriture} \\
\text{Non-nutritional disease}
\begin{cases}
\text{Congenital} \\
\text{Acquired by}
\begin{cases}
\text{Disease} \\
\text{Accident}
\end{cases}
\begin{cases}
\text{Past malnutriture} \\
\text{Non-nutritional disease}
\end{cases}
\end{cases}
\end{cases} \\[2ex]
\text{Structural defects not affecting health}
\end{cases}
$$

This relation is important because, as we shall see, many attempts to assess nutriture in fact assess physical state. Consider two persons, one with a leg missing because of an accident and the other with carious teeth. In neither case could physical state be considered perfect, but health could be perfect if general function was not seriously impaired. My friend Professor John Ryle[12] used to maintain that if an old carious lesion in a tooth were unfilled health could not be perfect, whereas if the cavity were filled health could be perfect. I would regard inactive caries of any type in a tooth as not affecting health unless it seriously impairs general function; if full health could exist after extraction of the tooth nutriture is to be used in preventive medicine quality must be considered; assessing the quality of malnutriture is essential in defining its cause. Since it is important to distinguish between malnutriture produced by deficiency of aliments and that produced by deficiency of specific nutrients, it is useful to restrict the term "nourishment" to the former.

The severity of malnutriture is defined in terms of its second quality, namely, *degree*. I[10] have defined five degrees: excess nutriture, normal nutriture, poor nutriture, latent malnutriture and clinical malnutriture. We need not consider them now. But severity is more than degree; it includes the last property of nutriture, namely, *duration*. Duration is of course concerned with the time for which the state has existed; since malnutriture of short duration is in general more easily reversed than that of long duration, this affects the severity in population groups. I would call your attention to the confusion that exists in medical literature concerning the use of the terms "acute" and "chronic." The latter is often used to denote a condition that has existed for a long time, — and indeed this is the derivational and dictionary usage. But we find that the dictionary, — of course I speak of the Oxford English Dictionary, edited by my colleague Dr. Onions, — is inconsistent; it opposes "chronic" to "acute," and therefore we expect to find an acute disease defined as one that is of short duration. We do not. It is one that comes sharply to a point

or the filling of its cavity, full health can certainly exist while it is there unless some factor such as pain limits function; rats on a diet free of fluorine die of starvation because chewing is painful.[13] Obviously non-nutritional factors that affect health may affect nutriture; the cachexia of carcinoma is an example.

There are three properties of nutriture that are relevant in its assessment: quality, degree and duration. *Quality* expresses the types or kinds of deficiencies of other factors that are affecting nutriture. It therefore includes specificity and takes account of the unfortunate fact that deficiency of each of several nutrients can produce the same clinical sign; angular stomatitis can apparently arise from deficiency of iron or of riboflavin, and Dr. Ramalingaswami and I[14] have found (following earlier work of others on man) that this is an early and consistent sign of pyridoxine deficiency in rats. A clinical sign is often produced by a defect in the nutriture of a cell caused by a break in the metabolic chain in that cell, and deficiencies of one or more nutrients or the presence of one or more poisons may break the chain — even at the same link. One puzzling fact is that lesions produced by combined deficiencies may be very different from the sum of the lesions of the individual deficiencies: Follis[15] showed that deficiency of potassium or of thiamine causes myocardial necrosis, whereas deficiency of these two nutrients together is not accompanied by any pathologic change in the heart. If the assessment of or crisis of severity. We note therefore that a disease with a succession of crises of severity could be simultaneously both acute and chronic. I would suggest that we should regard processes as being short or long, and as having a degree that is changing rapidly or slowly. A rapid change of degree is acuteness; a slow or zero change is chronicity. These terms then do not denote duration, but the first differential coefficient of degree with respect to duration (I[10] have discussed this more fully in my paper in *Vitamins and Hormones*. Kruse[16-18] has, in his concept of the deficiency states, made much use of certain properties he ascribes to the pathologic process in the tissue in a deficiency disease. I think he has added confusion, and I have criticized him elsewhere[10,19]; I spare you further colloquy on this and pass to consider the third function of a nutritional survey, namely, the assessment of the distribution or incidence of malnutriture. We are here concerned with sampling and with the concept of normality.

Sampling is as important in nutritional surveys as in any other branch of epidemiology. Of the many pitfalls that occur I will select three as examples. In the Netherlands, houses were chosen on a sound statistical basis and all occupants were asked to attend a clinic. Had the apparent refusals not been followed up, the worst cases of malnutriture would have been missed because no person was unwilling to attend but some were too ill to do so.

When attending a committee in Germany I was shown the results of a nutritional survey in a certain town and was surprised at the gloomy picture presented. I was assured by the medical officer in charge that they were correct. I asked about the sampling and was shown the letter of instructions sent to the Oberburgermeister. I spoke to him, and he told me that as the army authorities were making a nutritional survey he supposed they would like to see cases of malnutrition, so he sent them as the sample those persons who were to be given extra rations on the grounds of malnutrition. In another town, I watched a street-corner weight survey carried out by an allied sergeant with two GIs and a German interpreter with scales and measuring-rod at a busy street junction. Here nothing but the errors was random. The GIs selected the pretty girls, the interpreter (and the onlookers) discouraged fat Germans and pushed forward thin ones; the bright-eyed healthy streetwalkers came up to see what was going on and were weighed; the listless malnourished passed by on the other side; the worst were at home in bed. Even if one gets a proper sample to examine, troubles are not ended. One may, for instance, be taking a venous blood sample from, and measuring the dark adaptation of, every tenth person clinically examined. A case of suspected vitamin-A deficiency is found and this is not a tenth person. One can scarcely refrain from making further tests to clinch the diagnosis — but one must refrain unless the results are rigidly excluded from the sample. The results obtained on our sample are used for comparison with other samples; we usually grade them as normal or abnormal. The concept of normality was excellently discussed by Ivy[20] and less satisfactorily by Ryle,[12] who was unaware of the former publication; I have commented upon their definitions elsewhere.[10] We must use two definitions of normality in different circumstances, one based on physiologic and the other on statistical concepts. We may say that the condition of the body or of some part or function of the body is (physiologically) normal when it allows the usual functions to be performed in the usual environment. We may also say that a value of a variate is statistically normal when it does not differ from the mean by more than the standard deviation. Unfortunately we have as yet insufficient data concerning the distribution of variates from which the range of normality can be assessed. In the course of the work of the Oxford Nutrition Survey we obtained a large number of such data. We are analyzing them in this way: We can exclude from our estimations all those subjects who were found to have any clinical, dietary or other abnormality that might affect values, and thereby we can obtain the distribution of the variate in normal members of the population.

The assessment of nutriture is a diagnosis and, like clinical diagnosis, it uses various disciplines.

The methods used may most conveniently be discussed under four headings: dietary, chemical, functional and clinical methods. These reflect the order in which deficiency is likely to arise; the person partakes of an inadequate diet, which may be detected by dietary methods. Later a change occurs in the concentration of nutrients (such as ascorbic acid) or of metabolites (such as pyruvate), or of compounds found through the agency of the nutrients (such as hemoglobin), and these may be detected by chemical methods. Later, changes in function may be found. Lastly, morphologic changes in tissues may occur, and these may be detected by what we shall for convenience call clinical methods.

I[10] have pointed out elsewhere that dietary methods cannot usually alone be used to assess nutriture, but a dietary history is an essential part of such an assessment and must be included even in rapid nutritional surveys. The purpose of including dietary information is to aid in interpreting the cause of, and suggesting a remedy for, any malnutriture that may be found. The difficulties of obtaining accurate dietary information have often been underestimated; for instance, the housewife may wish to show her efficiency, or she may hope to prove shortage of food, and not infrequently she cannot be bothered accurately to fill in the formidable array of forms that are presented to her. The errors involved in the computation of the nutrient content of diets are not always appreciated. We have spent much time in evolving a foolproof method of calculating dietaries; for this and for other purposes we have used a nutritional unit based partly on the share system of Dr. Mary Swartz Rose; this unit we call the Atwater.[21] Finally, even when the nutrient content of a diet is known difficulties arise in assessing availability and requirements.

Chemical methods are very valuable even in rapid field surveys. Their use has not been fully appreciated, — partly because they were introduced for public-health purposes before their interpretation had been investigated experimentally and partly because it is supposed that complicated and expensive apparatus is needed. We still know comparatively little about the interpretation of estimations of substances in the tissues, blood and urine. Since the composition of tissues alters before structural changes occur through malnutriture, estimations of vitamins of the B complex (thiamine, riboflavin, niacin and pyridoxine) and of ascorbic acid in tissues such as leukocytes are particularly useful. Compounds in plasma vary in their behavior: Some, such as carotenoids and ascorbic acid, tend to reflect the immediate past intake; others, such as niacin and riboflavin and, usually, protein, remain at concentrations that are within the statistically normal range even though clinical malnutriture may be present; therefore in these cases a single estimation that falls within the statistically

normal range gives little information. Estimations upon urine, with or without test doses, have not yet been proved valuable for assessing nutriture.

The only functional method that has so far proved valuable is the measurement of the rod threshold of the dark-adapted eye, provided that retesting after therapy with vitamin A is included. There are many factors other than nutrition, and there are nutritional factors other than vitamin A, that affect this measurement. Since endurance is affected by malnutriture, particularly by undernourishment, objective tests of this would be useful; unfortunately such a test, that would be applicable in a nutritional survey of a population group and that would scarcely be affected by variations in strength and motivation, has not yet been devised. We will, I think, all be agreed that clinical methods are essential in the proper assessment of nutriture, and we may here include somatometric methods, such as the use of sex-age-height-weight tables. These, and the calculation of various indices based on body measurements, are by themselves useless in the cross-sectional assessment of nutriture in an individual, although the measurements of height and weight and of the thickness of skin and subcutaneous tissue are useful in nutritional surveys. Although I speak with some hesitation about the growth of children in Boston, where Dr. Harold Stuart has carried out such valuable studies, I think we should bear in mind that the optimum rate of growth of children is not necessarily the maximum, and that harm may be done by excessive feeding of children with milk or school meals and now by medication with vitamin B_{12} and aureomycin, even though these activities make them grow more quickly and mature earlier. I have already mentioned the introduction of the general assessment of nutriture by inspection soon after the start of this century: In Great Britain it was adopted by the Board of Education in 1907, extended by the Dunfermline system a few years later[8] and revised in 1934; it is still used by many school medical officers. There is no doubt that the method assesses muscular development and physique more than general nutriture and, as shown admirably by Jones,[22] experienced school medical officers are not consistent with one another or even with themselves. Our Ministry of Health has continued such surveys with modification, and I have criticized these in detail elsewhere.[10, 11] The clinical assessment must include a medical history in which symptoms, though subjective, will be recorded. Examination of clinical signs must take into account all the non-nutritional factors that may produce them; the interpretation of clinical signs is exceptionally difficult because almost all the early signs are unspecific. The tissues that are most often affected structurally in malnutrition are the epidermis and the cornea; this statement is not derived entirely from the fact that few other tissues can be inspected but is based upon post-mortem examinations of man and lower animals.

It is by a combination of the methods just outlined that the diagnosis of the estimation of nutriture must be made; this is not necessarily a formidable procedure, as I hope to show you by the following brief account of our own work.

THE OXFORD NUTRITION SURVEY

The Oxford Nutrition Survey started early in 1941 at the request of the Ministry of Health and with the support of the Ministry of Food and the Medical Research Council. It began in a hurry, since the Ministry of Food happened to be conducting a dietary survey in Oxford and it was desirable to link with this the more detailed studies that were projected. Since governmental funds are slow to mobilize, it ran for its first few months without financial support; it was accommodated in the university's Department of Biochemistry through the kindness of professor R. A. Peters. Soon it was put on a secure basis by the generous financial assistance of the International Health Division of the Rockefeller Foundation.

At this time there was a tendency for those not immediately connected with nutritional research to believe that the methods of nutritional surveys had been established and that all that was required was a team to apply them. We did not accept this view. We therefore decided in our initial surveys to include all reasonable methods of assessing nutriture so that we could test their efficacy and select those most valuable. These initial surveys were made on working-class families in and around Oxford. Such surveys, essential through they were, did not give the rapid answers expected and required by governmental officials. Mobile surveys were therefore started by a group that consisted initially of a clinician (Dr. A. P. Meiklejohn), a dietitian and a biochemist. This team went to areas selected by the Ministry of Health, where families were investigated, samples of blood being dispatched to Oxford for some of the analyses. Initially this type of survey was also slower than required by the authorities; information was therefore obtained from two further sources: rapid clinical surveys and large-scale biochemical analyses upon samples of blood obtained by the Blood Transfusion Services.

The first rapid clinical survey was made in response to a specific problem: it was alleged that deficiency of riboflavin was prevalent in the Royal Navy, and we therefore organized early in 1942 a survey in which 1373 naval ratings in various parts of the country were examined clinically for the early signs of nutritional deficiency, were examined with a slit-lamp microscope for superficial corneal vascularisation and were tested for night vision. We were most fortunate in being joined by

Dr. Sydenstricker just as the survey was about to start; without his authoritative views upon the supposed early signs of deficiency of riboflavin, our survey would have given no trustworthy answer. Thereafter Dr. Sydenstricker conducted rapid nutritional surveys throughout the country; he[23] has described these in outline. His technic was essentially a rapid inspection of the skin, hair, tongue and mucous membranes of the mouth and a slit-lamp examination of the cornea and scleral conjunctiva; he also recorded certain data about the teeth, including the presence of dentures. In addition, he noted what he called "general physical status," dividing it into "good," "satisfactory" and "poor." He made no general physical examination and performed no laboratory tests. As I have mentioned elsewhere,[11] it was invaluable to have Dr. Sydenstricker in Great Britain during the war to reassure us about the nutriture of the population by showing that there was a low incidence of the clinical signs he sought; there was and is no greater authority upon those signs. I consider, however, that such surveys are by themselves inadequate for public-health purposes in England in peace, since they do not detect early malnutriture.

It was important during the war to be able to detect as early as possible any trends that might be occurring in the nutriture of the population, particularly concerning certain nutrients like ascorbic acid, of which the dietary levels were considerably affected by import policy. For this purpose we included large-scale analyses of blood samples obtained with the co-operation of a former Cutter Lecturer, Sir Lionel Whitby. The Army Blood Transfusion Service under him sent teams to various localities, where volunteers were bled. A small sample of this blood was at once sent to us at Oxford and analyzed for hemoglobin, protein, vitamin A and ascorbic acid. We fully appreciated from the experience of others and from our own investigations that samples of blood taken from a recumbent donor with a tourniquet on his arm, even if the first portion of the blood was reserved for us, would not give analyses that would approach the customary standard of accuracy. Our purpose, however, was not this. We wished to obtain a rough idea of trends that might be occurring in fairly comparable population groups at different seasons of the year, and of different mean levels between such population groups. We were able to obtain for each sample data about age, sex, locality and occupation of the donor.

While these four survey methods were being used in England, a limited amount of research was undertaken to examine some of the methods, and special groups of the population or of the Armed Forces were investigated as required. Toward the end of the war we turned our attention to the possibility of making surveys in liberated parts of Europe so that the distribution of food, whether carried out by the Army, UNNRA or, later, by the national governments, might be made according to certain priorities; since, in an area of malnutriture, the distribution of limited foodstuffs must be according to need, the problem is to discover who is in greatest need and the nature of that need. We set forth in 1944 the methods that should be used and received generous financial assistance from the Rockefeller Foundation for equipping two mobile laboratories for such work. In response to an invitation from the government of the Netherlands, these laboratories, with members of our staff, went shortly before the end of the war to study the acute famine in the Western Netherlands. In this work, of which a preliminary account has been published,[24] we were greatly assisted by the skilled help of colleagues in the Netherlands. Thereafter we conducted nutritional surveys in the British Zone of Germany. We hope shortly to publish in detail the results, in which we shall be able to compare and contrast the findings upon a comparatively well nourished population as in wartime Britain, a population subjected to an acute famine as in the Netherlands in April and May, 1945, and a population in which there was chronic malnutriture as in Germany immediately after the war.

We are convinced, as a result of our experience,

particularly in the Netherlands and in Germany, that a team consisting of a clinician, a biochemist, a dietist, a technician and a clerk can carry out rapid nutritional surveys of population groups by a combination of clinical, biochemical, functional and dietary methods and can report without delay their results to the relevant authorities. It remains to consider who those authorities are and the auspices under which such surveys should be conducted.

THE PLACE OF NUTRITIONAL SURVEYS IN THE UNIVERSITY AND IN PUBLIC-HEALTH PRACTICE

In this section I shall speak of my own country, for I do not presume to know — still less to understand — the relative responsibilities of state and Federal organizations here, — and if I did, it would be impertinent of me to comment.

The object of any science is the accumulation of systematized verifiable knowledge; this is achieved through observation, experiment and thought (which includes both criticism and imagination). Experiment is observation under controlled conditions. Observation (by which I mean bare observation) played an important part in primitive nutritional science: our ancestors observed that certain berries produced stomach-ache; we observe that they contain vitamin C. Progress only became rapid when medical science adopted the much more efficient tool of experiment. Though Galen devised brilliant physiologic experiments and almost discovered the circulation of the blood, his method was lost in the Dark Ages. The importance of Harvey is that he restored to medical science the study of function by deliberately planned experiment; that, no doubt, was why Sir Thomas Browne wrote that he "preferred the circulation of the blood to the discovery of America." We have seen earlier in this lecture the experimental work of Michael, Lind and Stark; we may also note that Lind's work was put into practice by the Admiralty after forty years and by the Mercantile Marine almost a century after his work. The use of the experimental method has brilliant discoveries to its credit, whereas the method of observation has achieved little. The observer must await the occurrence of the natural succession of events he wishes to study, and he is very apt to be misled by the fallacy of *post hoc ergo propter hoc* or by the existence of a correlation without causality. Nutritional science should be studied as a branch of experimental science. Human nutritional science is a branch of fundamental science that studies nutrition in man experimentally, whether on patients with developed malnutriture or on volunteers in whom malnutriture is induced. It has recourse to experiments upon lower animals to supplement experiments on man. The importance of the experimental approach I learned from the master at whose feet I sat — Charles Sherrington (still fortunately among us). When an Institute of Social Medicine was projected in Oxford I hoped the nutritional work we were doing would be included in it, for social medicine could scarcely be studied without reference to nutrition. The Regius Professor of Physics at Cambridge, John Ryle, was appointed its professor. He had, in several articles and in his research at Cambridge, extolled the observational as opposed to the experimental method, crossing swords with Sir Thomas Lewis. When Ryle told me that no experiments, either upon man or upon lower animals, would be done in the Institute of Social Medicine, I decided that human nutrition could not be usefully studied there. I have in consequence been said to have quarreled with Ryle. He assured me, and I know, that there was no quarrel. He was a great clinician whose wise counsel and inspiring friendship I treasured. His memory I treasure. We shared a desire to advance medicine and we differed seriously only in our approach to that end.

Medicine, like agriculture or engineering, rests with shifting emphasis upon a tripod, the three limbs of which are a practical art, a theoretical science and an applied science. The theoretical science derives from two sources. First, there are the fundamental sciences of human physiology, biochemistry and the like; secondly, there is clinical science. Both study man experimentally and both have recourse to experiments upon lower animals.

Both are proper subjects of study in a university. The main difference between the two is that in the fundamental sciences knowledge is pursued for its own sake, whereas the object of clinical science is to elucidate disease. The distinction between the two is often slight; the science of human nutrition may include either or both. For instance, we may study the human dietary requirements, metabolism and bodily functions of silver without supposing that silver deficiency ("anargyrosis") ever arises in man; that is in the realm of fundamental science. Or we may study the disease Kwashiorkor in the field, in volunteers and in lower animals to elucidate its etiology and establish its therapy; that is in the realm of clinical science. Both, we would all agree, are properly pursued in the university, which is concerned with the advancement of knowledge by research and teaching. We may well ask whether the applied science of nutrition belongs to the universities. I believe it does not. Lind was right in establishing the prevention and cure of scurvy by fruit juice; it was not for him to ensure that the juice was distributed.

In England, as a former Cutter Lecturer has stated elsewhere,[25] it is the duty of the Ministry of Health to advise on nutritional policy, and it is the business of the departments of food and agriculture to give effect, so far as is practicable, to this policy, either by importing or by producing the foodstuffs required. This was a wise Cabinet decision, since the primary object of agriculture is to produce human food, and the purpose of food is to maintain or promote health. Indeed, one may wonder whether ministries of agriculture and food are necessary. The latter is a recent outgrowth of two World Wars, disappearing in the peace between. The former was started by my great-grandfather, who was its first president, in 1793; with great filial impiety I should like to propose its abolition. If the Ministry of Health is responsible for advising on nutritional policy, it must be informed of the epidemiology of malnutrition in the country. As agreed in 1943 at the United Nations Conference on Food and Agriculture[25] at Hot Springs, "The first step in planning a nutrition program for any country should be to determine the needs of the population for health in terms of the more important nutrients" by making nutritional surveys. The countries that have accepted the unanimous recommendations of the Conference are pledged to undertake immediately such nutritional surveys, to deal by dietary and medical means with any malnutrition discovered and to take steps to prevent its recurrence. They are further pledged to report periodically to the permanent organization upon the results of their investigations into national food habits and nutriture. Two teams might be established by the Ministry of Health for this purpose, each consisting of clinician, biochemist, dietitian, technician and clerk. They would work through the local medical officers of health and be based on a central organization that could carry out certain chemical analyses and assist in computing the results on Hollerith cards so that they would be rapidly available; for this we introduced a punch-card that has been reproduced elsewhere[24] and has, I understand, proved satisfactory in Dr. Pett's surveys in Canada. Apart from the essential epidemiologic information that they provide, such nutritional surveys are very valuable in an educational program because they stimulate a personal interest in nutrition among both the population and the local practitioners of medicine. These teams should be available to undertake any special nutritional problem that arises in the field of public health. An alleged outbreak of scurvy in a town could be investigated at once, but their normal function would be to anticipate any such outbreak. Obviously only valid methods and the latest technics of proved worth must be used. This cannot be ensured by sending the teams out from an office in the Ministry in the center of London, but it can be achieved by locating them near a university institute where active research on human nutrition is in progress. A bureau of human nutrition might link the two, its object being to collect and integrate relevant knowledge to the subject so that its staff, its library and its museum might be consulted by medical officers, munity, as on an island. Such research would best be undertaken jointly by two or more university departments, preferably situated in different countries. Your penultimate Cutter Lecturer, Sir James Spence, once wrote as follows[27]:

The second aspect of the problem (of nutritional assessment) is the group investigation. By this I mean physical examinations or nutritional inquiries in various representative groups of the community. I admit that there is difficulty in getting this type of work carried out satisfactorily. It is clinical research. It requires trained clinical research workers. This in turn needs organization and financial support. The place for all this is surely in the universities and medical schools. It is a pressing need that in some of our universities there should be set up departments for the study of human nutrition. They would be primarily clinical departments, but, having contact with other scientific departments, they would serve as centers of research and training. They would test and apply the scientific knowledge of nutrition which comes from departments of physiology and biochemistry. If human nutrition is as important a subject as this meeting suggests it is, it is surely not too much to ask that it should have this recognition in our universities.

One should now, I believe, demur from this. A university institute of human nutrition should work on its own lines of research, using any methods that seem applicable; it should not be testing and applying scientific knowledge of nutrition so much as advancing such knowledge; the application should belong to the Ministry of Health. For there is a danger, even in a university, of dictated research;

hospital dietitians and others anxious to keep in close touch with the advancing front of research; it would exchange information with all the other countries that, by accepting the Final Act of the United Nations Conference on Food and Agriculture,[26] pledge themselves to such an interchange of information.

If we agree, then, that pure science, whether fundamental experimentation or clinical science, belongs in the universities, whereas the application of science does not, we may ask whether nutritional surveys of population groups should be made entirely by public-health agencies or can also find a place in the university. I believe that the function here is strictly limited. We would not deny to the Ministry's teams the right to do research; indeed, this should be encouraged, since they could not efficiently continue to make public-health surveys endlessly. Nor would we deny a university department the right to undertake field research if, — and only if, — new and useful knowledge is likely to result. Goldberger's brilliant research on the epidemiology of pellagra was work that any nutritional scientist might be proud to imitate. Research on similar lines has been and is being done on goiter, fluorosis and Kwashiorkor. It would be extended to study the distribution of disease in relation to diet in various parts of the world, and might well include the detailed and long-term study of improved diet upon a fairly circumscribed com-the clinician or the medical practitioner who approaches the scientist and asks him to work on a problem may be a menace, since few scientists are at a loss for interesting problems to pursue. And the university scientist will shun the routine organization needed in nutritional surveys.

We see, therefore, that there can be two types of nutritional surveys of population groups: those designed to study the epidemiology of malnutriture and conducted repeatedly by the Ministry of Health, — these are nutritional surveys proper, — and field research, which is used as an occasional tool by the nutritional scientist to advance fundamental knowledge. The two should not be entirely divorced, and indeed might usefully collaborate on occasion in studying malnourished populations abroad. But the observational method of nutritional surveys proper is very unlikely to advance fundamental knowledge of nutritional science and has no place in a university nutritional institute. We were faced in the Netherlands and Germany with the conduct of surveys for the specific and urgent public health purpose of establishing the incidence, cause and severity of malnutriture and then of recommending treatment.

There were therefore no opportunities for the detailed experimental study of a few cases, which is the method of choice for advancing fundamental knowledge. Had we gone with a government team whose function was the conduct of epidemiologic surveys, their normal routine work would have been enlivened by the new experience, and our research would have been possible — but no such team existed. Fortunately, controlled observations on a few persons were made by others on malnourished persons during and after the war.[28] In the future, we may hope that all countries will have such epidemiologic teams and such institutes for experimental research and that there may be constant interchange of information between them and the joint field studies, especially within their own country but also with the corresponding teams and institutes in other countries. There are, unfortunately, large areas of the world where gross malnutriture exists, and surely combined studies by experimental scientists and nutritional epidemiologists should be made. We need close collaboration between the pure and the applied science of human nutrition.

John Clarence Cutter, himself the son of a physiologist, combined a professorship of physiology with a consultant post in a government colonial department. I have tried to assess the role of nutritional surveys in establishing the epidemiology of malnutriture in population groups, such surveys being conducted by teams from the governmental agency responsible for nutritional policy but in close contact with those engaged upon basic research in human nutrition. There is perhaps at the present time no more important field of preventive medicine.

References

1. *The Works of the Late William Stark, M.D.* Revised by James C. Smyth. 190 pp. London: J. Johnson, 1788.

2. Gordon, J. E., and Le Riche, H. Epidemiologic method applied to nutrition. *Am. J. M. Sc.* 219:321-345, 1950.

3. Atwater, W. O. *Methods and Results of Investigations on the Chemistry and Economy of Food.* 222 pp. U. S. Department of Agriculture. Office of Experimental Stations. Bull. No. 21. Washington, D. C., 1895.

4. Koppe, O. *Der Schulartz* 3:219, 1905.

5. Gaspar, A. Die Beurteilung des Ernährungszustandes der Schulkinder. *Ztschr. f. Schulgesundhpflg.* 21:689-702, 1908.

6. Hogarth, A. H. *Medical Inspection of Schools.* 360 pp. London: Oxford University Press, 1909.

7. Howarth, W. J., and Kelynack, T. N. *Medical Examination of Schools and Scholars.* London: King and Son, 1910.

8. Mackenzie, A. *Seventh Annual Report, Dunfermline.* Edinburgh: King and Son, 1912-1913.

9. Bigwood, E. J. *Guiding Principles for Studies on the Nutrition of Populations.* 281 pp. League of Nations Health Organisation. New York: Columbia University Press, 1939.

10. Sinclair, H. M. Assessment of human nutriture. *Vitamins & Hormones* 6:101-162, 1948.

11. *Idem.* Clinical surveys and correlation with biochemical, somatometric and performance measurements. *Brit. J. Nutrition* 2:161-170, 1948.

12. Ryle, J. A. Meaning of normal. *Lancet* 1:1-5, 1947.

13. McClendon, J. F. Fluorine is necessary in diet of rat. *Federation Proc.* 3:94, 1944.

14. Ramalingaswami, V., and Sinclair, H. M. Lesions of muco-cutaneous junctions in pyridoxin deficiency in albino rats. *Brit. J. Nutrition* 4:xiii, 1950.

15. Follis, R. H., Jr. Myocardial necroses in rats on potassium low diet prevented by thiamine deficiency. *Bull. Johns Hopkins Hosp.* 71: 235-241, 1942.

16. Kruse, H. D. Concept of deficiency states. *Milbank Mem. Fund Quart.* 20:245-261, 1942.

17. *Idem.* Handbook of nutrition: medical evaluation of nutritional status. *J. A. M. A.* 121:584-591, and 669-677, 1943.

18. *Idem.* Concept of etiological complex of deficiency states with especial consideration of conditions. *Milbank Mem. Fund Quart.* 27:5-97, 1949.

19. Sinclair, H. M. Nutrition. *Ann. Rev. Biochem.* 19:339-370, 1950.

20. Ivy, A. C. What is normal or normality? *Quart. Bull. Northwestern Univ. M. School* 18:22-32, 1944.

21. Sinclair, H. M. Atwater — nutritional unit. *Brit. J. Nutrition* 3: x, 1949.

22. Jones, R. H. Physical indices and clinical assessments of nutrition of school children. *J. Roy. Stat. Soc.* 101:1-18, 1938.

23. Sydenstricker, V. P. Nutrition under wartime conditions. *Bull. New York Acad. Med.* 19:749-765, 1943.

24. Burger, G. C. E., Drummond, J. C., and Sandstead, H. R. *Malnutrition and Starvation in Western Netherlands September 1944–July 1945.* Part I. 187 pp. Part II. 299 pp. The Hague: General State Printing Office, 1948.

25. Jameson, W. War and advancement of social medicine. *Lancet* 2: 475-480, 1942.

26. Food and Agriculture Organisation. United Nations Conference on Food and Agriculture, Hot Springs, Virginia, 1943.

27. Spence, J. C. Discussion on difficulties of nutritional assessment. *J. Roy. San. Int.* 58:70-72, 1937.

28. Keys, A., et al. *The Biology of Human Starvation.* 158 pp. Minneapolis: University of Minnesota Press, 1950.

Additional Reference

Board of Education. *Admin. Mem.* No. 124, 1934.

APPENDIX D

324

[Reprinted with permission from the *Lancet*, i (1956), pp. 381–3 © by The Lancet Ltd (1956)]

Letters to the Editor

DEFICIENCY OF ESSENTIAL FATTY ACIDS AND ATHEROSCLEROSIS, ETCETERA

Sir,—Scant attention seems to be paid by the medical profession and by food administrators to a very important change in the dietaries of the more civilised countries that has been occurring over recent decades with increasing intensity. I refer to a chronic relative deficiency of the polyethenoid essential fatty acids (E.F.A.). It is true that the matter was raised in your columns a year ago, but then no less a person than a professor of nutrition stated [1] that such deficiency rarely if ever occurs in man ; Professor Yudkin, however, has the disadvantage of not having worked upon E.F.A. Our own experimental work, humble in scope, combined with a careful assessment of the literature, has led us to exactly the opposite conclusion. The causes of death that have increased most in recent years are lung cancer, coronary thrombosis, and leukæmia ; I believe that in all three groups deficiency of E.F.A. may be important. Your readers with stereotyped minds should stop reading at this point.

The most important E.F.A. is the vitamin arachidonic acid with four double bonds, tetraenoic. This can be formed in the body from the dienoic linoleic acid, also a vitamin, but vitamin B_6 is needed for the conversion.[2] The body probably has some slight ability to synthesise E.F.A., and these unsaturated fatty acids are of course easily oxidised. Oleic acid, with one double bond, can readily be synthesised in the body and for convenience I shall include it below in the term "saturated fat." For a reason that is obscure, the requirement of male animals for E.F.A. is at least five times that of females.

Eighteen months ago Dr. V. Basnayake and I made the surprising discovery that in the rat, fed a diet entirely free of fats and of sterols but adequate in other respects, there was a large deposition in the epidermis of cholesterol (or, more accurately, of Liebermann-Burchard-positive sterols) and this was mainly esterified ; after about 7 weeks of the deficient diet the esterified cholesterol was more than doubled. Alfin-Slater and others [3] had slightly earlier found accumulation of esterified cholesterol in the liver of rats on a fat-free diet, but their work had not then appeared in print ; and two groups of workers [4] [5] have shown that linoleic acid lowers serum-cholesterol in man. Obviously in our experiments the cholesterol was being synthesised in the body and its accumulation might have arisen for one of three reasons ; these we discussed when we reported [6] our observations

1. Yudkin, J. *Lancet*, 1955, i, 46.
2. Witten, P. W., Holman, R. T. *Arch. Biochem.* 1952, **41**, 266.
3. Alfin-Slater, R. B., Aftergood, L., Wells, A. F., Deuel, H. J. jun. *Ibid*, 1954, **52**, 180.
4. Friskey, R. W., Michaels, G. D., Kinsell, I. W. *Circulation Res.* 1955, **12**, 492.
5. Bronte-Stewart, B., Eales, L., Antonis, A., Brock, J. F. *Lancet*, Jan. 14, 1956, p. 101.
6. Basnayake, V., Sinclair, H. M. Proceedings of the 2nd International Conference on Lipids, Ghent, 1955.

to the 2nd International Conference on Lipids in Ghent last July.

First, there is an enormous increase in permeability of the skin in E.F.A.-deficiency[6-8] and there is an increase in capillary fragility,[9] and therefore cholesterol might pass more readily from blood as has recently been supposed to occur in atherogenesis[10]; this we rejected because the great increase is in the epidermis and not in the dermis.

Secondly, the epidermis might be synthesising unusual amounts of esterified cholesterol; but liver synthesis is normal in E.F.A.-deficiency.[3]

Thirdly, the cholesterol might be esterified with unusual fatty acids and might therefore be less readily disposed of from tissues.

We obtained evidence that this is true. Cholesterol is normally esterified mainly with unsaturated fatty acids[11] and when—as in our experiments—these are extremely deficient in the body it is esterified with much more saturated fatty acids synthesised in the body from carbohydrate. Thereby, I believe, abnormal esters are produced which tend to be deposited in tissues. Such deposition would be enhanced by a diet relatively high in saturated fat or high in cholesterol (which increases the dietary requirement of E.F.A.[12]). A diet deficient in vitamin B_6 and relatively low in arachidonic acid would act similarly, and this I believe may be the reason for the occurrence of atherosclerosis in pyridoxine-deficient monkeys[13] and possibly for the occasional hypertension in pyridoxine-deficient rats, which however we have never observed—probably because we give such rats large amounts of E.F.A. Atherosclerosis might therefore be regarded as a chronic deficiency of the vitamin arachidonic acid, this deficiency occurring in the body under the following conditions: (1) on diets very low in E.F.A. and adequate in aliments, but such diets would be exceedingly unusual; (2) on diets high in cholesterol and fairly low in E.F.A.; (3) on diets high in saturated or unnatural fats and fairly low in E.F.A. where the fats would be acting as an antivitamin; and (4) on diets deficient in vitamin B_6 and fairly low in arachidonic acid. Are such diets encountered?

DEFICIENCY OF E.F.A.

For many years and particularly in recent years animal fats and more especially vegetable fats have become increasingly oxidised and deprived of E.F.A. before being eaten. Vegetable oils, in many cases rich in E.F.A., are hardened by hydrogenation: margarine and shortenings are produced by hydrogenation of cotton-seed and soybean oil, some peanut-oil, and certain other oils; during this hydrogenation much of the E.F.A. are destroyed and unnatural *trans* fatty acids are formed. Unnatural fatty acids are formed not only during hydrogenation but also during the practice of deep-frying. Whale-oil is sometimes added to margarine, and this, like other marine oils of plant and animal origin which contain a variety of highly unsaturated fatty acids, is deodorised and hardened by hydrogenation. Some people who consume high-fat diets, such as some Norwegians and also Eskimos[14] uncontaminated by so-called civilisation, fare well by taking marine foods. The extraction-rate of bread is important in this context. The unsaturated fatty acids occur in the germ—wheat-germ oil is an excellent source—and therefore only

7. Sinclair, H. M. *Biochem. Soc. Symp.* 1952, **9**, 80.
8. Basnayake, V., Sinclair, H. M. *J. Physiol.* 1954, **126**, 55P.
9. Kramár, J., Levine, V. E. *J. Nutr.* 1953, **50**, 149.
10. Wang, C.-I., Schaefer, L. E., Adlersberg, D. *Circulation Res.* 1955, **3**, 293.
11. Kelsey, F. E., Longenecker, H. E. *J. biol. Chem.* 1941, **139**, 727.
12. Bromer, W. W., Day, H. G. *Chem. Engng. News,* 1953, **31**, 3964.
13. Rinehart, J. F., Greenberg, L. D. *Amer. J. Path.* 1949, **25**, 481.
14. Sinclair, H. M. *Proc. nutr. Soc.* 1953, **12**, 69.

small amounts survive in 70%-extraction flour. Of this remnant part is destroyed (as indeed is part of that in 80%-extraction flour) by the use of the so-called "improvers." Those that are bleaching agents, such as agene and chlorine dioxide, are all powerful oxidising agents, and the other improvers are also oxidisers. As well as the chemical improvers, substances with surface-active properties are added in baking, such as glyceryl monostearate, stearyl tartrate, and polyoxy-ethylene stearate; nothing is known of the effect of these upon the metabolism of E.F.A. Vitamin B_6 is mainly present in high-extraction flour, which is an important dietary source. Therefore, the use of 70%-extraction flour tends to produce E.F.A.-deficiency partly by having a low content of E.F.A. and partly by having only about a third of the vitamin B_6 present in wholemeal flour, this vitamin being needed for the conversion of linoleic to arachidonic acid. Further, vitamin E, which is a powerful antioxidant and therefore protects E.F.A., is largely removed from low-extraction flour and what remains is destroyed by the "improvers." The dietary intake of E.F.A. and of unnatural fats in this and other countries during the past years is at present being examined by Miss Dorothy Hollingsworth, Dr. J. N. Morris, and myself.

EFFECTS OF E.F.A.-DEFICIENCY

What are the consequences of deficiency of E.F.A. in man? We can only surmise by arguing from the effects of such deficiency in lower animals. First, deficiency would be likely to be at least five times commoner in males than in females. Secondly, we might expect deposition of cholesterol since cholesterol esterified with unusually saturated or with unnatural fatty acids is probably disposed of less readily; diets high in saturated or unnatural fats or cholesterol or deficient in vitamin B_6

would enhance this. Thirdly, E.F.A.-deficient animals are much more sensitive to noxious agents such as ultra-violet light (mild doses of which we found to produce complete necrosis of an irradiated paw [7]), to X-irradia-tion [15] including probably its carcinogenic action, and perhaps to chemical carcinogens.[16] In your columns Dr. Nyrop has argued [17] for E.F.A.-deficiency increasing the sensitivity, of males particularly, to a chemical carcinogen in tobacco-smoke and so contributing to bronchial carcinoma, and in general I endorse his views. Aortic atherosclerosis is unusually common in persons dying of bronchial carcinoma. Are there not other forms of cancer the incidence of which might be similarly increased? For instance, deaths from leukaemia have increased rapidly in the past two decades, are somewhat commoner in males than in females, show peaks of incidence (particularly in the case of lymphatic) in persons under 5 and over 45 years, are commoner in the U.S.A. and Denmark (a country with a highly developed medical service including radiography) than in England, are commoner in the rich than in the poor and (in the U.S.A.) in white than in coloured persons, and are nine times more frequent amongst U.S.A. radiologists than amongst other physicians.[18] If the more highly civilised and wealthier peoples are becoming increasingly sensitive to the carcinogenic effects of X rays through deficiency of E.F.A., the above facts become explicable provided those young children who died of leukaemia were irradiated when their pregnant mothers had diagnostic radiography; a chemical carcinogen could hardly be responsible for their deaths. But a chemical carcinogen in food, susceptibility

15. Cheng, A. L. S., Kryder, G. D., Bergquist, L., Deuel, H. J. jun. *J. Nutr.* 1952, **48**, 161. Deuel, H. J. jun., et al. *Science*, 1953, 117, 254.
16. Dickens, F., Weil-Malherbe, H. *Cancer Res.* 1946. **6**, 161.
17. Nyrop, J. E. *Lancet*, 1954, ii, 1288.
18. Hewitt, D. *Brit. J. prev. soc. Med.* 1955, **9**, 81. March, H. C. *Amer. J. med. Sci.* 1950, **220**, 282.

to which has been increased by E.F.A.-deficiency, might account for the susceptibility of males in particular to carcinoma of the stomach.

EPITHELIUM AND NERVOUS SYSTEM

Seborrhoeic eczema is a disease predominantly of males and appears to be increasing in incidence. It resembles histologically the lesion found in E.F.A.-deficient rats; some cases are alleged to respond to vitamin B_6 by inunction but not by injection, and inunction of an antagonist of vitamin B_6 produces dermatitis[19]; it is not impossible that this therapeutic effect is achieved by local conversion of linoleic to arachidonic acid, and that some cases of seborrhoeic dermatitis are E.F.A.-deficiency. Similarly, the dermatitis that sometimes follows the local application of detergents may arise through these extracting E.F.A. from the skin. In lower animals, in which we have carefully studied the skin lesion of E.F.A.-deficiency[20] and found a dramatic increase in permeability of the epidermis,[7,8] we believe there is a structural fault perhaps through failure of phospholipids containing E.F.A. to polymerise and form the impermeable barrier in the stratum granulosum. Phospholipids are rich in unsaturated fatty acids (though not so rich as cholesteryl esters), and although we found them to be normal in amount in the epidermis of E.F.A.-deficient rats they were abnormal in composition since the epidermal content of dienes and tetraenes had been enormously lowered. In the E.F.A.-deficient dog[21] the iodine number of serum phospholipid fatty acids is lowered, as is that of serum cholesteryl ester fatty acids, whereas the iodine number of serum glyceride fatty acids is normal. So, as in the case of esterified cholesterol, we have an abnormal type of phospholipid being formed which may not only cause a structural defect in the skin which is responsible for the great increase in permeability but may also be out-standingly important for four reasons.

First, the nervous system is rich in phospholipids containing polythenoid fatty acids, and these are found together with highly unsaturated cholesteryl esters in myelin; the presence of abnormal types of the compounds that are known to be important to it would be unlikely to leave function undisturbed: disseminated sclerosis is a disease of highly civilised countries being almost unknown in India and China, and other diseases in which the ectodermal neuroglia is affected may be relevant; since serum-E.F.A. fall in acute infections, it is not impossible that postinfective encephalo-myelitis should be considered in this context; early mild dementia appears to be becoming commoner in males.

Secondly, duodenal ulcers are increasing particularly in males, and in lower animals ulcers of the gut can be produced by feeding saturated fat.

Thirdly, the mitochondrial membrane probably contains phospholipids, and derangement of this through deficiency of E.F.A. might be responsible for the uncoupling of oxidative phosphorylation found in such deficiency.[22]

Fourthly, it is known that after a meal rich in fat the blood is more coagulable,[23] and recently it has been shown[24] that the substance concerned is almost certainly phosphatidyl ethanolamine.

CORONARY THROMBOSIS

Phospholipids are increased in the serum of persons with acute myocardial infarction, and if this compound contained unusually saturated or abnormal fatty acids, it would be expected to persist in serum for a longer time

19. Schreiner, A. W., Slinger, W. N., Hawkins, V. R., Vilter, R. W. *J. Lab. clin. Med.* 1952, **40**, 121.
20. Ramalingaswami, V., Sinclair, H. M. *Brit. J. Derm.* 1953, **65**, 1.
21. Hansen, A. E., Wiese, H. F. *Texas Rep. Biol. Med.* 1951. 9, 491.

22. Klein, P. D., Johnson, R. M. *J. biol. Chem.* 1954, **211**, 103.
23. Fullerton, H. W., Davie, W. J. A., Anastasopoulos, G. *Brit. med. J.* 1953, ii, 250.
24. Robinson, D. S., Poole, J. C. F. *Quart. J. exp. Physiol.* 1956, **41**, 36.

and might therefore be a factor in causing thrombosis, whether coronary or cerebral, in persons with atherosclerosis which I have concluded above may well be caused by cholesterol esterified with abnormal fatty acids. Eighteen months ago, when we had obtained some evidence of the production of abnormal compounds, I called attention [25] to the correlation in this country between the dietary intake of vegetable fat and deaths from coronary thrombosis : " I believe the answer may be that the level of total dietary fat is related to atheroma, but that the level of dietary *vegetable* fat is related to clotting upon atheromatous plaques, and hence to coronary thrombosis." In general it is the vegetable fats that are hardened and may give rise to abnormal phospholipids. But at that time we had been fortunate in having Dr. Ancel Keys recently to work with us and I had been impressed in general with his belief that total dietary fat, of whatever kind, was related to atheroma though we differed about its relation to coronary thrombosis and about Eskimos who have high dietary fat and allegedly little atheroma. Since then I have come to realise that what matters in atheroma is, I believe, the amount and structure of the dietary fatty acids.

NEED FOR RESEARCH

You will now conclude, Sir, if indeed you have not done so long ago, that I have strayed into the realms of fantasy. You will remind me that long ago Sir Thomas Browne wrote : " Imagination is apt to rove, and conjecture to keep no bounds." My inclusion of " etcetera," in the title invites the scorn we so readily pour on vendors of patent cure-alls. I have no wares to sell. The experiments to test my hypotheses are simple

25. Sinclair, H. M. *Advanc. Sci.* 1955, **12**, 115.

to devise, and for eighteen months I have hoped to be able to perform them. But at that time when Dr. Basnayake and I observed as we think the deposition of an unusual cholesteryl ester and the formation of unusual phospholipids we were engaged upon the ninth move of our laboratory; he did the estimations in another department while I did plumbing and carpentry. Now we are engaged upon our tenth move which will be followed later by our eleventh. Nomadism is incompatible with experimental science. And all the time we note that these diseases of civilisation have increased, are increasing, and ought to be diminished—diseases which I believe are caused by a dietary high in saturated fats and unnatural fats, and relatively low linoleic and arachidonic acids so that abnormal esters of cholesterol and abnormal phospholipids are formed in the body. I have not in this letter been concerned with conditions I believe to be caused by a pure dietary deficiency of essential fatty acids, without any relative increase in saturated or unnatural fat : these I consider to include phynoderma [19] and also vesical calculus so common in the Far East, the latter particularly affecting children under the age of 10 and being far more common in males than in females [26] (7 : 1, which is about the ratio of the requirement for E.F.A.). My purpose in sending you these speculations, which are supported by experimental evidence, is to try to help to prevent the devastating and usually increasing mortality in highly civilised countries from atherosclerosis and its sequelae, from cancer of the lung and stomach, from leukaemia, from disseminated sclerosis, and possibly other diseases of civilisation. There is a particular urgency in that if any part of my thesis is right it would be unfortunate to adopt flour of a low extraction-rate for three reasons : first, the practice

226. Ramalingaswami, V. Personal communication.

would reduce the dietary content of E.F.A. partly because flour of low extraction has most of these removed, much of the rest being destroyed by the so-called improvers; secondly, it would remove an important dietary source of vitamin B_6, this vitamin being needed to convert linoleic to arachidonic acid; thirdly, it would remove, in presence of "improvers," all the vitamin E which protects E.F.A., and indeed this protection might account for the unsubstantiated claims of Shute and others that administration of vitamin E is useful in certain of the conditions in which I suggest deficiency of E.F.A. is important. You, Sir, have consistently opposed any such retrograde step[27]; when Professor McCance and Dr. Widdowson on the basis of their short-term experiment upon children in Germany pronounced[28] that "general information and also scientific information had been wrong in laying so much emphasis on the value of high extraction [bread] for all persons under all conditions," you rightly concluded: "we maintain unrepentantly that, when it is a question of altering what has been a staple food for a very long time, the onus of proving advantage lies on those who wish to make the alteration not on those who oppose it." To revert to flour of a lower extraction-rate fortified with only three nutrients would I fear be likely to increase these diseases of civilisation in this country. And let us not rush into "fortifying" low-extraction flour with arachidonic acid until we are sure that this will not produce cataract or have other disadvantages. With no sympathy for long-haired naturalism, I humbly plead that we should give more thought and perform more research before extracting and "improving" wheat, manufacturing unnatural fats that then appear in margarine (which could otherwise be an excellent food), and before foisting upon the public sophisticated fodder which it unsuspectingly accepts. "Truth," wisely wrote Leonardo, "is the daughter of Time and not of Authority." But time is fast carrying away—unnecessarily—increasing numbers of the greatest persons. I am sure you, Sir, will agree that research upon the problems I have mentioned is worth while, particularly at a time when we are trying to determine correct policy regarding the desirable extraction-rate of flour.

CONCLUSIONS

In summary, my thesis is that the dietaries of the more highly civilised countries are becoming increasingly deficient in the essential polyethenoid fatty acids (the vitamin arachidonic acid which in presence of vitamin B_6 can be formed in the body from the vitamin linoleic acid) and increasingly rich in more saturated fatty acids and unnatural fatty acids which may act as antivitamins. This disastrous change arises partly from processes of manufacture such as the hardening of fats, and partly from the use of low-extraction flour "improved" with oxidising agents. Since the requirement of males for E.F.A. is much greater than that of females, the consequences are more serious in males. These consequences may include the following:

1. Cholesterol becomes esterified with abnormal or unusually saturated fatty acids, and these abnormal esters are less readily disposed of and so cause atheroma.

2. Phospholipids (such as phosphatidyl ethanolamine) contain abnormal or unusually saturated fatty acids, and these abnormal phospholipids, being less readily disposed of, are retained in plasma and increase the coagulability of blood thereby contributing to coronary and cerebral thrombosis.

3. The deficiency of normal phospholipids in the epidermis and gut makes these structures faulty, and this may contribute to the incidence of seborrhoeic eczema and peptic ulcer.

27. Lancet, 1955, ii, 235.
28. McCance, R. A., Widdowson, E. M. Ibid, p. 205.

4. The similar deficiency of normal phospholipids (or presence of abnormal phospholipids) in the nervous system causes defective structure including demyelination which may cause disseminated sclerosis and possibly mental disease.

5. Deficiency of polyethenoid fatty acids may increase susceptibility to noxious agents including the carcinogenic effect of X-irradiation and chemical carcinogens; the former, taken in conjunction with increased presentation of the carcinogen, may account for the increased incidence of leukæmia and the latter for the increased incidence of carcinoma of the bronchus and for the predominance of carcinoma of the stomach in males.

6. If further experimental work gives any support to these broad conclusions, steps should be taken to increase the dietary content of polyethenoid fatty acids, and perhaps to decrease the content of unnatural fatty acids; bread and flour of high extraction without added oxidising agents, and margarine and shortenings made with less hardening of fats, would be important factors in thus decreasing some very serious diseases of civilisation.

H. M. SINCLAIR.

Oxford.

Prologue

[On the afternoon of Tuesday 16 July the Conference adjourned to Stratford-on-Avon to see a performance of William Shakespeare's *Julius Caesar*.]

ACT III SCENE 2

Enter Mark Antony contemplating Caesar's atheromatous aorta

ANTONY: If you have plaques, prepare to shed them now.
    You all do know this corn-oil: I remember
    The first time Larry Kinsell gave a dose;
    'Twas on a summer's evening, in his ward;
    That day we overcame the Keysii.
    Look, in this place ran chylomicrons through;
    See what a plaque cholesterol esters made,
    Esterified with saturated fat.
    This plaque the well-belovéd butter caused
    Through Hilditch's cis-trans milk isomers,
    For Hilditch, just like Klenk here, studies lipids—
    Judge, O you gods, how dearly these two love them—
    This was the most unkindest plaque of all,
    For when the butter lipids get laid down
    Corónary clots, more strong than traitors' arms,
    Quite vanquish us. Then burst our mighty hearts.

    But if we take a tot or two of corn-oil—
    E'en in the face of Mead's silicic columns,
    Or alkali ísomerisation,
    Or counter-current distributíon,
    Reversed-phase chromatography as well
    As James's gas-phase—watch the plaques dissolve.
    Oh what a find is here for middle-aged men!
    Then I and you and all of us drink down
    Polyethenoid acids with our meals—
    Essential fatty acids sometimes called—
    While lipoproteins sludge along our veins.
    If not, you ail; and I perceive you need
    A pint of corn-oil full of double bonds:
    This great elixir with B_6 as well
    Dissolves our clots and makes our plaques dispel.

                    H. S.

Index

The index covers the main text, but not preliminary or end pages. Terms categorize the life of Hugh Macdonald Sinclair in letter-by-letter order (spaces and punctuation have no filing values). A page number in **bold** type indicates a more significant section; '(*fig.*)' after a page number indicates inclusion of an illustration.